Lecture Notes in Mathematics

Edited by A. Dold, F. Takens and B. Teissier

Editorial Policy
for the publication of monographs

1. Lecture Notes aim to report new developments in all areas of mathematics – quickly, informally and at a high level. Monograph manuscripts should be reasonably self-contained and rounded off. Thus they may, and often will, present not only results of the author but also related work by other people. They may be based on specialized lecture courses. Furthermore, the manuscripts should provide sufficient motivation, examples and applications. This clearly distinguishes Lecture Notes from journal articles or technical reports which normally are very concise. Articles intended for a journal but too long to be accepted by most journals, usually do not have this "lecture notes" character. For similar reasons it is unusual for doctoral theses to be accepted for the Lecture Notes series.

2. Manuscripts should be submitted (preferably in duplicate) either to one of the series editors or to Springer-Verlag, Heidelberg. In general, manuscripts will be sent out to 2 external referees for evaluation. If a decision cannot yet be reached on the basis of the first 2 reports, further referees may be contacted: the author will be informed of this. A final decision to publish can be made only on the basis of the complete manuscript, however a refereeing process leading to a preliminary decision can be based on a pre-final or incomplete manuscript. The strict minimum amount of material that will be considered should include a detailed outline describing the planned contents of each chapter, a bibliography and several sample chapters.
Authors should be aware that incomplete or insufficiently close to final manuscripts almost always result in longer refereeing times and nevertheless unclear referees' recommendations, making further refereeing of a final draft necessary.
Authors should also be aware that parallel submission of their manuscript to another publisher while under consideration for LNM will in general lead to immediate rejection.

3. Manuscripts should in general be submitted in English.
Final manuscripts should contain at least 100 pages of mathematical text and should include
– a table of contents;
– an informative introduction, with adequate motivation and perhaps some historical remarks: it should be accessible to a reader not intimately familiar with the topic treated;
– a subject index: as a rule this is genuinely helpful for the reader.

Continued on back inside cover

Lecture Notes in Mathematics

1745

Editors:
A. Dold, Heidelberg
F. Takens, Groningen
B. Teissier, Paris

Springer
Berlin
Heidelberg
New York
Barcelona
Hong Kong
London
Milan
Paris
Singapore
Tokyo

V. D. Milman   G. Schechtman (Eds.)

Geometric Aspects of Functional Analysis

Israel Seminar 1996-2000

1996-2000

Springer

Editors

Vitali D. Milman
Department of Mathematics
Tel Aviv University
69978 Tel Aviv, Israel

E-mail: vitali@math.tau.ac.il

Gideon Schechtman
Department of Mathematics
Weizmann Institute of Science
76100 Rehovot, Israel

E-mail: gideon@wisdom.weizmann.ac.il

Cataloging-in-Publication Data applied for

Die Deutsche Bibliothek - CIP-Einheitsaufnahme

Geometric aspects of functional analysis / Israel Seminar (GAFA) 1996
- 2000. V. D. Milman ; G. Schechtman (ed.). - Berlin ; Heidelberg ;
New York ; Barcelona ; Hong Kong ; London ; Milan ; Paris ; Singapore
; Tokyo : Springer, 2000
  (Lecture notes in mathematics ; Vol. 1745)
  ISBN 3-540-41070-8

Mathematics Subject Classification (2000): 40-06, 46B07, 52-06, 60-06

ISSN 0075-8434
ISBN 3-540-41070-8 Springer-Verlag Berlin Heidelberg New York

Springer-Verlag Berlin Heidelberg New York
a member of BertelsmannSpringer Science+Business Media GmbH

© Springer-Verlag Berlin Heidelberg 2000
Printed in Germany

Typesetting: Camera-ready TeX output by the author
SPIN: 10724232   41/3142-543210 - Printed on acid-free paper

Preface

During the last two decades the following volumes containing papers presented at the Israel Seminar in Geometric Aspects of Functional Analysis appeared

1983-84 Published privately by Tel Aviv University
1985-86 Springer Lecture Notes, Vol. 1267
1986-87 Springer Lecture Notes, Vol. 1317
1987-88 Springer Lecture Notes, Vol. 1376
1989-90 Springer Lecture Notes, Vol. 1469
1992-94 Operator Theory: Advances and Applications, Vol. 77, Birkhauser
1994-96 MSRI Publications, Vol. 34, Cambridge University Press.

The first six were edited by Lindenstrauss and Milman while the last, which also contains material from the program in Convex Geometry and Geometric Analysis held at MSRI in 1996, was edited by Ball and Milman.

The current volume reflects some of the new directions in Banach Space Theory in the last few years. These include the tighter connection with classical convexity and as a result the added emphasis on convex bodies which are not necessarily centrally symmetric. Initially, emerging from the functional analysis point of view, symmetric convex bodies were the natural object of investigation but, as it becomes more and more clear, a large portion of the theory carries over to the non-symmetric case and this sometimes sheds new light even on the symmetric case. A similar situation, which is also reflected in some of the articles of this volume, is the treatment of bodies which have only very weak convex–like structure - they are only p-convex for some $0 < p < 1$. Another topic which is represented here is the use of some new probabilistic tools; in particular transportation of measures methods and new inequalities emerging from Poincare–like inequalities. Finally, several of the papers here deal with improving and finding the best, or best order, constants in several results. This is another topic which has received considerable attention recently.

All the papers here are original research papers and were subject to the usual standards of refereeing.

As in previous volumes of the GAFA Seminar, we also list here all the talks given in the seminar as well as talks in related workshops and conferences. We believe this gives a sense of the main directions of research in our area.

We are grateful to Ms. Diana Yellin for taking excellent care of the typesetting aspects of this volume.

Vitali Milman
Gideon Schechtman

Contents

The Transportation Cost for the Cube

M. Anttila

Department of Mathematics, University College London, Gower Street, London WC1E 6BT, UK

Abstract. The transportation method for proving concentration of measure results works directly for the cube. Here we find the best constant that can be found using this method which turns out to be better than those obtained by previous methods and which cannot be far from that which is best possible.

1 Introduction

In this paper we prove a deviation inequality for the cube using a method developed by K. Marton to show similar results for Markov chains. Talagrand named this method the transportation method when simplifying Marton's arguments for certain product spaces.

Let us consider the cube, $[0,1]^n \subset \mathbf{R}^n$, and denote by P the n-dimensional Lebesgue measure on it. If B is any measurable subset of the cube, let B_t be its expansion,

$$B_t = \{x \, \epsilon \, [0,1]^n : d(x, B) \leq t\},$$

where $d(x, B)$ denotes the Euclidean distance from x to B. We shall prove a deviation inequality of the form

$$1 - P(B_t) \leq e^{-ct^2}, \tag{1}$$

provided B does not have too small probability, where c is a constant dependent on $P(B)$.

Concentration results of this form have been known for the cube for some time. Indeed, it was pointed out in [TIS] that inequality (1), with bound $\frac{1}{P(B)}e^{-\frac{\pi}{2}t^2}$, can be obtained directly from concentration in Gauss space via a measure preserving Lipschitz map. Here our objective is to point out that the transportation method works directly for the cube and, more importantly, to ascertain the best constant that can be found using this method. This constant is better than those previously obtained and cannot be far from best possible. Finding this constant gives rise to a "text-book example" of a variational problem which has a surprisingly neat solution.

Marton's original method uses an inequality bounding the so-called $\bar{d}$-distance by informational divergence to prove a concentration of measure result for certain Markov chains (see [M] for definitions and a detailed account).

Supported by EPSRC-97409672.

The important thing in her method is that her one-dimensional inequality can be inducted on dimension and quickly implies a concentration of measure result. Marton's method certainly works for product spaces. However, Talagrand simplifies it and strengthens the result for certain product spaces in [T], by considering l_2, rather than l_1, distance in the inequality. More precisely, Talagrand's inequality bounds something called transportation cost, with the square of the l_2-distance as the cost function. The definition of transportation cost now follows.

Suppose we have two probability measures μ_1 and μ_2 on measurable spaces Ω_1 and Ω_2 respectively. The basic idea is to look at all bijections $b : \Omega_1 \to \Omega_2$ which transport μ_1 to μ_2, i.e. for which

$$\mu_1(A) = \mu_2\big(b(A)\big) \quad \text{whenever } A \subset \Omega_1.$$

For a given function $C : \Omega_1 \times \Omega_2 \to \mathbf{R}^+ \cup \{\infty\}$, $(C(x, b(x))$ measures the cost of moving a unit mass from x to $b(x)$), we seek to minimise

$$\int_{\Omega_1} C\big(x, b(x)\big) \, d\mu_1(x).$$

If μ_1 or μ_2 has atoms then there may be no such function b. So, formally, the transportation cost is defined in terms of an integral over the product space $\Omega_1 \times \Omega_2$ with respect to a probability measure with marginals μ_1 and μ_2. However, in our case $\Omega_1 = \Omega_2 = [0, 1]^n$ and our measures will be the Lebesgue measure on the cube itself and a weighted Lebesgue measure on one of its subsets, so no such formality is needed here.

As already mentioned, we shall use the square of the Euclidean distance as our "cost function", C, just as Talagrand did for Gaussian measure. So now we can define the transportation cost, $\tau(\mu_1, \mu_2)$, to be the minimum, over all functions b as above, of

$$\int_{\Omega_1} |x - b(x)|_2^2 \, d\mu_1(x).$$

The main result of this article is the following:

Theorem (Bound on Transportation Cost) *If A is a subset of $[0, 1]^n$ and μ is the normalised restriction of the Lebesgue measure, P, to A (i.e. has density $1_A/P(A)$ with respect to P), then*

$$\tau(\mu, P) \leq \frac{2}{\pi^2} \log \frac{1}{P(A)}.$$

From this it is easy to get a concentration estimate using the following short argument. Let $B \subset [0, 1]^n$. The cost of transporting $[0, 1]^n$ to the complement of the expanded B, B_t^c, is clearly greater than that of transporting B (a subset of $[0, 1]^n$) to B_t^c. The Theorem gives an upper bound on the former and the latter is greater than $P(B)t^2$. So

$$P(B)t^2 \leq \frac{2}{\pi^2} \log \frac{1}{P(B_t^c)}.$$

Rearranging this we have

$$P(B_t^c) \leq e^{\frac{-\pi^2}{2}P(B)t^2}.$$

However, this bound can be improved by applying the Theorem to B as well as to B_t^c as in [M] and [T]. This gives the slightly better estimate

$$P(B_t^c) \leq \exp\left\{\frac{-\pi^2}{2}\left(t - \sqrt{\frac{2}{\pi^2}\log\frac{1}{P(B)}}\right)^2\right\}.$$

As already mentioned, we will see, in the proof of the Theorem, that $\frac{\pi^2}{2}$ is the best constant that we can find using this method. Before we begin the proof, however, we observe that our constant is not far from best possible.

The following tells us that c in (1) cannot be greater than 6. Let K be the cube of volume 1, now centered at zero. We regard K as a probability space and define on it the random variable $X_\theta : x \mapsto \langle x, \theta \rangle$, where $\theta = \left(\frac{1}{\sqrt{n}}, \ldots, \frac{1}{\sqrt{n}}\right)$, so that the density of X_θ is obtained by scanning across K with hyperplanes perpendicular to θ. Since

$$X_\theta(x) = \frac{1}{\sqrt{n}}\sum_{i=1}^{n} X_i(x),$$

where $X_i : x \mapsto x_i \in [-\frac{1}{2}, \frac{1}{2}]$ are random variables with zero mean and variance $\rho^2 = \frac{1}{12}$, the Central Limit Theorem tells us that as $n \to \infty$

$$\text{Prob}(X_\theta > t) \to \frac{1}{\sqrt{2\pi}\rho}\int_t^\infty e^{-\frac{y^2}{2\rho^2}}\,dy$$

$$\geq \frac{\rho}{\sqrt{2\pi}t}e^{-\frac{t^2}{2\rho^2}}$$

$$= \frac{\text{const}}{t}e^{-6t^2}.$$

Now we need only notice that the left hand side is precisely $P(H_t^c)$ for $H \subset [0, 1]^n$ given by the intersection of the cube with the halfspace through zero perpendicular to θ:

$$H = \{x \in [0, 1]^n : \langle x, \theta \rangle \leq 0\}. \qquad \square$$

2   Proof of the Theorem

We wish to show by induction that

$$\tau(\mu, P) \leq \frac{2}{\pi^2} \log \frac{1}{P(A)} \quad \text{for } A \subset [0,1]^n,$$

where the left hand side is the minimum cost of transporting A to $[0,1]^n$.

We show in Section 2.1 that a result of the form

$$\tau(\mu, P) \leq c \log \frac{1}{P(A)} \quad \text{for } A \subset [0,1]^n, \tag{2}$$

can be obtained from the following one-dimensional inequality for absolutely continuous $f : [0,1] \to [0,1]$,

$$\int_0^1 (f(t) - t)^2 f'(t) \, dt \leq c \int_0^1 f'(t) \log f'(t) \, dt, \tag{3}$$

where c is the same constant in both inequalities.

We continue in the ensuing sections to use the Calculus of Variations to show that (3) holds for all appropriate f if (and only if) $c \geq \frac{2}{\pi^2}$. More precisely we show that there is an optimising function satisfying an appropriate Euler-Lagrange equation and then we analyse the solutions of this equation.

2.1   The Inductive Step

Let us choose one of the n coordinate directions, e_1 say. We denote by $g(t)$ the $(n-1)$-dimensional volume of A intersected with the "slice of the cube at $t \epsilon [0,1]$":

$$\{x \epsilon [0,1]^n : \langle x, e_1 \rangle = t\}.$$

The idea is that we transport in the e_1 direction, the $(n-1)$-dimensional slices via an increasing function, f, such that the proportion of A between the slices at t and $t + \delta$ is equal to the proportion of $[0,1]^n$ between $f(t)$ and $f(t + \delta)$:

$$\frac{\delta g(t)}{P(A)} = f(t + \delta) - f(t). \tag{4}$$

The weighted cost of transporting in this way in one dimension is clearly

$$\int_0^1 (f(t) - t)^2 \frac{g(t)}{P(A)} \, dt. \tag{5}$$

We then use the inductive hypothesis, (2), to transport in each $(n-1)$-dimensional slice. The total of the transportation costs in all of the slices is at most

$$\int_0^1 c \log \frac{1}{g(f^{-1}(s))} \, ds.$$

After substituting $s = f(t)$, this is

$$-\int_0^1 cf'(t) \log g(t) \, dt. \tag{6}$$

We see from (4) that $f'(t) = \frac{g(t)}{P(A)}$. So to complete the inductive step we combine (5) and (6) and ask whether

$$\int_0^1 (f(t) - t)^2 f'(t) \, dt - \int_0^1 cf'(t) \log\left[f'(t) P(A)\right] \, dt \;\leq\; c \log \frac{1}{P(A)}.$$

When rearranged, this is

$$\int_0^1 (f(t) - t)^2 f'(t) \, dt - \int_0^1 cf'(t) \log f'(t) \, dt \;\leq\; c \log \frac{1}{P(A)} \int_0^1 (1 - f'(t)) \, dt,$$

which simplifies to

$$\int_0^1 (f(t) - t)^2 f'(t) \, dt \;\leq\; c \int_0^1 f'(t) \log f'(t) \, dt, \tag{7}$$

since $f(0) = 0$ and $f(1) = 1$.

The same inequality handles the one-dimensional case because we can transport in exactly the same way in dimension one, where $g(t) = 1_A(t)$ and where clearly we will not be required to transport further within $(n - 1)$-dimensional sheets. So the transportation cost is at most (5). Further, since $f'(t) = \frac{1_A(t)}{P(A)}$, we have

$$c \log \frac{1}{P(A)} = c \int_0^1 f'(t) \log f'(t) \, dt. \quad \square$$

It is not difficult to find some c for which (7) holds (and hence such that (2),

$$\tau(\mu, P) \leq c \log \frac{1}{P(A)},$$

is true). For example, if we rewrite the left hand side of (7) as below, we see that (7) holds with $c = 2$ by using standard methods from information theory and the Csiszár-Kullback-Pinsker inequality. We mention here also that the logarithmic Sobolev inequality for the cube implies (2) with $c = \frac{1}{\pi}$, see [OV]. However we wish to find the smallest c.

We begin by rewriting (7). Notice that

$$\int_0^1 (f(t) - t)^2 (f'(t) - 1) \, dt = 0, \quad \text{since } f(0) = 0 \text{ and } f(1) = 1.$$

So we can rewrite the left hand side of (7) as

$$\int_0^1 (f(t) - t)^2 \, dt.$$

If we consider instead the deviation of f from t, $h(t) = f(t) - t$, then (7) becomes

$$\int_0^1 h^2(t)\, dt \le c \int_0^1 \left(1 + h'(t)\right) \log \left(1 + h'(t)\right)\, dt. \qquad (8)$$

Our problem is to find the smallest constant, c, such that the functional in (8),

$$\mathcal{F}_c(h) = \int_0^1 \left[c\left(1 + h'(t)\right) \log \left(1 + h'(t)\right) - h^2(t) \right] dt, \qquad (9)$$

is non-negative for all h in the admissible class of functions given by

$$\mathcal{C} = \left\{ h \text{ absolutely continuous} : h(0) = h(1) = 0, \ h' > -1 \right\}.$$

This variational problem is the subject of the following sections.

2.2 The Variational Problem

Recall that our aim is to find the smallest c such that the functional $\mathcal{F}_c$ is non-negative for all functions in $\mathcal{C}$. First we will show that for all $c > 0$, a minimiser of $\mathcal{F}_c$ exists and satisfies the Euler-Lagrange equation:

$$\left(1 + h'(t)\right) h(t) + \frac{c}{2}\, h''(t) = 0. \qquad (10)$$

Then we will find that if $c > \frac{2}{\pi^2}$, the only solution of (10) which satisfies the boundary conditions of $\mathcal{C}$, is the trivial one, $h = 0$. Hence $\mathcal{F}_c \ge 0$ for such c. To show that $c = \frac{2}{\pi^2}$ is the smallest constant for which $\mathcal{F}_c$ is non-negative, we will consider specific functions in our admissible class.

A classical theorem of Tonelli on the existence of minimisers of a one-dimensional variational integral,

$$\mathcal{F}(v) = \int_I F(x, v, v')\, dx,$$

can be found, for example, in [BGH]. The standard conditions are that the Lagrangian, $F(x, v, p)$, is continuous, convex in p and has superlinear growth in p at ∞ (i.e. is such that there exists a function $\theta(p)$ such that

$$F(x, v, p) \ge \theta(p) \quad \text{for all } (x, v, p) \ \epsilon \ I \times \mathbf{R} \times \mathbf{R}$$

$$\text{and } \frac{\theta(p)}{|p|} \to \infty \quad \text{as } |p| \to \infty).$$

The superlinearity condition clearly does not hold for our Lagrangian,

$$F_c(x, v, p) = c(1 + p)\log(1 + p) - v^2,$$

because the "v" term could make $\mathcal{F}_c$ very small. However, it is not hard to see that the standard arguments can be adapted to demonstrate the existence of minimisers in our case. In fact, our Lagrangian has certain invariance properties which, if anything, make our problem easier than the general one. We include here a very rough explanation.

We wish to show that there exists a function $u \in C$ such that

$$\mathcal{F}_c(u) = \inf\{\mathcal{F}_c(v) : v \in C\}.$$

Call this infimum λ, say. From the boundary conditions on C, we have that $|v| \leq 1$ for $v \in C$, so $\mathcal{F}_c$ is bounded below. Hence we can find a minimising sequence, $\{u_k\} \subset C$, such that $\mathcal{F}_c(u_k) \to \lambda$.

The following properties of our Lagrangian allow us to take the functions in this minimising sequence to be positive and concave. Since F_c comprises only the square of the function, v, and its derivatives, rotating a negative section of the function by 180° leaves the functional unaltered. Further, if we approximate any positive function in the minimising sequence by a piecewise linear function and make this concave in steps, it is clear that in doing so the functional, $\mathcal{F}_c$, decreases. This follows since v increases and since

$$(1 + p) \log(1 + p) \quad \text{is convex for } p > -1.$$

We can use the Ascoli-Arzelà Theorem to show that a subsequence of $\{u_k\}$ converges uniformly to a continuous function u, say. Equiboundedness is clear. To prove equicontinuity, we need to show that u'_k cannot get too large on $[0, 1]$. But since we restricted u_k to being positive and concave, we need only show that u'_k is not too large near zero.

Notice that since $u_k(0) = u_k(1) = 0$, we can write $\mathcal{F}_c(x, u_k, u'_k)$ as

$$\int_0^1 c\big[(1 + u'_k) \log[1 + u'_k] - u'_k\big] - u_k^2 \, dx.$$

So if, for $\varepsilon > 0$, $u_k(\varepsilon) = L\varepsilon$, it is not hard to see, using the restriction $|u_k| < 1$ and that $(1 + p) \log[1 + p] - p > 0$ for all positive p, that

$$\mathcal{F}_c(x, u_k, u'_k) \geq \varepsilon c\big[(1 + L) \log[1 + L] - L\big] - 1.$$

This in turn gives us an upper bound on $L\varepsilon$ which tends to zero as $\varepsilon \to 0$.

To show that $u' > -1$, and hence $u \in C$, requires noticing that $(1 + p) \log(1 + p)$ has infinite derivative at $p = -1$ and so a minimiser will not have derivative equal to -1, except possibly at 1. Finally, the concavity of the functions in the minimising sequence ensures that $u'_k \to u'$ a.e. Then $\mathcal{F}_c(u_k) \to \mathcal{F}_c(u)$ dominatedly. □

That any minimiser satisfies the Euler-Lagrange equation (10) is standard, see e.g. [BGH]. The only possible issue in our case is that the functional must be defined for all functions in a neighbourhood of the minimiser, u. But since we just saw that our Lagrangian forces $u' > -1$, this does not pose a problem.

2.3 Periodicity Analysis

It remains to show that for $c > \frac{2}{\pi^2}$, the only solution of the Euler-Lagrange equation is the trivial one (hence $\mathcal{F}_c \geq 0$ for such c) and that conversely there are functions in our admissible class for which $\mathcal{F}_c < 0$ if $c < \frac{2}{\pi^2}$.

Recall that the Euler-Lagrange equation is given by

$$\left(1 + h'(t)\right) h(t) + \frac{c}{2} h''(t) = 0. \tag{11}$$

If we rearrange and multiply both sides by $h'(t)$, (11) becomes

$$h(t)h'(t) = -\frac{c\, h''(t)}{2(1 + h'(t))} h'(t)$$

and this integrates to

$$-h^2(t) + M^2 = c\left[h'(t) - \log\left(1 + h'(t)\right)\right], \quad \text{where } M = \sup |h|. \tag{12}$$

If we define the function $\Omega : (-1, \infty) \to [0, \infty)$ to be

$$\Omega(s) = s - \log(1 + s),$$

then (12) can be written in terms of Ω as

$$\frac{1}{c}(-h^2 + M^2) = \Omega(h'). \tag{13}$$

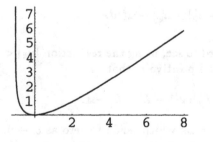 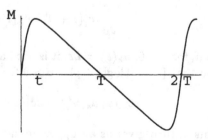

Fig. 1. The function Ω, (left), and a solution, h, of (12), (right).

It is not difficult to see that a solution of (12) either increases to M or is periodic. Since we have the restriction that any function in our admissible class is zero at 1, we need only consider periodic solutions. So if for $c > \frac{2}{\pi^2}$ every non-trivial solution has period greater than 2, then we know that there is no non-trivial solution in our admissible class for such c. Hence we will have $\mathcal{F}_c \geq 0$ for $c > \frac{2}{\pi^2}$.

Let $2T$ denote the period of a solution, h, of (12). Suppose that h attains its maximum at the point $t \in (0, T)$. Then we can express t as an integral over h between 0 and M:

$$t = \int_0^t ds = \int_0^M \frac{1}{h'} \, dh. \tag{14}$$

Similarly, for the second section of the semiperiod, on which $h' \leq 0$ we have

$$T - t = \int_M^0 \frac{1}{h'} \, dh. \tag{15}$$

So if we denote the two branches of Ω^{-1}, Ω_+^{-1} and Ω_-^{-1}, using (13), we can express h' in terms of the inverses Ω_+^{-1} and Ω_-^{-1}, depending on the sign of h'.

Hence from (14) and (15), we know the semiperiod of a periodic solution of (12) to be

$$T = \int_0^M \frac{1}{\Omega_+^{-1}(\frac{1}{c}(M^2 - h^2))} \, dh - \int_0^M \frac{1}{\Omega_-^{-1}(\frac{1}{c}(M^2 - h^2))} \, dh. \tag{16}$$

We shall see below that

$$\frac{1}{\Omega_+^{-1}(x)} - \frac{1}{\Omega_-^{-1}(x)} \geq \frac{\sqrt{2}}{\sqrt{x}} \quad \text{for } x \geq 0. \tag{17}$$

Applying this to (16) we have

$$T \geq \int_0^M \frac{\sqrt{2}}{\sqrt{\frac{1}{c}(M^2 - h^2)}} \, dh = \frac{\pi}{\sqrt{2}} \sqrt{c}.$$

Hence for $c > \frac{2}{\pi^2}$ the return time, T, is strictly greater than 1 and we are done.

To prove (17) we fix $x \in [0, \infty)$ and define $s, t \geq 0$ by

$$\Omega_+^{-1}(x) = t \quad \text{and} \quad \Omega_-^{-1}(x) = -s.$$

Then

$$x = t - \log(1 + t) = -s - \log(1 - s) \tag{18}$$

and we need to show that

$$\frac{1}{s} + \frac{1}{t} \geq \frac{\sqrt{2}}{\sqrt{x}},$$

i.e.

$$\frac{1}{2}\left(\frac{1}{s} + \frac{1}{t}\right) \geq \frac{1}{\sqrt{2x}}.$$

By the AM/GM inequality, the left hand side is at least $\frac{1}{\sqrt{st}}$, so it suffices to show that under (18),

$$st \leq 2x.$$

By (18) this will follow if we show that for any $s, t \geq 0$,

$$st \leq t - \log(1 + t) - s - \log(1 - s),$$

i.e.

$$\log(1 + t) + \log(1 - s) \leq t - s - st.$$

But the left hand side is

$$\log\big((1 + t)(1 - s)\big) = \log(1 + t - s - st)$$
$$\leq t - s - st. \quad \square$$

Finally, to show that $\frac{2}{\pi^2}$ is the best constant, we find that there are specific admissible functions for which the inequality, (8),

$$\int_0^1 h^2(t) \, dt \leq c \int_0^1 \big(1 + h'(t)\big) \log\big(1 + h'(t)\big) \, dt,$$

does not hold, if $c < \frac{2}{\pi^2}$.

Let $j(t) = \delta \sin \pi t$ where, among other things, δ is sufficiently small to ensure that $j \in C$. Substituting this function into (8) we have

$$\int_0^1 \delta^2 (\sin \pi t)^2 \, dt \leq c \int_0^1 (1 + \delta \pi \cos \pi t) \log(1 + \delta \pi \cos \pi t) \, dt. \qquad (19)$$

For small δ, the right hand side is

$$c \int_0^1 \delta \pi \cos \pi t + \frac{(\delta \pi \cos \pi t)^2}{2} \, dt \, + O(\delta^3).$$

So we can rewrite (19) as

$$\int_0^1 \delta^2 (\sin \pi t)^2 \, dt \leq c \int_0^1 \frac{(\delta \pi \cos \pi t)^2}{2} \, dt \, + O(\delta^3).$$

Dividing both sides by δ^2 and letting $\delta \to 0$, we get

$$\int_0^1 (\sin \pi t)^2 \, dt \leq c \frac{\pi^2}{2} \int_0^1 (\cos \pi t)^2 \, dt$$

which does not hold if $c < \frac{2}{\pi^2}$. $\qquad\qquad \square$

After this article had already been circulated, M. Ledoux communicated an alternative method of finding the same constant, c, in (2). His argument depends upon a reflection trick to modify the logarithmic Sobolev inequality on the interval to the periodic case for which it is known that spectral methods yield the log-Sobolev constant. This can then be transferred to a transportation constant using the methods of Otto and Villani [OV].

This work will form part of a Ph.D. thesis which is being supervised by Keith Ball.

References

[BGH] Buttazzo G., Giaquinta M., Hildebrandt S. (1998) One-Dimensional Variational Problems. Oxford Lecture Series in Mathematics and its Applications, 15, Clarendon Press, Oxford

[M]   Marton K. (1996) Bounding $\bar{d}$-distance by informational divergence: a method to prove measure concentration. The Annals of Probability 24(2):857–866

[OV]   Otto F., Villani C. (1999) Generalization of an inequality by Talagrand, and links with the logarithmic Sobolev inequality. Preprint

[T]   Talagrand M. (1996) Transportation cost for Gaussian and other product measures. Geometric and Functional Analysis 6(3):587–599

[TIS] Tsirel'son B.S., Ibragimov I.A., Sudakov V.N. (1976) Norms of Gaussian sample functions. Proceedings of the Third Japan-USSR Symposium on Probability Theory, Lecture Notes in Mathematics, 550

The Uniform Concentration of Measure Phenomenon in ℓ_p^n $(1 \leq p \leq 2)$

J. Arias-de-Reyna and R. Villa

Dpto. Análisis Matemático, Facultad de Matemáticas, Universidad de Sevilla, c/ Tarfia, s/n, 41012 Sevilla, Spain

Abstract. We prove the uniform concentration of Lebesgue measure phenomenon on the ball of ℓ_p^n for $1 \leq p \leq 2$. In particular, we give a first concentration inequality for Lebesgue measure on the ball of ℓ_1^n. An application is the lower exponential bound on the dimension of ℓ_∞ admitting an isomorphic embedding of ℓ_1^n and on the distortion of such those embeddings, proved in [L].

1 Introduction and Previous Results

It is known (see [GM2] or [ABV]) that a concentration of measure phenomenon is verified in the unit ball B of any n-dimensional uniformly convex normed space X: if

$$\delta(\varepsilon) = \inf\left\{1 - \left\|\frac{x+y}{2}\right\| : \|x - y\| \geq \varepsilon,\ \|x\| \leq 1,\ \|y\| \leq 1\right\}$$

and σ denotes the normalized Lebesgue measure restricted to B, then

$$\sigma(A_\varepsilon) > 1 - 2e^{-2\delta(\varepsilon)n}$$

for any Borel set $A \subset B$ with $\sigma(A) \geq 1/2$ and any $\varepsilon > 0$. This gives a concentration phenomenon for the space ℓ_p^n when $1 < p < \infty$, but the exponent in the right side of the inequality goes to zero when $p \to 1^+$. And, of course, this result can't be applied to the space ℓ_1^n.

In Theorem 1 we prove a uniform concentration phenomenon on the unit ball B_p^n of all the spaces ℓ_p^n $(1 \leq p \leq 2)$. In particular, we give a concentration phenomenon for the space ℓ_1^n. This fact was unknown at the present.

Related problems have been studied recently by G. Schechtman and J. Zinn. In a private communication, they gave a deviation inequality for any Lipschitz function with respect to the Euclidean metric. More precisely, they proved

$$\sigma\left(\left\{x : \left|f(x) - \int f\, d\sigma\right| > t\right\}\right) \leq C \exp(-ctn)$$

Research supported by DGICYT grant #PB96-1327. This paper includes part of the Ph.D. thesis of the second author.

for any $t > 0$ and $f : B_1^n \to \mathbb{R}$ satisfying $|f(x) - f(y)| \leq \|x - y\|_2$ for all $x, y \in B_1^n$ (c, C are absolute constants).

In the last section, we apply Theorem 1 to embedding problems into ℓ_∞^n.

To prove Theorem 1 we need two technical lemmas. Both of them are easily checked. A proof of Lemma 2 can be found in [K].

Lemma 1. *Let (X, Σ, μ) be a probability space, let f be a density on X, and let ν be the measure with density f with respect to μ. If $t > 0$ and $L_t = \{x \in X : f(x) \leq t\}$, then for any $A \in \Sigma$,*

$$\nu(A) \geq \nu(X \setminus L_t) \Longrightarrow \mu(A) \geq \mu(X \setminus L_t).$$

Lemma 2.

$$\lim_{b \to +\infty} \frac{1}{\Gamma(b)} \int_b^{+\infty} t^{b-1} e^{-t} \, dt = \frac{1}{2}.$$

Let $1 \leq p \leq 2$, let B_p^n be the unit ball of ℓ_p^n, let γ_p be the measure on $\mathbb{R}$ with density $c_p^{-1} e^{-|x|^p/p}$ with respect to Lebesgue measure ($c_p = 2\Gamma(1 + 1/p)p^{1/p}$) and let γ_p^n be the product measure on $\mathbb{R}^n$, where each factor is endowed with γ_p. Of course γ_p^n is a measure on $\mathbb{R}^n$ with density $c_p^{-n} e^{-\|x\|_p^p/p}$ with respect to Lebesgue measure. For $p = 2$, the measure γ_2^n is just the Gaussian measure.

The following result can be found essentially in a paper of Talagrand [T1] concerning Gauss space. He obtains an isoperimetric inequality for the measure γ_1^n from which it can be deduced a known inequality for the measure γ_2^n (see [P1], [B] or [P2]). A simplification of his argument was found and used by Maurey [M] to get the result. The statement given here can be found in Maurey's paper.

Theorem. *There exists a constant $\alpha > 0$ such that*

$$\gamma_1^n(A + \sqrt{t}B_2^n + tB_1^n) \geq 1 - \frac{1}{\gamma_1^n(A)} e^{-\alpha t}$$

for all $t > 0$ and all Borel subset $A \subset \mathbb{R}^n$ with $\gamma_1^n(A) > 0$.

From this theorem, and following the argument used in [T2], the following isoperimetric inequality for the measure γ_p^n can be drawn: there exists an absolute constant $K > 0$ such that, for any p, $1 \leq p \leq 2$, and any Borel subset $A \subset \mathbb{R}^n$ with $\gamma_p^n(A) > 0$

$$\gamma_p^n \left(A + \left(t^{1/2} n^{1/p-1/2} + t^{1/p} \right) B_p^n \right) \geq 1 - \frac{e^{-t/K}}{\gamma_p^n(A)}. \tag{1}$$

The main result of this paper, the uniform concentration of the normalized Lebesgue measure restricted to the unit ball of ℓ_p^n, is a consequence of the concentration of measure property for the measure γ_p^n stated in (1). One could expect this fact if realizes that both measures σ_p' (defined in the proof below) and γ_p^n are connected, in some sense. It is easy to see that, for any measurable subset $A \subset \mathbb{R}^n$,

$$\lim_{N \to \infty} \pi_N \sigma_p'(A) = \gamma_p^n(A),$$

where $\pi_N : \mathbb{R}^N = \mathbb{R}^n \times \mathbb{R}^{N-n} \to \mathbb{R}^n$ is the projection onto the first n coordinates. This fact, for the particular case $p = 2$, the Gaussian measure, is known as Poincaré's remark.

2 Uniform Concentration of Measure Phenomenon for ℓ_p^n $(1 \leq p \leq 2)$

Theorem 1. *Let $1 \leq p \leq 2$, let B_p^n be the closed unit ball of ℓ_p^n and let σ_p be the normalized Lebesgue measure restricted to B_p^n. There exist two constants $\alpha, c > 0$ such that, for any Borel set $A \subset B_p^n$ with $\sigma_p(A) \geq 1/2$, and any $0 < \varepsilon < 1$*

$$\sigma_p(A + \varepsilon B_p^n) \geq 1 - cne^{-\alpha \varepsilon^2 n}.$$

Proof. Consider the probability γ_p^n on $\mathbb{R}^n$ with density $c_p^{-n} e^{-\|x\|_p^p / p}$ with respect to Lebesgue measure ($c_p = 2\Gamma(1 + 1/p)p^{1/p}$).

For any measurable positive function $f : [0, +\infty) \to \mathbb{R}$ we have

$$c_p^{-n} \int_{\mathbb{R}^n} f(\|x\|_p)\, dx = \frac{n}{\Gamma(1 + \frac{n}{p})p^{n/p}} \int_0^\infty f(r) r^{n-1}\, dr. \qquad (2)$$

Let σ_p' be the probability on the Borel subsets of $\mathbb{R}^n$ defined by

$$\sigma_p'(A) = \frac{1}{v_{p,n} n^{n/p}} m(A \cap n^{1/p} B_p^n),$$

where $v_{p,n} = m(B_p^n) = \dfrac{(2\Gamma(1 + 1/p))^n}{\Gamma(1 + \frac{n}{p})}$ and m denotes the usual Lebesgue measure on $\mathbb{R}^n$. This probability σ_p' has a density with respect to γ_p^n, given by

$$\rho_n(x) = \begin{cases} \dfrac{\Gamma(1 + \frac{n}{p})}{(n/p)^{n/p}} e^{\|x\|_p^p / p} & \text{if } \|x\|_p \leq n^{1/p}, \\[2mm] 0 & \text{if } \|x\|_p > n^{1/p}. \end{cases}$$

For this function, the level sets L_t considered in Lemma 1 are

$$L_t = \{x \in \mathbb{R}^n : \rho_n(x) \leq t\} = r(t) B_p^n \cup (\mathbb{R}^n \setminus n^{1/p} B_p^n)$$

for each $t > 0$, where $r(t)$ satisfies $e^{r(t)^p/p} = t(n/p)^{n/p}/\Gamma(1 + \frac{n}{p})$.

Fix a Borel set $A \subset B_p^n$ such that $\sigma_p(A) \geq 1/2$. Choose $t_0 > 0$ so that $\sigma_p'(r(t_0)B_p^n) = 1/2$ (i. e., so that $r(t_0) = 2^{-1/n}n^{1/p}$). Then

$$\sigma_p'(n^{1/p}A) = \sigma_p(A) \geq 1/2 = \sigma_p'(\mathbb{R}^n \setminus L_{t_0}).$$

By Lemma 1, $\gamma_p^n(n^{1/p}A) \geq \gamma_p^n(\mathbb{R}^n \setminus L_{t_0})$. From (2) we have

$$\gamma_p^n(\mathbb{R}^n \setminus L_{t_0}) = \frac{n}{\Gamma(1 + \frac{n}{p})p^{n/p}} \int_{r(t_0)}^{n^{1/p}} r^{n-1}e^{-r^p/p} \, dr$$

$$\geq \frac{n^{n/p}e^{-n/p}}{2\Gamma(1 + \frac{n}{p})p^{n/p}} \sim \frac{1}{2}\sqrt{\frac{p}{2\pi n}}$$

and then, for large n,

$$\gamma_p^n(n^{1/p}A) > \frac{c_1}{\sqrt{n}}$$

for some absolute constant $c_1 > 0$. From the inequality given in (1), we obtain

$$\gamma_p^n\left(n^{1/p}A + \left(t^{1/2}n^{1/p-1/2} + t^{1/p}\right)B_p^n\right) \geq 1 - \frac{e^{-t/K}}{\gamma_p^n(n^{1/p}A)} > 1 - c_2\sqrt{n}e^{-t/K}$$

for any $t > 0$. Fix any $0 < \delta < 1$ and take $t = Kn\delta$. Then

$$\gamma_p^n\left(n^{1/p}\left(A + K'\delta^{1/2}B_p^n\right)\right) > 1 - c_2\sqrt{n}e^{-\delta n} \tag{3}$$

for some absolute constant $K' > 0$. Let M denote the set $n^{1/p}\left(A + K'\delta^{1/2}B_p^n\right)$. We claim that for $n \geq n_0$

$$c_3 e^{-\delta n} < 1 \implies \sigma_p'(M) \geq (1 - c_3 e^{-\delta n})^{n/p} \tag{4}$$

for some absolute constants $c_3 > 0$ and $n_0 \in \mathbb{N}$ (actually, we can take $c_3 = 8\sqrt{\pi}c_2$). Indeed, we can assume that $c_3 e^{-\delta n} < 1 - 2^{-p/n}$ since $\sigma_p'(M) \geq 1/2$. Then for large n_0 and $n \geq n_0$, we have

$$c_2\sqrt{n}e^{-\delta n} < c_3\sqrt{n}e^{-\delta n} < \sqrt{n}(1 - 2^{-p/n}) < 1/4.$$

From Lemma 2, there exists an $s(\delta)$ so that

$$c_2\sqrt{n}e^{-\delta n} = \frac{n}{p^{n/p}\Gamma(1 + \frac{n}{p})} \int_{s(\delta)}^{n^{1/p}} r^{n-1}e^{-r^p/p} \, dr$$

$$= \frac{n/p}{\Gamma(1 + \frac{n}{p})} \int_{s(\delta)^p/p}^{n/p} s^{n/p-1}e^{-s} \, ds.$$

Now we can reformulate (3) as follows

$$\gamma_p^n(M) > \gamma_p^n(L_u),$$

where $u = (n/p)^{-n/p}\Gamma(1 + \frac{n}{p})e^{s(\delta)^p/p}$. Then $\gamma_p^n(\mathbb{R}^n \setminus M) < \gamma_p^n(\mathbb{R}^n \setminus L_u)$ and therefore Lemma 1 implies $\sigma_p'(\mathbb{R}^n \setminus M) < \sigma_p'(\mathbb{R}^n \setminus L_u)$. Hence

$$\sigma_p'(M) > \sigma_p'(L_u) = \frac{s(\delta)^n}{n^{n/p}}. \tag{5}$$

We have $s(\delta)^p \geq n - c_3 n e^{-\delta n}$, for otherwise, for large n,

$$c_2\sqrt{n}e^{-\delta n} = \frac{n/p}{\Gamma(1 + \frac{n}{p})}\int_{s(\delta)^p/p}^{n/p} s^{n/p-1}e^{-s}\,ds$$

$$> \frac{n/p}{\Gamma(1 + \frac{n}{p})}\int_{\frac{n}{p} - c_3\frac{n}{p}e^{-\delta n}}^{\frac{n}{p}} s^{n/p-1}e^{-s}\,ds$$

$$> c_3\frac{n}{p}e^{-\delta n}\frac{n/p}{\Gamma(1 + \frac{n}{p})}\left(\frac{n}{p} - c_3\frac{n}{p}e^{-\delta n}\right)^{n/p-1}e^{-n/p}$$

$$> \frac{c_3\sqrt{n}e^{-\delta n}}{8\sqrt{\pi}}$$

which is impossible for $c_3 = 8\sqrt{\pi}c_2$. From (5) we obtain, for some n_0,

$$\sigma_p'(M) > \left(1 - c_3 e^{-\delta n}\right)^{n/p}$$

for any $n \geq n_0$ so that $c_3 e^{-\delta n} < 1 - 2^{-p/n}$. This proves (4). Consequently, for $n \geq n_0$

$$\sigma_p'(M) > 1 - \frac{c_3 n}{p}e^{-\delta n}$$

and finally

$$\sigma_p\left(A + K'\delta^{1/2}B_p^n\right) = \sigma_p'\left(n^{1/p}(A + K'\delta^{1/2}B_p^n)\right) \geq 1 - c_3 n e^{-\delta n}$$

for $n \geq n_0$, which implies the desired inequality for some $\alpha, c > 0$.   □

3 Application to Embedding Problems into ℓ_∞^n

Let X be an n-dimensional normed space. Let B denote the unit ball of X.

For a Borel probability measure μ defined on B, the *concentration function* φ is defined for any $\varepsilon > 0$ by

$$\varphi(\varepsilon) = \sup\left\{1 - \mu(A + \varepsilon B) : A \subseteq B \text{ Borel with } \mu(A) \geq 1/2\right\}.$$

The argument used in [GM1] allows one to prove the next theorem. Similar results can be found in [P1].

Theorem 2. *Let X be an n-dimensional normed space, B its closed unit ball and φ the concentration function of a symmetric Borel probability μ defined* on B. If there exists a d-embedding of X into ℓ_∞^N then, for any $0 < \varepsilon < 1/d$ such that $\varphi(\varepsilon) > 0$,

$$N \geq \frac{1}{2}\varphi(\varepsilon)^{-1}\left(1 - \mu(d\varepsilon B)\right).$$

Using the uniform concentration of measure property for ℓ_p^n $(1 \leq p \leq 2)$ given in Theorem 1, we can state that there exists no d-isomorphic embedding of ℓ_p^n into ℓ_∞^N for $N < \frac{1}{4cn}e^{\alpha n/4d^2}$. From this, if we define, following Larsson [L],

$$d_p(N,k) =$$

$$\inf\left\{\|T\|\,\|T^{-1}\| : T \text{ is an isomorphic embedding of } \ell_p^{[k\log N]} \text{ into } \ell_\infty^N\right\}$$

for $1 \leq p \leq 2$, then

$$\liminf_{N\to\infty} d_p(N,k) \geq \frac{\sqrt{\alpha k}}{2}.$$

This argument, simpler than the one used in [L], gives the right order of growth $\sqrt{k}$ of $d_p(N,k)$ with respect to k. For $p = 1$, the sharp lower bound $\sqrt{\frac{k}{\log 4}}$ was given in [L].

References

[ABV]   Arias-de-Reyna J., Ball K., Villa R. (1998) Concentration of the distance in finite dimensional normed spaces. Mathematica 45:245–252

[B]   Borell C. (1975) The Brunn-Minkowski inequality in Gauss space. Inventiones Math. 30:205–216

[GM1]   Gromov M., Milman V.D. (1983-1984) Brunn theorem and a concentration of volume phenomena for symmetric convex bodies. Geometric Aspects of Functional Analysis, Seminar Notes, Tel-Aviv

[GM2]   Gromov M., Milman V.D. (1987) Generalization of the spherical isoperimetric inequality to uniformly convex Banach spaces. Compositio Math. 62:263–282

[K]   Knuth D.E. (1968) The Art of Computer Programming (Vol. I) Fundamental Algorithms, Addison-Wesley, Reading Mass

[L]   Larsson J. (1986) Embeddings of ℓ_p^m into ℓ_∞^n, $1 \leq p \leq 2$. Israel J. Math. 55: 94–124

[M]   Maurey B. (1991) Some deviation inequalities. Geometric and Functional Analysis 1:188–197

[P1]   Pisier G. (1980-1981) Remarques sur un Résultat non publié de B. Maurey. Séminaire d'Analyse Fonctionnelle, Exp. V

[P2]   Pisier G. (1989) The Volume of Convex Bodies and Banach Spaces Geometry. Cambridge University Press, Cambridge

[T1]   Talagrand M. (1991) A new isoperimetric inequality and the concentration of measure phenomenon. Geometric Aspects of Functional Analysis (Israel Seminar, 1989-1990), Lecture Notes in Math. 1469:94–124

[T2]   Talagrand M. (1994) The supremum of some canonical processes. Amer. J. Math. 116:283–325

An Editorial Comment on the Preceding Paper

G. Schechtman

Department of Mathematics, Weizmann Institute of Science, Rehovot, Israel

I would like to present a more direct proof of Theorem 1 of the preceding paper [AV] of Arias-de-Reyna and Villa. I shall give the details of the proof for the most interesting case of $p = 1$ and remark at the end how to prove in a similar way the case $1 < p < 2$. I follow the notations of [AV]. Recall first a theorem of Talagrand [Tal], an equivalent form of which is also used in [AV].

Theorem. *Let $f : \mathbb{R}^n \to \mathbb{R}$ be a function satisfying*

$$|f(x) - f(y)| \leq \alpha \|x - y\|_2 \quad and \quad |f(x) - f(y)| \leq \beta \|x - y\|_1$$

Then

$$\gamma_1^n \left(|f(x) - \mathbf{E}f| > r \right) \leq C \exp(-\delta \min(r/\beta, r^2/\alpha^2)).$$

In particular,

$$\gamma_1^n \left(\left| \frac{\sum |x_i|}{n} - 1 \right| > r \right) \leq C \exp(-\delta n \min(r, r^2)).$$

Now if $f : \partial B_1^n :\to \mathbb{R}$ satisfy $|f(x) - f(y)| \leq \|x - y\|_1$ extend it to a function F on $\mathbb{R}^n$ with the same Lip constant with respect to $\|\cdot\|_1$ and note that $|F(x) - F(y)| \leq \sqrt{n}\|x - y\|_2$. Put $S = \sum |x_i|, T = \sum |y_i|$. Then, considering (x, y) as an element of $\mathbb{R}^{2n}$,

$$\gamma_1^{2n} \left(\left| F\left(\frac{x}{S}\right) - F\left(\frac{y}{T}\right) \right| > 3r \right)$$

$$\leq 2\gamma_1^n \left(\left| F\left(\frac{x}{S}\right) - F\left(\frac{x}{n}\right) \right| > r \right) + \gamma_1^{2n} \left(\left| F\left(\frac{x}{n}\right) - F\left(\frac{y}{n}\right) \right| > r \right).$$

By the $\|\cdot\|_1$-Lipschitsity of F, we get from the Theorem above that, for all $0 < r < 1$,

$$\gamma_1^n \left(\left| F\left(\frac{x}{S}\right) - F\left(\frac{x}{n}\right) \right| > r \right) \leq \gamma_1^n \left(\left| 1 - \frac{S}{n} \right| > r \right) \leq C \exp(-\delta n r^2).$$

While

$$\gamma_1^{2n} \left(\left| F\left(\frac{x}{n}\right) - F\left(\frac{y}{n}\right) \right| > r \right) \leq C \exp(-\delta n r^2)$$

since for $F(\frac{\cdot}{n})$ $\beta = 1/n$ and $\alpha = 1/\sqrt{n}$.

Supported in part by ISF.

Now, if x is distributed according to γ_1^n then x/S is distributed according to the normalized surface measure on the sphere of ℓ_1^n. This is an easy and known fact. The papers [MP] and [SZ] contain this and also a similar fact for γ_p^n (in this case the relevant measure is not the surface measure but the one induced fron the Lebesgue measure on the full ball - the measure of a set A on the sphere is the normalized Lebesgue measure of $[0,1] \times A$). In [SZ] this fact is used in a similar way to the one here. It follows that if X and Y are independent random variables distributed uniformly on the sphere of ℓ_1^n then for all $r \leq 2$,

$$\mathrm{Prob}(|f(X) - f(Y)| > r) \leq C' e^{-\delta' r^2 n}$$

from which the analog of Theorem 1 of [AV] for the sphere of ℓ_1^n easily follows. Going from the sphere to the ball is again easy. The proof for $1 < p < 2$ is very similar: use the relation, mentioned above, between γ_p^n and the normalized Lebesgue measure on the ball of ℓ_p^n and replace the use of the Theorem above with another theorem of Talagrand also used in [AV] (see (1) there). Again it is more convenient to state this theorem in its concentration form: *There are positive constants C and δ such that if $f : \mathbb{R}^n \to \mathbb{R}$ has Lipschits constant 1 with respect to $\|\cdot\|_p$, $1 \leq p \leq 2$, then*

$$\gamma_p^n(|f(x) - \mathbf{E}f| > r) \leq C \exp(-\delta \min(r^p, r^2 n^{1-2/p})).$$

References

[AV]   Arias-de-Reyna J., Villa R. The uniform concentration of measure phenomenon in ℓ_p^n ($1 \leq p \leq 2$). This volume.

[MP]   Meyer M., Pajor A. (1988) Sections of the unit ball of L_p^n. J. Funct. Anal. 80:109–123

[SZ]   Schechtman G., Zinn J. (1990) On the volume of the intersection of two L_p^n balls. Proc. A.M.S. 110:217–224

[Tal]   Talagrand M. (1991) A new isoperimetric inequality and the concentration of measure phenomenon. Geometric Aspects of Functional Analysis (1989–90), Lecture Notes in Math., 1469, Springer, 94–124

A Remark on the Slicing Problem

K. Ball

Department of Mathematics, University College London, Gower Street, London
WC1E 6BT, UK

1 Introduction

Over the last decade or so, quite a lot of effort has been expended on the
so-called slicing problem in convex geometry, which asks whether there is a
constant $\delta > 0$ independent of dimension, so that every (symmetric) convex
body of volume 1, in $\mathbf{R}^n$, has a slice of $(n-1)$-dimensional volume at least δ.
The problem has many equivalent formulations and a positive answer would
have many interesting consequences: not least, it would immediately imply a
version of the reverse Brunn-Minkowski inequality of Milman, [M]. A survey
of these reformulations can be found in the article of Milman and Pajor,
[MP].

A symmetric convex body K in $\mathbf{R}^n$ is called *isotropic* if its inertia tensor
is a multiple of the identity: that is, for some constant L,

$$\int_K \langle x, y \rangle^2 \, dx = L^2 |y|^2$$

for every $y \in \mathbf{R}^n$ (where $|y|$ is the Euclidean length of y). It is easy to see that
every convex body has an affine image which is isotropic. If K has volume 1,
the number L is called the isotropic constant of K. It is known (see [MP] or
[B1]) that the slicing problem is equivalent to the question of whether there is
a uniform upper bound, independent of dimension, on the isotropic constants
of isotropic bodies of volume 1.

To date, uniform bounds have been established on the isotropic constants
of various families of convex bodies in [MP], [B2] and [J]. Recently, [K],
the conjecture was proved for the 'natural' family of counterexamples: the
unit balls of the non-commutative ℓ_p spaces. The best estimate available for
arbitrary bodies is $n^{\frac{1}{4}}$ obtained by Bourgain in [Bou].

It was shown in the author's article [B1] that any estimate for the isotropic
constants of convex bodies in $\mathbf{R}^n$ implies an estimate on the ratio of the
volumes of n-codimensional slices of isotropic bodies (in any dimension). The
precise result is as follows.

Theorem 1. *Fix a natural number n and suppose that every isotropic symmetric convex body in $\mathbf{R}^n$, of volume 1 satisfies*

$$\int_K |x|^2 \, dx \leq nM^2.$$

Then for every $m > n$, every isotropic convex body C in $\mathbf{R}^m$ and any two n-codimensional subspaces H_1 and H_2 of $\mathbf{R}^m$,

$$\frac{vol(H_1 \cap C)}{vol(H_2 \cap C)} \leq (12M)^n.$$

(There is nothing special about the constant 12 here.)

At the time of this result, several people raised the question of whether the reverse statement might hold: whether an upper estimate on the ratios of volumes of n-codimensional slices would automatically provide an estimate of the isotropic constants for arbitrary bodies in $\mathbf{R}^n$. One reason for this question is that the problem of estimating ratios of volumes of slices looks as though it can be approached through more 'isomorphic' means than the original slicing problem. To see why this might be, observe first that there is indeed a constant which works for 1-codimensional slices. (This was proved by Hensley in [H]. It also follows from the theorem above since there is only one symmetric convex body in the real line.) Now, to pass from 1-codimensional sections to n-codimensional ones, it would be nice to use induction. Any two n-codimensional subspaces can be linked by a chain of n-codimensional subspaces, with n terms in the chain, in which adjacent subspaces belong to a common $(n-1)$-codimensional subspace. Clearly, the problem with this approach is that the sections defined by the larger subspaces need not be isotropic. But this problem looks much more amenable to 'isomorphic' methods than the original slicing problem.

The purpose of this note is to give a proof of the reverse statement. The Theorem proved here is the following.

Theorem 2. *Fix n and let M be the least constant so that for any isotropic symmetric convex body C, in any dimension, and any pair of n-codimensional subspaces H_1 and H_2,*

$$vol(H_1 \cap C) \leq M^n vol(H_2 \cap C).$$

Then for any n-dimensional isotropic symmetric convex body K of volume 1,

$$\int_K |x|^2 \, dx \leq \frac{1}{2\pi} n M^2.$$

This result is proved in Section 3 below. The fact that the proof is very short might suggest that the information provided by the result is not strong enough to be helpful in tackling the slicing problem. However, the proof makes essential use of the Central Limit Theorem (for measures on $\mathbf{R}^n$) and there are some reasons to believe that the kind of 'symmetrisation' involved in the Central Limit Theorem is very much what is needed to understand the distribution of mass in high-dimensional bodies. Section 2 of this article provides a brief discussion of the form of the Central Limit Theorem needed for the proof.

2 Pointwise Convergence in the Central Limit Theorem

Let K be an isotropic symmetric convex body of volume 1 in $\mathbf{R}^n$ and let σ^2 be the 'variance' defined by

$$\int_K |x|^2 \, dx = n\sigma^2.$$

Let (X_i) be IID random vectors, each one uniformly distributed on K. The normalised sums

$$\frac{1}{\sqrt{m}} \sum_1^m X_i$$

converge in distribution to the Gaussian on $\mathbf{R}^n$ with density g given by

$$g(x) = \frac{1}{(2\pi\sigma^2)^{\frac{n}{2}}} e^{-\frac{|x|^2}{2\sigma^2}}.$$

Let $f_m : \mathbf{R}^n \to \mathbf{R}$ be the density of

$$\frac{1}{\sqrt{m}} \sum_1^m X_i$$

so that f_m is a renormalised convolution of m copies of the indicator function of K. The convergence in distribution guarantees that if A is (say) a ball in $\mathbf{R}^n$ then

$$\int_A f_m \to \int_A g$$

as $m \to \infty$. For the purposes of the proof below it is necessary to check that there is pointwise convergence, at least at 0:

$$f_m(0) \to g(0)$$

as $m \to \infty$.

The classical approach to pointwise convergence of densities in the Central Limit Theorem, depends upon estimates for the L_p norms of the Fourier transform of the density. (See for example Feller [F] Chapter XVI.) It is intuitively obvious that the characteristic function of a convex body is smooth enough to have a Fourier transform which decays well at infinity. But it is natural to expect that more direct methods are available for such well-behaved densities. The point of this section is to observe that this is indeed the case. For the sake of clarity, the argument will only be sketched.

For K as before, the renormalised convolutions f_m have two properties that will be the important ones. Firstly, each f_m is log-concave: that is $\log f_m$ is a concave function (with the usual convention regarding $-\infty$). This is a consequence of the functional form of the Brunn-Minkowski inequality, the

Prékopa-Leindler inequality, which guarantees that the convolution of a pair of log-concave functions is again log-concave. (See eg. [B3] for a proof of the Prékopa-Leindler inequality.) Secondly, each f_m is isotropic with the same variance as the original body K: that is

$$\int_{\mathbf{R}^n} \langle x, y \rangle^2 f_m(x)\, dx = \sigma^2 |y|^2$$

for all $y \in \mathbf{R}^n$. This is just the usual remark that the variance of a sum of independent random variables is the sum of their variances.

For large values of m the average of f_m over a small ball centred at 0 will be close to the average of g over this ball, which in turn is close to $g(0)$. So, in order to show that $f_m(0) \to g(0)$ it suffices to check that $f_m(0)$ cannot be very different from the average of f_m over a small ball centred at 0. Since f_m is an even log-concave function, it attains its maximum at 0. So $f_m(0)$ certainly cannot be smaller than the average of f_m over a ball. The problem is to show that $f_m(0)$ cannot be too large (and this is what is needed later).

Suppose then, that $f_m(0)$ is significantly larger than the average of f_m over a small ball. Then there is a point e, close to zero, with $f_m(e)$ significantly less than $f_m(0)$. Since f_m is log-concave, this forces f_m to be small at all points λe with $\lambda \geq 1$. Since f_m is even, the same is true if $\lambda \leq -1$. As a result, the integral of f_m along the line spanned by e,

$$\int_{-\infty}^{\infty} f_m \left(r \frac{e}{|e|} \right) dr$$

will be small.

To be precise for a moment, the above remarks show that, if $f_m(0)$ does not converge to $g(0)$, then for every $\epsilon > 0$ there is an m and a unit vector u for which

$$\int_{-\infty}^{\infty} f_m (ru)\, dr < \epsilon.$$

Let's see why this can't happen. Let H be the subspace of $\mathbf{R}^n$ orthogonal to u. Define a new function h on H by integrating f_m along the fibres parallel to u:

$$h(x) = \int_{-\infty}^{\infty} f_m (x + ru)\, dr.$$

Then h is an even, log-concave density on H which is isotropic with the same variance as f_m, in the sense that

$$\int_H \langle x, y \rangle^2 h(x)\, dx = \sigma^2 |y|^2$$

for all y in H. Therefore

$$\int_H |x|^2 h(x)\, dx = (n-1)\sigma^2 \tag{1}$$

and the statement to be contradicted is that

$$h(0) < \epsilon.$$

The argument is now clear: because h attains its maximum at zero, it is easy to see that the expression

$$h(0)^{\frac{2}{n-1}} \int_H |x|^2 h(x)\, dx$$

is at least as large as it would be if h were replaced by the characteristic function of the Euclidean ball of volume 1, centred at 0. So equation (1) gives a lower bound for $h(0)$, depending only upon n and σ.

3 The Proof of Theorem 2

Assume that for any isotropic symmetric convex body C, in any dimension, and any pair of n-codimensional subspaces H_1 and H_2,

$$\text{vol}(H_1 \cap C) \leq M^n \text{vol}(H_2 \cap C).$$

Let K be an n-dimensional isotropic symmetric body of volume 1 and set

$$\int_K |x|^2\, dx = n\sigma^2.$$

For each m regard $\mathbf{R}^{mn}$ as $(\mathbf{R}^n)^m$ and consider the body $C = C_m$ in $\mathbf{R}^{mn}$ consisting of the m-fold Cartesian product of K with itself:

$$C = K \times K \times \ldots K.$$

Clearly C is isotropic and of volume 1. C has an n-codimensional section of volume 1 obtained by dropping one factor of K. Therefore, every n-codimensional section of C has volume at least $\frac{1}{M^n}$. Now consider the section of C by the n-codimensional space consisting of sequences $(x_1, x_2, \ldots, x_m)$, of vectors in $\mathbf{R}^n$, satisfying $x_1 + x_2 + \ldots + x_m = 0$. The volume of this section is the value

$$f_m(0)$$

where f_m is the renormalised convolution of m copies of the characteristic function of K, as in the preceding section. By the pointwise Central Limit Theorem explained in that section, $f_m(0)$ is approximately

$$\frac{1}{(2\pi\sigma^2)^{\frac{n}{2}}}$$

when m is large. Since

$$f_m(0) \geq \frac{1}{M^n}$$

for every m, this implies that

$$\frac{1}{(2\pi\sigma^2)^{\frac{n}{2}}} \geq \frac{1}{M^n}$$

and hence that

$$\sigma^2 \leq \frac{1}{2\pi} M^2,$$

which is what was wanted.

References

[B1]   Ball K.M. (1988) Logarithmically concave functions and sections of convex sets in $\mathbf{R}^n$. Studia Mathematica 88:69–84

[B2]   Ball K.M. (1991) Normed spaces with a weak Gordon-Lewis property. In: Odell, Rosenthal (Eds.) U.T. Functional Analysis Seminar (1987-88), Lecture Notes 1470, Springer-Verlag, 36–47

[B3]   Ball K.M. (1997) An elementary introduction to modern convex geometry. In: Levy S. (Ed.) Flavors of Geometry, Mathematical Sciences Research Institute Publications 31, Cambridge University Press

[Bou]   Bourgain J. (1991) On the distribution of polynomials on high-dimensional convex sets. In: Lindenstrauss J., Milman V.D. (Eds.) Israel Seminar on GAFA, Lecture Notes 1469, Springer-Verlag, 127–137

[F]   Feller W. (1966) An Introduction to Probability Theory and its Applications, Vol. 2, Wiley

[H]   Hensley D. (1980) Slicing convex bodies-bounds for slice area in terms of the body's covariance. Proc. Amer. Math. Soc. 79:619–625

[J]   Junge M. (1994) On the hyperplane conjecture for quotient spaces of L_p. Forum Math. 6:617–635

[K]   König H., Meyer M., Pajor A. (1998) The isotropy constants of the Schatten classes are bounded. Math. Annalen 312:773–783

[M]   Milman V.D. (1986) Inégalité de Brunn-Minkowski inverse et applications à la théorie locale des éspaces normées. C. R. Acad. Sci. Paris 302:25–28

[MP]   Milman V.D., Pajor A. (1989) Isotropic position and inertia ellipsoids and zonoids of the unit ball of an n-dimensional normed space. In: Lindenstrauss J., Milman V.D. (Eds.) Israel Seminar on GAFA, Lecture Notes 1376, Springer-Verlag, 107–131

Remarks on the Growth of L^p-norms of Polynomials

S.G. Bobkov

Department of Mathematics, Syktyvkar University, 167001 Syktyvkar, Russia

Abstract. We study the behaviour of constants in Khinchine-Kahane-type inequalities for polynomials in random vectors which have logarithmically concave distributions.

It is well known that polynomials on $\mathbf{R}^n$ of bounded degree satisfy dimension free Khinchine-Kahane-type inequalities with respect to the uniform distribution on convex sets. The best known result, which is due to J. Bourgain [Bou] and which gave an affirmative answer to a conjecture of V. D. Milman, indicates the following: there exist universal constants $t_0 > 0$ and $c \in (0,1)$ such that, for every convex set $K \subset \mathbf{R}^n$ of volume one, every polynomial $f = f(x_1, \ldots, x_n)$ of degree $d \geq 1$, we have a distribution inequality

$$\mu_K\{|f| \geq \|f\|_1 t\} \leq \exp\{-t^{c/d}\}, \quad t \geq t_0, \tag{1}$$

where μ_K is the Lebesgue measure on K, and $\|f\|_1$ is L^1-norm of f with respect to μ_K. As usual, for $p > 0$, one denotes $\|f\|_p = (\int |f|^p \, d\mu)^{1/p}$ which also refers to some probability measure μ on a space where the function f is defined. The inequality (1) may also be written in terms of a suitable Orlicz norm. For $\alpha \geq 1$, set $\psi_\alpha(t) = \exp\{t^\alpha\} - 1, t \geq 0$, and introduce the associated norm

$$\|f\|_{\psi_\alpha} = \inf\left\{\lambda > 0 : \int \psi_\alpha(|f|/\lambda) \, d\mu \leq 1\right\}.$$

Then, (1) is equivalent to the inequality

$$\|f\|_{\psi_{c/d}} \leq C^d \|f\|_1, \tag{2}$$

with some universal $c \in (0,1)$ and $C > 0$. In particular, this yields the equivalence between L^p and L^1-norms in the form of the Khinchine-Kahane-type inequality

$$\|f\|_p \leq C(d,p) \|f\|_1, \tag{3}$$

where $C(d,p)$ depends on d and p, only. The case $d = 1$, with the best constant $c = 1$ in (2), was settled before by M. Gromov and V. D. Milman [G-M1] (cf. also [M-S]). It is therefore interesting to know whether or not the inequality

The work was partially supported by a grant of the Russian Foundation for Fundamental Research.

(2) with $c = 1$ holds true in the general case $d \geq 1$. The latter is equivalent to the statement that the inequality (3) holds in the range $p \geq 1$ with constants $C(d,p) = (Cdp)^d$, for some universal C. In this note, we would like to show, following R. Kannan, L. Lovász and M. Simonovits [K-L-S], a short proof and a refinement of such a statement involving more general classes of probability measures on $\mathbf{R}^n$.

Theorem 1. *With respect to an arbitrary log-concave probability measure μ on $\mathbf{R}^n$, for every polynomial f on $\mathbf{R}^n$ of degree $d \geq 1$, we have, for some universal C,*

$$\|f\|_{\psi_{1/d}} \leq C^d \|f\|_0. \tag{4}$$

Here, $\|f\|_0 = \lim_{p \to 0} \|f\|_p = \exp\left\{\int \log|f|\, d\mu\right\}$. The log-concavity of μ means (cf. [Bor]) that μ is supported by some affine subspace E of $\mathbf{R}^n$ where it has a logarithmically concave density with respect to Lebesgue measure on E, i.e., a density $u : E \to [0, +\infty)$ such that, for all $x, y \in E$, and all $t, s \geq 0$ with $t + s = 1$,

$$u(tx + sy) \geq u(x)^t u(y)^s.$$

As a main statement (Theorem 1.3 in [Bou]), J. Bourgain formulated the inequality (2) with, however, a constant C instead of C^d. As we believe, the power d was lost when deriving (2) from (1). That the power of a constant cannot be omitted can be seen on the example of the set $K = [0,1]$ with the function $f = 1$ and with growing d. On the other hand, the power can be hidden, if one wishes to rewrite for example the inequality (4) in the equivalent form as

$$\left\| |f|^{1/d} \right\|_{\psi_1} \leq C \left\| |f|^{1/d} \right\|_0.$$

(with the same constant). This inequality is very much similar to what is known for norms on $\mathbf{R}^n$ in the place of $|f|^{1/d}$ (cf. [L]).

To prove (4), we use Theorem 2.7 from [K-L-S] which is based on the localization lemma of L. Lovász and M. Simonovits [L-S]. It reduces multi-dimensional Khinchine-Kahane-type inequalities with respect to log-concave measures to dimension one that was illustrated in [K-L-S] on the example of inequalities of the form (3) in the linear case $d = 1$. It was also mentioned there that the method applies as well to the general case $d \geq 1$ and allows one to recover Bourgain's theorem. This is completely true up to a remark that the one-dimensional case in (3) still requires and deserves a special careful consideration, since it determines behaviour of constants as a function of d and p. The localization lemma of [L-S] has also applications to other types of inequalities. In connection with isoperimetric inequalities on the sphere, a kind of localization technique was developed by M. Gromov and V. D. Milman in [G-M2], cf. also [A].

For the Gaussian measure on the real line $\mathbf{R}$, using expansions over Hermit polynomials, the inequality (3) with $p = 2$ was studied by Yu. V. Prokhorov

[P1]; he obtained as well similar inequalities when μ belongs to the family of $\Gamma(\alpha)$-distributions with the parameter α growing with the degree d, cf. [P2]. As shown in these works, in both cases, the constants $C(d, 2)$ grow exponentially, and this fact cannot be recovered by an inequality such as (2) or (4). Combining Prokhorov's approach with the localization method, one can prove:

Theorem 2. *With respect to an arbitrary log-concave probability measure μ on $\mathbf{R}^n$, for every polynomial f on $\mathbf{R}^n$ of degree $d \geq 1$, the Khinchine-Kahane-type inequality (3) holds with*

$$C(d, p) = p^{Cd}, \quad p \geq 2,$$

where C is universal.

Consider the example of the exponential measure $\mu = \nu$ on $\mathbf{R}$ with density $\frac{d\nu(x)}{dx} = e^{-x}$, $x > 0$, and take $f(x) = x^d$. Then, $\|f\|_p = \Gamma(dp + 1)^{1/p}$ and, by Stirling's formula,

$$c_1 \frac{p^d}{d^{(p-1)/(2p)}} \|f\|_1 \leq \|f\|_p \leq c_2 \frac{p^d}{d^{(p-1)/(2p)}} \|f\|_1,$$

for some $c_1, c_2 > 0$. Therefore, when p is fixed and d grows, the constant $C(d, p) = p^{Cd}$ gives a correct exponential rate of increase.

To compare with Theorem 1, note that, up to a universal constant, the inequality (4) is equivalent to

$$\|f\|_p \leq (Cdp)^d \|f\|_0, \quad p \geq 1/d. \tag{5}$$

Here, since we have replaced $\|f\|_1$ with a smaller quantity $\|f\|_0$, the constants have a different order. Again, for $f(x) = x^d$, we have $\|f\|_0 = \|x\|_0^d$ with respect to ν, so

$$(c_1 dp)^d \|f\|_0 \leq \|f\|_p \leq (c_2 dp)^d \|f\|_0,$$

for some $c_1, c_2 > 0$. One could also test sharpness of (5) on the example of the uniform distribution on the ℓ^1-ball in $\mathbf{R}^n$ for $f(x) = x_1^d$ and with growing dimension n.

Proof of Theorem 1. First let us comment on the one-dimensional case in (5). Motivated by the results of [P1][P2], inequalities of the form (5), with respect to an arbitrary probability measure μ on $\mathbf{R}$, were studied in [B-G]. As was observed there, given $p > 0$ and $d \geq 1$, the optimal constant $C = C(d, p; \mu)$ in

$$\|f\|_p \leq C \|f\|_0, \tag{6}$$

where f is an arbitrary polynomial of degree d on $\mathbf{R}$ with complex coefficients, is given by

$$C^{1/d} = \sup_{z \in \mathbf{C}} \frac{\|x - z\|_{dp}}{\|x - z\|_0}. \tag{7}$$

Since the argument is straightforward, let us recall it. A remarkable feature of the functional $\|f\|_0$ is its multiplicativity property: $\|f_1 \ldots f_d\|_0 = \|f_1\|_0 \ldots \|f_d\|_0$. Therefore, writing $f(x) = A(x - z_1) \ldots (x - z_d)$ and applying Hölder's inequality, we get

$$\|f\|_p \leq A \prod_{k=1}^{d} \|x - z_k\|_{pd} \leq CA \prod_{k=1}^{d} \|x - z_k\|_0 = C\|f\|_0,$$

where C is defined according to (7). This proves (6) with this constant which cannot be improved as the example of the polynomials $f(x) = x - z$ shows.

It might be helpful to note that the sup in (7) can be restricted to the real line $\mathbf{R}$. Indeed, write $z = a + bi$ so that $|x - z|^2 = \xi + t$ with $\xi = (x - a)^2 \geq 0$ and $t = b^2 \geq 0$. As can easily be verified by differentiation, for any $p > q$, the function of the form $g(t) = \|\xi + t\|_p / \|\xi + t\|_q$ is non-increasing in $t \geq 0$, hence it is maximized at $t = 0$. Therefore, as a function of b, the value of

$$\frac{\|x - z\|_p}{\|x - z\|_q} = \frac{\|\xi + t\|_{p/2}^{1/2}}{\|\xi + t\|_{q/2}^{1/2}}$$

is maximized at $b = 0$. So, we may apply this observation with $q = 0$ to (7).

In the particular case, where μ is log-concave, the right hand side of (7) can be bounded by a quantity which is independent of μ and grows like Cdp. Indeed, after shifting, one needs to estimate an optimal constant C in

$$\| |x| \|_{dp} \leq C^{1/d} \| |x| \|_0, \quad p \geq 1/d, \tag{8}$$

with respect to a log-concave measure on $\mathbf{R}$. The fact that such an inequality holds for an arbitrary log-concave measure μ on $\mathbf{R}^n$, and for an arbitrary norm $\|x\|$ instead of $|x|$ was established by R. Latala [L] (cf. also [B] and [Gu] for different proofs). More precisely, he showed that, for some universal C_1, we always have

$$\| \|x\| \|_1 \leq C_1 \| \|x\| \|_0.$$

On the other hand, it had been known, as an application of Borell's lemma [Bor], that, for $p \geq 1$,

$$\| \|x\| \|_p \leq C_2 p \| \|x\| \|_1.$$

These two inequalities give $\| \|x\| \|_p \leq C_0 p \| \|x\| \|_0$. Thus, the constant in (8), $C^{1/d}$, is bounded by $C_0 dp$, for some universal C_0. This proves (5) with $C = C_0$.

To treat the multidimensional case in (5), one may assume that μ is absolutely continuous. As proved in [K-L-S], Theorem 2.7, given non-negative continuous functions f_1, f_2, f_3 and f_4 on $\mathbf{R}^n$, and the numbers $\alpha, \beta > 0$, the following two properties are equivalent:

(a) for every absolutely continuous log-concave probability measure μ on $\mathbf{R}^n$,

$$\left(\int f_1 \, d\mu\right)^\alpha \left(\int f_2 \, d\mu\right)^\beta \leq \left(\int f_3 \, d\mu\right)^\alpha \left(\int f_4 \, d\mu\right)^\beta ; \qquad (9)$$

(b) for every non-degenerate interval $\Delta \subset \mathbf{R}^n$ with directional vector $v \in \mathbf{R}^n$, and for every $\lambda \in \mathbf{R}$,

$$\left(\int_\Delta e^{\lambda\langle v,x\rangle} f_1(x)\, dx\right)^\alpha \left(\int_\Delta e^{\lambda\langle v,x\rangle} f_2(x)\, dx\right)^\beta$$
$$\leq \left(\int_\Delta e^{\lambda\langle v,x\rangle} f_3(x)\, dx\right)^\alpha \left(\int_\Delta e^{\lambda\langle v,x\rangle} f_4(x)\, dx\right)^\beta,$$

where dx stands for the Lebesgue measure on Δ.

Thus, (9) reduces to the case where the measure μ is supported by a line and, moreover, represents there the restriction of the exponential measure $e^{\lambda\langle v,x\rangle}\, dx$ to the interval Δ. Applying the above equivalence to $\alpha = 1/p$, $\beta = 1/q$ and to the functions $f_1 = |f|^p$, $f_4 = C^q|f|^q$ ($C > 0$), and $f_2 = f_3 = 1$, R. Kannan, L. Lovász and M. Simonovits made the following striking conclusion which we state here as a lemma.

Lemma 1. *Let $p > q > 0$ and $C \geq 1$. Given a continuous function f on $\mathbf{R}^n$, the inequality*

$$\|f\|_p \leq C\|f\|_q \qquad (10)$$

holds true with respect to all log-concave probability measures μ on $\mathbf{R}^n$ if and only if it holds on all intervals in $\mathbf{R}^n$ with respect to the normalized exponential measures.

By continuity, one can clearly consider in (10) the case $q = 0$, as well.

Now, if f is polynomial, its restriction to every line is again a polynomial of the same degree but of one variable. Since the restrictions of the exponential measures are log-concave, the inequality (10) thus reduces to the one-dimensional case.

One may therefore conclude that the inequality (5) holds with $C = C_0$ for all polynomials f on $\mathbf{R}^n$ of degree d in the range $p \geq 1/d$. Applying (5) to $p = k/d$, $k = 1, 2, \ldots$, we get

$$\| |f|^{1/d} \|_k \leq C_0 k \|f\|_0^{1/d}.$$

Finally, by Taylor's expansion and using $k^k \leq e^k k!$, we obtain that

$$\int \psi_{1/d}\left(\frac{f}{(2eC_0)^d \|f\|_0}\right) d\mu \leq 1.$$

Proof of Theorem 2. First, we consider the growth of the constants $C(d, 2)$ in (3) in the case $p = 2$. We will now use Lemma 1, with $p = 2$ and $q = 1$, in full volume that gives more than just a reduction of the multidimensional inequality (10) to dimension one. Let Δ be a non-degenerate interval in $\mathbf{R}^n$ with endpoints a, b and with the directional vector $v = (b-a)/|b-a|$. Then, with respect to the normalized exponential measures on Δ, the norms in (10) are given by

$$\|f\|_p^p = \frac{1}{\int_0^{|b-a|} e^{\lambda x}\, dx} \int_0^{|b-a|} |f(a+xv)|^p e^{\lambda x}\, dx.$$

When f is a polynomial on $\mathbf{R}^n$, $f(a+xv)$ represents a polynomial in $x \in \mathbf{R}$ of the same degree. Moreover, after rescaling, it suffices to consider the case $\lambda = -1$. Therefore, $C(d, 2)$ is the optimal constant C in the inequality

$$\left(\int |f|^2\, d\nu_u \right)^{1/2} \leq C \int |f|\, d\nu_u \tag{11}$$

for the class of all polynomials f on $\mathbf{R}$ of degree d with respect to all measures ν_u on $(0, +\infty)$ with densities

$$\frac{d\nu_u(x)}{dx} = \frac{e^{-x}}{1-e^{-u}}\, 1_{(0,u)}(x), \quad u > 0.$$

The limit case represents the exponential measure $\nu_{+\infty} = \nu$ on $(0, +\infty)$ with density e^{-x}, $x > 0$. For a related family of densities, $x^\alpha e^{-x}/\Gamma(\alpha+1)$, the inequality (11), with exponentially increasing constants, was proved by Yu. V. Prokhorov [P2]. He assumed that $\alpha \geq c_0 d$ (for a numerical c_0), but his approach proposed before in [P1] actually works in a more general situation and can be applied in particular to the measures ν_u. Below, to prove (11), we follow Prokhorov's scheme of the proof and simplify his argument about Laguerre's polynomials.

Step 1: $0 < u < 8d$.

Let f be an arbitrary polynomial on $\mathbf{R}$ of degree d (with real coefficients) such that $\|f\|_2 = 1$ where L^2-norm is understood with respect to ν_u. Let $x_0 \in [0, u]$ be such that $|f(x_0)| = \|f\|_\infty = \max_{x \in [0,u]} |f(x)|$, and assume, without loss of generality, that $f(x_0) > 0$. By Taylor's expansion and by Markov's inequality $\|f'\|_\infty \leq \frac{2d^2}{u} \|f\|_\infty$, we get, for every point $x \in [0, u]$,

$$f(x) \geq f(x_0) - \|f'\|_\infty |x - x_0|$$
$$\geq f(x_0) - \frac{2d^2}{u} \|f\|_\infty |x - x_0| = \left(1 - \frac{2d^2}{u} |x - x_0| \right) \|f\|_\infty.$$

Therefore, in the interval $\delta = [x_1, x_2] \equiv [x_0 - u/(4d^2), x_0 + u/(4d^2)] \cap [0, u]$, we have $f(x) \geq \frac{1}{2} \|f\|_\infty$ so that

$$\|f\|_1 \geq \int_\delta f(x)\, d\nu_u(x) \geq \frac{1}{2} \|f\|_\infty\, \nu_u(\delta).$$

In addition, since $x_2 - x_1 \geq u/(4d^2)$, for some middle point $x_3 \in [x_1, x_2]$, we get

$$\nu_u(\delta) = \frac{e^{-x_1} - e^{-x_2}}{1 - e^{-u}} = \frac{x_2 - x_1}{1 - e^{-u}} e^{-x_3} \geq \frac{1}{4d^2} \frac{u}{1 - e^{-u}} e^{-8d} \geq \frac{1}{4d^2} e^{-8d}.$$

Hence, $8d^2 e^{8d} \|f\|_1 \geq \|f\|_\infty \geq \|f\|_2$ so that (11) is fulfilled with $C = 8d^2 e^{8d}$.

The second step requires some preparation.

Lemma 2. *For every polynomial f on $\mathbf{R}$ of degree $d \geq 1$,*

$$\int_{8d}^{+\infty} |f(x)|^2 e^{-x} \, dx \leq \frac{1}{2} \int_0^{+\infty} |f(x)|^2 e^{-x} \, dx.$$

Proof. Assume that $\int_0^{+\infty} |f(x)|^2 e^{-x} \, dx = \|f\|_2^2 = 1$ (with respect to ν) and introduce the Laguerre polynomials

$$L_k(x) = \frac{e^x}{k!} \frac{d^k}{dx^k} (x^k e^{-x}) = \sum_{j=0}^k (-1)^j C_k^j \frac{x^j}{j!}, \quad k = 0, 1, \dots. \qquad (12)$$

They form a complete orthonormal system of functions in $L^2(\nu)$ so that there exists a representation $f = \sum_{k=0}^d a_k L_k$ with $\sum_{k=0}^d |a_k|^2 = 1$. Hence, $|f|^2 \leq \sum_{k=0}^d |L_k|^2$ so that

$$\|f\|_4^2 = \| |f|^2 \|_2 \leq \sum_{k=0}^d \| |L_k|^2 \|_2 = \sum_{k=0}^d \|L_k\|_4^2. \qquad (13)$$

According to (12) and since $(4j)! \leq 4^{4j} j!^4$, we get

$$\|L_k\|_4 \leq \sum_{j=0}^k C_k^j \frac{\|x^j\|_4}{j!} = \sum_{j=0}^k C_k^j \frac{(4j)!^{1/4}}{j!} \leq \sum_{j=0}^k C_k^j 4^j = 5^k.$$

Thus, by (13), $\|f\|_4^2 \leq \sum_{k=0}^d 25^k < \frac{25^{d+1}}{24}$. Now, by Cauchy-Schwarz inequality,

$$\int_{8d}^\infty |f(x)|^2 e^{-x} \, dx \leq \|f\|_4^2 \, \|1_{[8d,+\infty)}\|_2 \leq \frac{25^{d+1}}{24} e^{-4d} < \frac{1}{2}.$$

Step 2: $u \geq 8d$.

Again, let f be an arbitrary polynomial on $\mathbf{R}$ of degree d such that $\|f\|_2 = 1$ where L^2-norm is with respect to ν_u. Now, our basic interval will be $[0, 8d]$. Let x_0 be a point of maximum of $|f|$ on this interval, and assume that $f(x_0) > 0$. Again, by Taylor's expansion and by Markov's inequality, for every point $x \in [0, 8d]$,

$$f(x) \geq f(x_0) - \|f'\|_\infty |x - x_0| \geq \left(1 - \frac{d}{4} |x - x_0|\right) \|f\|_\infty$$

(L^∞-norm is taken on $[0, 8d]$). Therefore, in the interval

$$\delta = [x_1, x_2] \equiv [x_0 - 2/d, x_0 + 2/d] \cap [0, 8d]$$

we have $f(x) \geq \frac{1}{2} \|f\|_\infty$ so that

$$\|f\|_1 \geq \int_\delta f(x)\, d\nu_u(x) \geq \frac{1}{2} \|f\|_\infty\, \nu_u(\delta). \tag{14}$$

On the other hand, since $u \geq 8d$ and by Lemma 2,

$$\int_0^{8d} |f|^2\, d\nu_u = \frac{1}{1 - e^{-u}} \int_0^{8d} |f(x)|^2\, e^{-x}\, dx \geq \frac{1}{2(1 - e^{-u})} \int_0^{+\infty} |f(x)|^2\, e^{-x}\, dx$$

$$\geq \frac{1}{2(1 - e^{-u})} \int_0^u |f(x)|^2\, e^{-x}\, dx = \frac{1}{2}.$$

Therefore, $\frac{1}{2} \leq \|f\|_\infty^2\, \nu_u([0, 8d]) \leq \|f\|_\infty^2$. Combining with (14), we get $\|f\|_1 \geq \frac{1}{2\sqrt{2}} \nu_u(\delta)$. Using $x_2 - x_1 \geq 2/d$, we obtain that, for some middle point $x_3 \in [x_1, x_2]$,

$$\nu_u(\delta) = \frac{e^{-x_1} - e^{-x_2}}{1 - e^{-u}} = \frac{x_2 - x_1}{1 - e^{-u}} e^{-x_3} \geq \frac{2}{d} e^{-8d}.$$

Hence, $\sqrt{2}d e^{8d} \|f\|_1 \geq \|f\|_2$ so that (11) is fulfilled with $C = \sqrt{2}d e^{8d}$. Note that this constant is majorized by the constant $8d^2 e^{8d}$ ($\leq e^{11d}$) obtained on the first step.

Thus, for every polynomial f on $\mathbf{R}^n$ of degree d, with respect to an arbitrary log-concave probability measure μ on $\mathbf{R}^n$,

$$\|f\|_2 \leq e^{11d} \|f\|_1. \tag{15}$$

It remains to consider the general case $p \geq 2$, in order to complete the proof of Theorem 2. One can iterate an inequality of the form (15), $\|f\|_2^2 \leq A^d \|f\|_1^2$, starting from f and successively applying it to the polynomials $f, f^2, f^4, \ldots, f^{2^k}$. This yields

$$\|f\|_{2^k}^{2^k} \leq A^{k2^{k-1}d} \|f\|_1^{2^k}, \quad k \geq 1.$$

Assume $\|f\|_1 = 1$ and pick up $k \geq 1$ such that $2^k \leq p < 2^{k+1}$. Then,

$$\|f\|_p \leq \|f\|_{2^{k+1}} \leq A^{(k+1)d/2} \leq A^{kd} \leq p^{d \log A / \log 2}.$$

According to (15), we can apply these estimates with $A = e^{22}$ and thus get the statement of Theorem 2 with $C = 22/\log 2$.

Acknowledgment

I am grateful to V.D. Milman and Yu.V. Prokhorov for stimulating discussions.

References

[A]   Alesker S. (1999) Localization technique on the sphere and the Gromov-
         Milman theorem on the concentration phenomenon on uniformly convex
         sphere. Convex Geometric Analysis (Berkeley, CA, 1996), Math. Sci. Res.
         Inst. Publ., 34, Cambridge Univ. Press, Cambridge

[B]   Bobkov S.G. (1998) Isoperimetric and analytic inequalities for log-concave
         probability measures. Preprint. Ann. Probab., to appear

[B-G]   Bobkov S.G., Götze F. (1997) On moments of polynomials. Probab. Theory
         Appl. 42(3):518–520

[Bor]   Borell C. (1974) Convex measures on locally convex spaces. Ark. Math.
         12:239–252

[Bou]   Bourgain, J. (1991) On the distribution of polynomials on high dimensional
         convex sets. Lecture Notes in Math. 1469:127–137

[Gu]   Guédon O. (1998) Kahane-Khinchine type inequalities for negative expo-
         nents. Preprint. Mathematika, to appear

[G-M1]   Gromov M., Milman V.D. (1983-84) Brunn theorem and a concentration
         of volume phenomena for symmetric convex bodies. GAFA Seminar Notes,
         Tel Aviv University, Israel

[G-M2]   Gromov M., Milman V.D. (1987) Generalization of the spherical isoperi-
         metric inequality to uniformly convex Banach spaces. Composition Math.
         62:263–282

[K-L-S] Kannan R., Lovász L., Simonovits M. (1995) Isoperimetric problems for
         convex bodies and a localization lemma. Discrete and Comput. Geom.
         13:541–559

[L]   Latala R. (1996) On the equivalence between geometric and arithmetic
         means for logconcave measures. Convex Geometric Analysis, Berkeley, CA,
         123–127

[L-S]   Lovász L., Simonovits M. (1993) Random walks in a convex body and an
         improved volume algorithm. Random Structures and Algorithms 4(3):359–
         412

[M-S]   Milman V.D., Schechtman G. (1986) Asymptotic theory of finite dimen-
         sional normed spaces. Lecture Notes in Math. 1200, Springer, Berlin

[P1]   Prokhorov Yu.V. (1992) On polynomials in normally distributed random
         variables. Probab. Theory Appl. 37(4):692–694

[P2]   Prokhorov Yu.V. (1993) On polynomials in random variables that have the
         gamma distribution. Probab. Theory Appl. 38(1):198–202

Positive Lyapounov Exponents for Most Energies

J. Bourgain

Institute for Advanced Study, Princeton, NJ 08540, USA

0   Summary

Consider the ID lattice Schrödinger operator

$$H = \lambda \cos(2\pi n^\rho \omega + \theta)\delta_{nn'} + \Delta \tag{1}$$

with $1 < \rho < 2$ and define

$$\gamma(E, \lambda) = \lim_{N \to \infty} \frac{1}{N} \int_{\mathbb{T}} \log \left\| \prod_N^0 \begin{pmatrix} E - \lambda\cos(2\pi n^\rho \omega + \theta) & -1 \\ 1 & 0 \end{pmatrix} \right\| d\theta .$$

If $\lambda > 2$, M. Herman's [H] argument implies that $\gamma(E, \lambda) \geq \log \frac{\lambda}{2} > 0$, for all E. We are interested here in small λ and show that for all $E \in \mathcal{E}_\lambda \subset [-2, 2]$,

$$\mathrm{mes}\,([-2, 2]\backslash\mathcal{E}_\lambda) \xrightarrow{\lambda \to 0} 0 \tag{2}$$

we have that $\gamma(E, \lambda) > 0$. See Proposition 4.

Considering the skew shift on $\mathbb{T}^2$

$$T(x, y) = (x + y, y + \omega) \tag{3}$$

and the Hamiltonian

$$H = \lambda \cos\left(\pi_1 T^n(x, y)\right)\delta_{nn'} + \Delta \tag{4}$$

where

$$\pi_1 T^n(x, y) = x + ny + \frac{n(n-1)}{2}\omega$$

we show that the Lyapounov exponent

$$\gamma(E, \lambda) \overset{\text{a.e.}}{=} \lim_{N \to \infty} \frac{1}{N} \log \left\| \prod_N^0 \begin{pmatrix} E - \lambda\cos \pi_1 T^n(x, y) & -1 \\ 1 & 0 \end{pmatrix} \right\|$$

is strictly positive for $E \in \mathcal{E}_\lambda \subset [-2, 2]$ satisfying (2), provided we assume in (3) that

$$|\omega| < \varepsilon(\lambda).$$

See Proposition 5.

The method is based on a local approximation of (1), (4) by the almost Mathieu model

$$H_{\alpha,\lambda,\theta} = \lambda \cos(2\pi\alpha + \theta)\delta_{nn'} + \Delta \tag{5}$$

and uses the fact (see Corollary 3) that for λ small and all $E \in \mathcal{E}_\lambda \subset [-2, 2]$ satisfying (2),

$$\int_{\mathbb{T}} \gamma(\alpha, \lambda, E)d\alpha > 0 \tag{6}$$

where $\gamma(\alpha, \lambda, E)$ refers to the Lyapounov exponents of (5). The proof of (6) does rely on the Aubry duality, [A-A], [La]).

Added in Proof. Concerning lattice Schrödinger operators of the form (1), related references were pointed out to the author by Y. Last. First, it is shown in the paper [L-S] that $H = \lambda \cos(n^\rho) + \Delta$ on $\mathbb{Z}_+$ has no absolutely continuous spectrum for $\lambda > 2$, $\rho > 1$. In fact, Theorem 1.4 from [L-S] provides an alternative proof of Proposition 4 in this paper. Other numerical and heuristic studies appear in [G-F],[B-F]. The particular case $1 < \rho < 2$ was studied in [Th]. See [L-S] for further details.

1 Spectrum of Almost Mathieu Operator with Small Disorder

Denote Δ the lattice Laplacian

$$\Delta_{n,n'} = \delta_{|n-n'|=1}.$$

Lemma 1. *Assume $\lambda \neq 0$ small enough and $\omega \in \mathbb{T}$ s.t.*

$$\|k\omega\| \equiv \text{dist}(k\omega, 2\pi\mathbb{Z}) > \lambda^{\frac{1}{100}} \text{ for } 0 < |k| < 10^6. \tag{0.1}$$

Define

$$E(\omega) = \cos\left(\omega - \frac{\lambda^2}{\sin\omega}\left(\frac{1}{1-\cos\omega} + \frac{1}{\cos 2\omega - \cos\omega}\right)\right). \tag{0.2}$$

Then there is the following bound on the Green's function

$$\|[(\cos n\omega - E(\omega) + io)\delta_{n,n'} + \lambda\Delta]^{-1}\| < C\lambda^{-5}. \tag{0.3}$$

Proof.

(1) Consider the finite matrix A_σ with index set $|k| \leq 10^5$ given by

$$A_\sigma = \left(\cos(\sigma + k\omega) - E\right)\delta_{k,k'} + \lambda(\delta_{k+1,k'} + \delta_{k-1,k'}).$$

Then, we claim that for

$$|\sigma| < \lambda^{\frac{1}{10}} \tag{1.1}$$

and $E = E(\omega)$ given by (0.2),

$$\|A_\sigma^{-1}\| < C\lambda^{-5} \tag{1.2}$$

and

$$|A_\sigma^{-1}(k, k')| < \lambda^{10^2} \text{ for } |k - k'| > 10^4. \tag{1.3}$$

Write

$$E = \cos\alpha \text{ where } \alpha = \omega + 0(\lambda^2) \tag{1.4}$$

and

$$\Lambda = \{1, -1\}$$
$$\Lambda^c = \{k \in \mathbb{Z} \mid |k| \le 10^5, |k| \ne 1\}.$$

For $k \in \Lambda^c$ we have by assumption (0.1), (1.1), (1.4)

$$\left| \sin\frac{1}{2}(\sigma + k\omega \pm \alpha) \right| \sim \text{dist}\left(\frac{1}{2}(\sigma + k\omega \pm \alpha), \pi\mathbb{Z}\right)$$
$$\sim \|\sigma + k\omega \pm \alpha\|$$
$$\ge \|(k \pm 1)\omega\| - 2\lambda^{\frac{1}{10}}$$
$$> \lambda^{\frac{1}{100}} - 2\lambda^{\frac{1}{10}} > \frac{1}{2}\lambda^{\frac{1}{100}}.$$

Hence, for $k \in \Lambda^c$

$$|\cos(\sigma + k\omega) - E| = |\cos(\sigma + k\omega) - \cos\alpha|$$
$$\sim \left| \sin\frac{1}{2}(\sigma + k\omega + \alpha) \right| \left| \sin\frac{1}{2}(\sigma + k\omega - \alpha) \right|$$
$$> \frac{1}{4}\lambda^{\frac{1}{50}} \tag{1.5}$$

and

$$\|(R_{\Lambda^c} A_\sigma R_{\Lambda^c})^{-1}\| < 5\lambda^{-\frac{1}{50}}$$

$$(R_{\Lambda^c} A_\sigma R_{\Lambda^c})^{-1}(k, k') = \frac{1}{\cos(\sigma + k\omega) - E}\delta_{k,k'} + 0\left(\lambda^{-\frac{1}{10}}\lambda^{\frac{1+|k-k'|}{2}}\right). \tag{1.6}$$

It follows that

$$\|A_\sigma^{-1}\|$$
$$\lesssim \|(R_{\Lambda^c} A_\sigma R_{\Lambda^c})^{-1}\|^2 \|(R_\Lambda A_\sigma R_\Lambda - R_\Lambda A_\sigma R_{\Lambda^c}(R_{\Lambda^c} A_\sigma R_{\Lambda^c})^{-1} R_{\Lambda^c} A_\sigma R_\Lambda)^{-1}\|$$
$$\lesssim \lambda^{-\frac{1}{25}} |\det[R_\Lambda A_\sigma R_\Lambda - \lambda^2 R_\Lambda \Delta R_{\Lambda^c}(R_{\Lambda^c} A_\sigma R_{\Lambda^c})^{-1} R_{\Lambda^c} \Delta R_\Lambda]|^{-1}. \tag{1.7}$$

Clearly, from (1.6),(1.1),(1.4)

$$R_A A_\sigma R_A - \lambda^2 R_A \Delta R_{A^c} (R_{A^c} A_\sigma R_{A^c})^{-1} R_{A^c} \Delta R_A =$$

$$
\begin{bmatrix}
\cos(\sigma+\omega)-E & \lambda^2\left(\frac{1}{\cos\sigma-E}+0(\lambda^{\frac{1}{8}})\right) \\
-\lambda^2\left(\frac{1}{\cos\sigma-E}+\frac{1}{\cos(\sigma+2\omega)-E}+0(\lambda^{\frac{1}{8}})\right) & \\
& \cos(\sigma-\omega)-E \\
\lambda^2\left(\frac{1}{\cos\sigma-E}+0(\lambda^{1/3})\right) & -\lambda^2\left(\frac{1}{\cos\sigma-E}+\frac{1}{\cos(\sigma-2\omega)-E}+0(\lambda^{\frac{1}{8}})\right)
\end{bmatrix}
$$

$$
=
\begin{bmatrix}
\cos(\sigma+\omega)-E & \lambda^2\left(\frac{1}{1-\cos\omega}+0(\lambda^{\frac{1}{10}})\right) \\
-\lambda^2\left(\frac{1}{1-\cos\omega}+\frac{1}{\cos 2\omega-\cos\omega}+0(\lambda^{\frac{1}{10}})\right) & \\
& \cos(\sigma-\omega)-E \\
\lambda^2\left(\frac{1}{1-\cos\omega}+0(\lambda^{\frac{1}{10}})\right) & -\lambda^2\left(\frac{1}{1-\cos\omega}+\frac{1}{\cos 2\omega-\cos\omega}+0(\lambda^{\frac{1}{10}})\right)
\end{bmatrix}
$$
$$(1.8)$$

Thus

$$\det(1.8) =$$
$$\left[\cos(\sigma+\omega)-E-\lambda^2\left(\frac{1}{1-\cos\omega}+\frac{1}{\cos 2\omega-\cos\omega}+0(\lambda^{\frac{1}{10}})\right)\right]$$
$$\left[\cos(\sigma-\omega)-E-\lambda^2\left(\frac{1}{1-\cos\omega}+\frac{1}{\cos 2\omega-\cos\omega}+0(\lambda^{\frac{1}{10}})\right)\right]$$
$$-\lambda^4\left(\frac{1}{(1-\cos\omega)^2}+0(\lambda^{\frac{1}{10}})\right).$$
$$(1.9)$$

We have again by (1.1),(1.4)

$$\cos(\sigma+\omega)-E = 2\sin\frac{1}{2}(\sigma+\omega+\alpha)\sin\frac{1}{2}(\sigma+\omega-\alpha)$$
$$= (\sigma+\omega-\alpha)[\sin\omega+0(\lambda^{\frac{1}{10}})]$$
$$\cos(\sigma-\omega)-E = 2\sin\frac{1}{2}(\sigma-\omega+\alpha)\sin\frac{1}{2}(\sigma-\omega-\alpha)$$
$$= (\sigma-\omega+\alpha)[-\sin\omega+0(\lambda^{\frac{1}{10}})]$$

hence

$$(1.9) \sim \left[\omega - \alpha + \sigma - \frac{\lambda^2}{\sin\omega}\left(\frac{1}{1-\cos\omega} + \frac{1}{\cos 2\omega - \cos\omega} + 0(\lambda^{\frac{1}{10}})\right)\right]$$

$$\left[\omega - \alpha - \sigma - \frac{\lambda^2}{\sin\omega}\left(\frac{1}{1-\cos\omega} + \frac{1}{\cos 2\omega - \cos\omega} + 0(\lambda^{\frac{1}{10}})\right)\right]$$

$$-\frac{\lambda^4}{\sin^2\omega}\left(\frac{1}{(1-\cos\omega)^2} + 0(\lambda^{\frac{1}{10}})\right) =$$

$$\left[\omega - \alpha - \frac{\lambda^2}{\sin\omega}\left(\frac{1}{1-\cos\omega} + \frac{1}{\cos 2\omega - \cos\omega}\right)\right]^2$$

$$-\sigma^2 - \frac{\lambda^4}{\sin^2\omega(1-\cos\omega)^2} + 0\{\lambda^{4+\frac{1}{10}} + \lambda^{2+\frac{1}{10}}|\sigma|\}. \tag{1.10}$$

For the choice

$$\alpha = \omega - \frac{\lambda^2}{\sin\omega}\left(\frac{1}{1-\cos\omega} + \frac{1}{\cos 2\omega - \cos\omega}\right)$$

it follows that

$$(1.10) = -\sigma^2 - \frac{\lambda^4}{\sin^2\omega(1-\cos\omega)^2} + 0(\lambda^{4+\frac{1}{10}} + \lambda^{2+\frac{1}{10}}|\sigma|$$

$$< -\frac{\lambda^4}{2\sin^2\omega(1-\cos\omega)^2}.$$

Therefore

$$|\det(1.8)| \gtrsim \lambda^4 \tag{1.11}$$

and substituting (1.11) in (1.7) gives

$$\|A_\sigma^{-1}\| \lesssim \lambda^{-4-\frac{1}{25}}.$$

This establishes the claim (1.2).

We verify the off-diagonal estimate (1.3). Let thus $|k|, |k'| \leq 10^5, |k-k'| > 10^4$. Then clearly

$$\max\left(\text{dist}(k, \Lambda), \text{dist}(k', \Lambda)\right) > \frac{1}{3}10^4.$$

Assume

$$\text{dist}(k, \Lambda) > \frac{1}{3}10^4. \tag{1.12}$$

From the resolvent identity

$$A_\sigma^{-1}(k, k') = (R_{\Lambda^c}A_\sigma R_{\Lambda^c})^{-1}(k, k') - ((R_{\Lambda^c}A_\sigma R_{\Lambda^c})^{-1}R_{\Lambda^c}A_\sigma R_\Lambda A_\sigma^{-1})(k, k').$$

By (1.6)

$$|(R_{\Lambda^c}A_\sigma R_{\Lambda^c})^{-1}(k, k')| < \lambda^{10^3}$$

and from (1.12), (1.6), (1.2)

$$|((R_{\Lambda^c} A_\sigma R_{\Lambda^c})^{-1} R_{\Lambda^c} A_\sigma R_\Lambda A_\sigma^{-1})(k,k')| \le$$
$$\lambda \sum_{\substack{k_1 \in \Lambda^c, k_2 \in \Lambda \\ |k_1 - k_2| = 1}} |R_{\Lambda^c} A_\sigma R_{\Lambda^c})^{-1}(k,k_1)| \, |A_\sigma^{-1}(k_2, k')| < \lambda . \lambda^{\frac{1}{6} 10^4} . \lambda^{-5} < \lambda^{10^3}.$$

Thus $|A_\sigma^{-1}(k,k')| < \lambda^{10^3}$, proving (1.3).

(2) We establish (0.3).

Denote

$$H = (\cos n\omega)\delta_{nn'} + \lambda\Delta = D + \lambda\Delta.$$

Write

$$\mathbb{Z} = S + S^c$$

where

$$S = \{n \in \mathbb{Z} | \, |\cos n\omega - E(\omega)| < \lambda^{1/2}\}$$

is the set of singular sites.

Denote $D_{S^c} = R_{S^c} D R_{S^c}$, $H_{S^c} = R_{S^c} H R_{S^c}$ and $E = E(\omega)$. We have thus

$$\|(D_{S^c} - E)^{-1}\| < \lambda^{-1/2}$$

and

$$(H_{S^c} - E)^{-1} = (I + \lambda(D_{S^c} - E)^{-1} R_{S^c} \Delta R_{S^c})^{-1}(D_{S^c} - E)^{-1}$$
$$(H_{S^c} - E)^{-1}(n, n') = \frac{1}{\cos n\omega - E}\delta_{n,n'} + 0\left(\lambda^{-\frac{1}{2}} \lambda^{\frac{|n-n'|v_1}{2}}\right)$$

controlled by a Neumann series.

Assume $n \in S$, i.e.

$$|\cos n\omega - E| \le \lambda^{\frac{1}{2}}.$$

Since by (0.2)

$$\cos n\omega - E = 2\sin\frac{1}{2}(n+1)\omega \sin\frac{1}{2}(n-1)\omega + 0(\lambda)$$

it follows from (1.5) that

$$\min(\|(n+1)\omega\|, \|(n-1)\omega\|) < \lambda^{\frac{1}{6}}.$$

Thus for some $n_0 \in \{n+1, n-1\}$,

$$n_0\omega = \sigma(\bmod 2\pi) \text{ with } |\sigma| < \lambda^{\frac{1}{6}}.$$

Denoting

$$\Lambda = \Lambda(n_0) = \{n \in \mathbb{Z} | \, |n - n_0| \le 10^5\}$$

we have

$$|n - n_0| = 1$$

and

$$H_\Lambda - E = A_{\sigma \equiv n_0 \omega}$$

satisfying from (1.2), (1.3)

$$\|(H_\Lambda - E)^{-1}\| < C\lambda^{-5} \tag{2.1}$$

$$|(H_\Lambda - E)^{-1}(k, k')| < \lambda^{10^8} \text{ if } |k - k'| > 10^4. \tag{2.2}$$

Thus each element n in the singular set S is at distance 1 from a site n_0 with $\Lambda = \Lambda(n_0)$ satisfying (2.1), (2.2). The estimate (0.3), i.e.

$$\|(H - E)^{-1}\| < C\lambda^{-5}$$

for the full Green's function is then a routine application of the resolvent identity and the covering of $\mathbb{Z}$ obtained above.   □

(3) Observe that condition (0.1) restricts ω to a union of a bounded number of intervals $I_s \subset \mathbb{T}(s < 10^{12})$ with

$$\text{mes}\,(\mathbb{T} \backslash \bigcup I_s) < \delta(\lambda) \xrightarrow{\lambda \to 0} 0. \tag{3.1}$$

For $\omega \in I_\alpha$, (0.3) implies that $[E(\omega) - \lambda^6, E(\omega) + \lambda^6]$, $E(\omega)$ given by (0.2) is disjoint from the spectrum of the operator $\cos n\omega.\delta_{nn'} + \lambda\Delta$.

2   Almost Mathieu Operator with Varying Frequency

(4) Define

$$H_{\alpha,\lambda,\theta}\psi = \lambda \cos(2\pi\alpha n + \theta)\psi_n + (\psi_{n-1} + \psi_{n+1})$$

and

$$\sigma(\alpha, \lambda, \theta) = \text{Spec}\, H_{\alpha,\lambda,\theta}.$$

If α is irrational, $\sigma(\alpha, \lambda, \theta) = \sigma(\alpha, \lambda)$ does not depend on θ. We also recall Aubry's duality property

$$\sigma(\alpha, \lambda) = \frac{\lambda}{2}\sigma\left(\alpha, \frac{4}{\lambda}\right) \tag{4.1}$$

(for all λ), permitting to describe the spectrum for $\lambda < 2$ from the case $\lambda > 2$.

Remark. Recall that if ω is diophantine, then θ a.e.

For $\lambda > 2$: $H_{\alpha,\lambda,\theta}$ has pure point spectrum.

For $\lambda = 2$: $H_{\alpha,\lambda,\theta}$ has purely singular continuous spectrum.

For $\lambda < 2$: $H_{\alpha,\lambda,\theta}$ has purely absolutely continuous spectrum.

See [La], [Ji] (and further references).

The (perturbative) case of large λ was settled earlier in the work of Fröhlich-Spencer-Wittwer [F-S-W] and, independently, Sinai [S]. In this case, the closure $\sigma(\alpha, \lambda)$ of the point spectrum is a Cantor set.

For almost all α and all λ, one has

$$\operatorname{mes} \sigma(\alpha, \lambda) = |4 - 2|\lambda||$$

(see [La]).

It follows from section I that for small λ and

$$\alpha \in \bigcup I_s'$$

$$\left[\frac{1}{\lambda} E(2\pi\alpha) - \lambda^5, \frac{1}{\lambda} E(2\pi\alpha) + \lambda^5\right] \cap \sigma\left(\alpha, \frac{1}{\lambda}\right) = \phi. \tag{4.2}$$

Here

$$I_s' = \frac{1}{2\pi} I_s \cap \{\text{irrationals}\}$$

and $\{I_s\}$ is the set of intervals introduced above. The energy $E(\omega)$ is given by (0.2), i.e.

$$E(2\pi\alpha) = \cos\left(2\pi\alpha - \frac{\lambda^2}{\sin 2\pi\alpha}\left(\frac{1}{1 - \cos 2\pi\alpha} + \frac{1}{\cos 4\pi\alpha - \cos 2\pi\alpha}\right)\right). \tag{4.3}$$

From (4.1), (4.2)

$$[2E(2\pi\alpha) - 2\lambda^6, 2E(2\pi\alpha) + 2\lambda^6] \cap \sigma(\alpha, 4\lambda) = \phi. \tag{4.4}$$

Hence, the set

$$\mathcal{E}_{4\lambda} = \bigcup_{s, \alpha \in I_s'} [2E(2\pi\alpha) - 2\lambda^6, 2E(2\pi\alpha) + 2\lambda^6]$$

is disjoint from

$$\bigcap_{\alpha \in [0,1]} \sigma(\alpha, 4\lambda).$$

Notice that by (3.1), (4.3), the sets $\mathcal{E}_\lambda$ contain 10^{12} intervals that fill up $[-2, 2]$ except for a subset of small measure. It also follows from (4.4) and Kotani's theorem that the Lyapounov exponent $\gamma(\alpha, 4\lambda, E) > 0$ for almost all E in $[2E(2\pi\alpha) - 2\lambda^6, 2E(2\pi\alpha) + 2\lambda^6]$. Hence

Lemma 2. *For small $\lambda > 0$, there is a subset $\mathcal{E}_\lambda$ of $[-2, 2]$ obtained as union of 10^{12} intervals and of small complementary measure, s.t. for almost all $E \in \mathcal{E}_\lambda$*

$$\operatorname{ess\,sup}_{\alpha} \gamma(\alpha, \lambda, E) > 0.$$

Corollary 3. *For small λ and almost all $E \in \mathcal{E}_\lambda$*

$$\int_{\mathbb{T}} \gamma(\alpha, \lambda, E) d\alpha > 0. \tag{4.5}$$

3 Sequences $\{n^\rho\}$, $1 < \rho < 2$

(5) Consider the Schrödinger operator

$$H = \lambda \cos(2\pi n^\rho \omega + \theta)\delta_{nn'} + \Delta \tag{5.1}$$

where $1 < \rho < 2$. Denote

$$\gamma(E, \lambda) = \lim_{n \to \infty} \frac{1}{N} \int_{\mathbb{T}} \log \left\| \prod_N^0 \begin{pmatrix} E - \lambda \cos(2\pi n^\rho \omega + \theta) & -1 \\ 1 & 0 \end{pmatrix} \right\| d\theta.$$

We prove the following

Proposition 4. *Given $\omega \in \mathbb{T}$ and small $\lambda \neq 0$, there is a subset $\mathcal{E}$ of $[-2, 2]$ of small complementary measure, such that*

$$\gamma(E, \lambda) > 0 \text{ for all } E \in \mathcal{E}. \tag{5.2}$$

Writing

$$2\pi\omega(n_0 + k)^\rho = 2\pi\omega n_0^\rho + (2\pi\omega\rho n_0^{\rho-1})k + 0(|n_0|^{\rho-2}k^2)$$

and assuming

$$|k| < |n_0|^{\frac{2-\rho}{10}}$$

we see that

$$2\pi\omega(n_0 + k)^\rho = 2\pi\omega n_0^\rho + (2\pi\omega\rho n_0^{\rho-1})k + 0(|n_0|^{-\frac{2-\rho}{2}}). \tag{5.3}$$

This permits us to view (5.1) locally as a shift over a variable angle $2\pi\omega\rho n_0^{\rho-1}$ $= 2\pi\alpha$. As a consequence, one may show that

$$\gamma(E, \lambda) \leq \int_{\mathbb{T}} \gamma(\alpha, \lambda, E) d\alpha$$

where $\gamma(E, \lambda)$ is the exponent associated to (5.1) and $\gamma(\alpha, \lambda, E)$ denote the exponent for the almost Mathieu operator $H_{\alpha,\lambda,\theta} = \lambda \cos(2\pi n\alpha + \theta) + \Delta$, considered previously.

We don't explicit details, since we are interested in the converse inequality.

Since for $E \in \mathcal{E}_\lambda$, by (4.5)

$$\int \gamma(\alpha, \lambda, E) d\alpha > 0$$

there is $c_1 = c_1(E, \lambda) > 0$ such that

$$\gamma(\alpha, \lambda, E) > c_1 \qquad (5.4)$$

for all $\alpha \in \mathcal{R} = \mathcal{R}_E \subset \mathbb{T}$ and

$$|\mathcal{R}| > c_1. \qquad (5.5)$$

We restrict moreover $\mathcal{R}$ to α's satisfying a diophantine condition

$$\|k\alpha\| = \mathrm{dist}(k\alpha, \mathbb{Z}) > |k|^{-2} \text{ for } k \text{ large enough.}$$

It follows then from [B-G] and [G-S] that

$$\mathrm{mes}\,[\theta \in \mathbb{T}|\,\left|\frac{1}{m}\log\|M_m(\alpha, \theta, E)\| - L_m(\alpha, E)\right| > m^{-\sigma}] < e^{-m^\sigma} \qquad (5.6)$$

and

$$|\gamma(\alpha, \lambda, E) - L_m| < m^{-\sigma} \qquad (5.7)$$

for m large enough. Here $\sigma > 0$ is a fixed constant,

$$M_m(\alpha, \theta, E) = \prod_1^m \begin{pmatrix} E - \lambda \cos(2\pi j\alpha + \theta) & -1 \\ 1 & 0 \end{pmatrix}$$

$$L_m(\alpha, E) = \frac{1}{m}\int \log\|M_m(\alpha, \theta, E)\|$$

and

$$\gamma(\alpha, \lambda, E) = \lim L_m(\alpha. E).$$

(In fact, [G-S] provides also more precise estimates).

We may clearly restrict further the set $\mathcal{R}$ to ensure that for some $m_0 = m_0(E)$ and all $\alpha \in \mathcal{R}$, inequalities (5.6), (5.7) hold for $m > m_0(E)$.

Recall that from Cramer's rule, for $0 < k_1 \leq k_2 \leq m$

$$[R_{[0,m]}(H_{\alpha,\lambda,\theta} - E)R_{[0,m]}]^{-1}(k_1, k_2)| =$$
$$\frac{|\det[H_{k_1-1}(\theta) - E]|.|\det[H_{m-k_2-1}(\theta + 2\pi(k_2+1)\alpha) - E]|}{|\det[H_m(\theta) - E]|}$$
$$\leq \frac{M_{k_1-1}(\alpha, \theta, E)M_{m-k_2}(\alpha, \theta + 2\pi(k_2+1)\alpha, E)}{|\det[H_m(\theta) - E]|} \qquad (5.8)$$

where we denoted $R_{[0,k]}H_{\alpha,\lambda,\theta}R_{[0,k]}$ by $H_k(\theta)$.

Since also

$$\|M_m(\alpha, \theta, E)\| \sim \max\{|\det[H_m(\theta) - E]|, |\det[H_{m-1}(\theta) - E]|,$$
$$|\det[H_{m-1}(\theta + 2\pi\alpha) - E]|,$$
$$|\det[H_{m-2}(\theta + 2\pi\alpha) - E]|\}$$

there is a set

$$\Theta = \Theta_{\alpha,E} \subset \mathbb{T} \tag{5.9}$$

$$\mathrm{mes}\,(\mathbb{T}\backslash\Theta) < e^{-\frac{1}{2}m^{\sigma/2}}$$

such that for $\theta \in \Theta$ and one of the sets

$$\Lambda \in \{[0, m], [0, m-1], [1, m], [1, m-1]\}$$

(depending on θ), the Green's function

$$G_\Lambda(\alpha, \theta) = [R_\Lambda(H_{\alpha,\lambda,\theta} - E)R_\Lambda]^{-1} \tag{5.10}$$

satisfies the estimate

$$|G_\Lambda(\alpha, \theta)(k_1, k_2)| < e^{(m-|k_1-k_2|)L_{\sqrt{m}}(\alpha,E) - mL_m(\alpha,E) + 0(m^{1-\frac{\varsigma}{2}})}$$
$$< e^{-|k_1-k_2|\gamma(\alpha,E) + 0(m^{1-\frac{\varsigma}{2}})}$$
$$< e^{-c_1|k_1-k_2| + 0(m^{1-\frac{\varsigma}{2}})}. \tag{5.11}$$

This follows mainly from (5.6), (5.7), (5.8) and (5.4) (cf. [B-G]). Returning to the Hamiltonian H in (5.1), restrict n to an interval $n \in [N_1, N_2], N_2 > 2N_1, \log N_1 \sim \log N_2$. Let $[n_0, n_0 + m] \subset [N_1, N_2]$ be an interval of length

$$m < N_1^{\delta(\rho)}. \tag{5.12}$$

It follows then from (5.3) that

$$H_{[n_0,n_0+m]} = R_{[0,m]}(H_{\rho\omega n_0^{\rho-1},\lambda,\theta+2\pi\omega n_0^\rho})R_{[0,m]} + 0(N_1^{-\frac{2-\rho}{2}}) \tag{5.13}$$

with on the right the Almost Mathieu operator.
Assume n_0 satisfies

$$\mathrm{dist}_{\mathbb{T}}(\rho\omega n_0^{\rho-1}, \mathcal{R}) = \|\rho\omega n_0^{\rho-1} - \alpha\| < N_1^{-\frac{(2-\rho)(\rho-1)}{5}} \tag{5.14}$$

for some $\alpha \in \mathcal{R}$. Then from (5.12), (5.13), (5.14)

$$H_{[n_0,n_0+m]} = R_{[0,m]}H_{\alpha,\lambda,\theta+2\pi\omega n_0^\rho}R_{[0,m]} + 0\left(N_1^{-\frac{(2-\rho)(\rho-1)}{6}}\right). \tag{5.15}$$

It is clear that the estimate (5.11) will be preserved if the matrix $R_\Lambda H_{\alpha,\lambda,\theta}R_\Lambda$ is perturbed by less than $e^{-m^{1-\frac{\varsigma}{3}}}$. In view of (5.15), take thus

$$m^{1-\frac{\varsigma}{3}} \ll \log N_1 \tag{5.16}$$

so that from (5.9), (5.11), if

$$\theta + 2\pi\omega n_0^\rho \notin \Theta_{\alpha,E} \tag{5.17}$$

then for one of the sets

$$\Lambda \in \{[n_0, n_0+m], [n_0, n_0+m-1], [n_0+1, n_0+m], [n_0+1, n_0+m-1]\} \tag{5.18}$$

there is the Green's function estimate

$$|(R_\Lambda(H - E)R_\Lambda)^{-1}(k_1, k_2)| < e^{-c_1|k_1-k_2|+0(m^{1-\frac{\epsilon}{2}})}. \tag{5.19}$$

Recalling (5.5), simple considerations on the distribution of $\{\rho\omega n^{\rho-1}\}_{n\in[N_1,N_2]}$ show that (5.14) holds for

$$n_0 \in J_E \subset [N_1, N_2]$$

where

$$|J_E| > \frac{c_1}{2}(N_2 - N_1).$$

If $n_0 \in J_E$, (5.17) and the preceding imply that for θ in a set $\Theta'_{(n_0)} \subset \mathbb{T}$

$$\mathrm{mes}\,(\mathbb{T}\backslash\Theta'_{(n_0)}) < e^{-\frac{1}{2}m^{\sigma/2}} \tag{5.20}$$

the estimate (5.19) will hold for one of the sets (5.18).
   Since

$$\{(n, \theta) \in J_E \times \mathbb{T} | \theta \notin \Theta'_{(n)}\}$$

is of measure $< (N_2 - N_1)e^{-\frac{1}{2}m^{\sigma/2}}$, by (5.20), there exists by Fubini a subset $\Theta' = \Theta'_E \subset \mathbb{T}$ such that

$$\mathrm{mes}\,\Theta' > 1 - o(1)$$

and for $\theta \in \Theta'$

$$\mathrm{card}\,(J_{E,\theta}) > \frac{c_1}{3}(N_2 - N_1)$$

with

$$J_{E,\theta} = \{n \in J_E | \theta \in \Theta'_{(n)}\}.$$

For $\theta \in \Theta'$, there is thus a system of disjoint intervals $\Lambda_s \subset [N_1, N_2]$ such that

$$|\Lambda_s| \in \{m, m - 1, m - 2\}$$
$$\sum |\Lambda_s| > \frac{c_1}{6}N_2 \tag{5.21}$$

each $\Lambda = \Lambda_s$ satisfies the Green's function estimate (5.19). (5.22)

Denote n_s an element in the center of Λ_s and let

$$N_1 = n_0 < n_1 < \cdots < n_{s_\bullet} < n_{s_\bullet+1} = N_2$$

providing a covering of $[N_1, N_2]$ as

$$[N_1, N_2] = \bigcup \Lambda_s \cup \bigcup_{s=0}^{s_*} [n_s, n_{s+1}]. \tag{5.23}$$

Observe that again (5.19), (5.22) remains valid if the energy E is perturbed by less than $e^{-m^{1-\frac{\sigma}{3}}}$.

Hence, there is a subset

$$\mathcal{E} = \mathcal{E}_{E,\theta} \subset [E - e^{-m^{1-\sigma/3}}, E + e^{-m^{1-\sigma/3}}]$$
$$\text{mes}\,\mathcal{E} > (2 - N_2^{-1}) e^{-m^{1-\sigma/3}} \tag{5.24}$$

such that for $E' \in \mathcal{E}$, any interval $I \subset [N_1, N_2]$

$$\|[R_I (H - E') R_I]^{-1}\| < N_2^3 e^{m^{1-\sigma/3}}$$

and for all $s = 1, \ldots, s_*$

$$|[R_{\Lambda_s}(H - E') R_{\Lambda_s}]^{-1}(k_1, k_2)| < e^{-c_1 |k_1 - k_2| + 0(m^{1-\frac{\sigma}{2}})}.$$

Choose m such that, cf. (5.16)

$$m^{1-\sigma/3} \ll \log N_1 \text{ and } m \gg \log N_1 \sim \log N_2.$$

Thus in particular for each s

$$|(R_{\Lambda_s}(H - E') R_{\Lambda_s})^{-1}(k_1, k_2)| < e^{-\frac{c_1}{8}m} \text{ for } |k_1 - k_2| > \frac{m}{4}$$

with

$$m \gg \max_s \log \|[R_{[n_s, n_{s+1}]}(H - E') R_{[n_s, n_{s+1}]}]^{-1}\|.$$

Application of the resolvent identity to the paving (5.23) shows that for $E' \in \mathcal{E}_{E,\theta}$

$$|[R_{[N_1,N_2]}(H - E') R_{[N_1,N_2]}]^{-1}(k_1, k_2)| < e^{-\frac{c_1}{20}d(k_1,k_2) + 0(m)}$$

where for $k_1 < k_2$ we denote

$$d(k_1, k_2) = \sum_s |[k_1, k_2] \cap \Lambda_s|. \tag{5.25}$$

In particular, by (5.21), (5.25)

$$\left\| \prod_{N_2}^{N_1} \begin{pmatrix} E' - \lambda \cos(2\pi n^\rho \omega + \theta) & -1 \\ 1 & 0 \end{pmatrix} \right\| \geq$$

$$|\det(R_{[N_1,N_2]}(H(\theta) - E') R_{[N_1,N_2]})| =$$

$$\{|[R_{[N_1,N_2]}(H(\theta) - E') R_{[N_1,N_2]}]^{-1}(N_1, N_2)|\}^{-1} >$$

$$e^{\frac{c_1^2}{200} N_2} \tag{5.26}$$

for all $\theta \in \Theta'_E$ and $E' \in \mathcal{E}_{E,\theta}$.

Since by (5.24), $\mathrm{mes}\,\mathcal{E}_{E,\theta} > (2 - \frac{1}{N_2})e^{-m^{1-\sigma/3}}$ in $[E - e^{-m^{1-\sigma/3}}, E + e^{-m^{1-\sigma/3}}]$ for each $\theta \in \Theta'_E$, there exists again by Fubini, a subset

$$\mathcal{E}_E \subset [E - e^{-m^{1-\sigma/3}}, E + e^{-m^{1-\sigma/3}}] \tag{5.27}$$

$$\mathrm{mes}\,\mathcal{E}_E > \left(2 - \frac{1}{\sqrt{N_2}}\right)e^{-m^{1-\sigma/3}} \tag{5.28}$$

such that if $E' \in \mathcal{E}_E$, then

$$\mathrm{mes}\,[\theta \in \Theta'_E | E' \in \mathcal{E}_{E,\theta}] > 1 - o(1).$$

Thus, for $E' \in \mathcal{E}_E$, (5.26) implies that

$$\int_{\mathbb{T}} \log \left\| \prod_{N_2}^{N_1} \left(\begin{matrix} E' - \lambda\cos(2\pi n^\rho\omega + \theta) & -1 \\ 1 & 0 \end{matrix} \right) \right\| d\theta > \frac{c_1(E,\lambda)^2}{300} N_2$$

and letting $N_1 \ll N_2$, also

$$\frac{1}{N_2} \int \log \left\| \prod_{N_2}^{0} \left(\begin{matrix} E' - \lambda\cos(2\pi n^\rho\omega + \theta) & -1 \\ 1 & 0 \end{matrix} \right) \right\| d\theta > \frac{c_1(E,\lambda)^2}{400}. \tag{5.29}$$

Property (5.29) holds for all $E' \in \mathcal{E}_E$, $E \in \mathcal{E}_\lambda$ (assuming N_2 sufficiently large). The set $\mathcal{E}_E$ depends on N_2 but satisfies (5.27), (5.28). One easily sees that one may therefore obtain a subset $\mathcal{E} \subset \mathcal{E}_\lambda$, $\mathrm{mes}\,\mathcal{E} > \mathrm{mes}\,\mathcal{E}_\lambda - o(1)$ and some $\delta = \delta_\lambda > 0$ s.t. the left side of (5.29) is at least δ, for all $E' \in \mathcal{E}$ and sufficiently large N_2. In particular, (5.2) holds for all $E \in \mathcal{E}$. This proves Proposition 4.

Remark. Refining a bit the previous considerations shows that in fact for $E \in \mathcal{E}$, the limit

$$\lim_{N \to \infty} \frac{1}{N} \log \left\| \prod_{N}^{0} \left(\begin{matrix} E - \lambda\cos(2\pi n^\rho\omega + \theta) & -1 \\ 1 & 0 \end{matrix} \right) \right\|$$

exists θ a.e. and equals

$$\int_{\mathbb{T}} \gamma(\alpha, \lambda, E) d\alpha.$$

6   Green's Function Estimate with Varying Energy

(6)

Lemma 6.1 *Let*

$$A = v_n \delta_{nn'} + \Delta \quad (1 \leq n, n' \leq N)$$

be an $N \times N$ matrix. Assume $|v_n| < C$. Let

$$M > \delta^{-2}, \delta \text{ small}.\tag{6.2}$$

Let $\mathcal{E} \subset \mathbb{R}$ be a set of positive measure

$$\text{mes}\,\mathcal{E} > c_0$$

such that for each energy $E \in \mathcal{E}$, there is a collection of disjoint intervals $\{I_\alpha\}$ in $[1, N]$ (depending on E), $|I_\alpha| = M$, s.t.

$$\sum |I_\alpha| > c_0 N\tag{6.3}$$

and for each α

$$\|(R_{I_\alpha}(A - E)R_{I_\alpha})^{-1}\| < e^{\delta M}\tag{6.4}$$

$$|(R_{I_\alpha}(A - E)R_{I_\alpha})^{-1}(k, k')| < e^{-c_1|k-k'|} \text{ if } k, k' \in I_\alpha, |k - k'| > \delta M.\tag{6.5}$$

Then there is a subset $\mathcal{E}' \subset \mathcal{E}$

$$\text{mes}\,(\mathcal{E} \backslash \mathcal{E}') < \frac{1}{M}$$

such that for $E \in \mathcal{E}'$, there is a restriction σ of the counting measure on $[1, N]$ (depending on E) such that

$$|(A - E)^{-1}(k, k')| < e^{-\frac{1}{2}c_1 d(k, k') + N^{1/2}}\tag{6.6}$$

where

$$d(k, k') = \sigma([k, k']) \text{ for } k < k'$$

and σ satisfies

$$\sigma([1, N]) > \frac{1}{2}c_0 N.\tag{6.7}$$

Proof. Fix $E \in \mathcal{E}$. Let $\{I_\alpha\}$ be the associated system of intervals. Denote for $\Lambda \subset [1, N]$

$$\sigma_0(\Lambda) = \sum |I_\alpha \cap \Lambda|$$

and for $k < k'$

$$d_0(k, k') = \sigma_0([k, k']).$$

Properties (6.4), (6.5) for the I_α-restrictions remain essentially preserved if E is perturbed to

$$E' \in \tau = \tau_E \equiv [E - e^{-3\delta M}, E + e^{-3\delta M}].$$

Define

$$M_1 = M^{100}\tag{6.8}$$

and partition $[1, N]$ in intervals $J_s = [sM_1, (s+1)M_1]$ of size M_1. Denote

$$S = \left\{ s < \frac{N}{M_1} \,\middle|\, \sigma_0(J_s) > \delta M_1 \right\} \tag{6.9}$$

so that by (6.3)

$$\sum_{s \in S} \sigma_0(J_s) > \sum |I_\alpha| - \delta N > (c_0 - \delta)N. \tag{6.10}$$

Fixing s and defining

$$\tau_s = \{ E' \in \tau \,|\, \mathrm{dist}(E', \,\mathrm{Spec}\, R_\Lambda A R_\Lambda) < M_1^{-5} e^{-3\delta M} \text{ for some interval } \Lambda \subset J_s \} \tag{6.11}$$

is of measure

$$|\tau_s| < M_1^3 M_1^{-5} e^{-3\delta M} < M_1^{-2}|\tau|.$$

Hence

$$\frac{1}{|\tau|} \int_\tau \sum_s \chi_{\tau_s}(E') dE' < M_1^{-2} \frac{N}{M_1}$$

and there is a subset $\tau' \subset \tau$,

$$|\tau'| < \frac{1}{M_1}|\tau| \tag{6.12}$$

so that if $E' \in \tau \backslash \tau'$ and

$$S' = S'_{E'} = \{ s \in S \,|\, E' \notin \tau_s \}$$

then

$$|S \backslash S'| < \frac{N}{M_1^2}.$$

Hence, from (6.10)

$$\sum_{s \in S'} \sigma_0(J_s) > (c_0 - \delta)N - \frac{N}{M_1} > (c_0 - 2\delta)N. \tag{6.13}$$

Fix $s \in S'$ and denote $J = J_s$. Thus from (6.9), (6.11)

$$\sigma_0(J_s) > \delta M_1 \tag{6.14}$$

and

$$\|(R_\Lambda(A - E')R_\Lambda)^{-1}\| < M_1^2 e^{3\delta M} \text{ for any interval } \Lambda \subset J. \tag{6.15}$$

Our aim is to estimate $(R_J(A - E')R_J)^{-1}$.
    Let thus $k_1, k_2 \in J, k_1 < k_2$

$$d_0(k_1, k_2) > 10M. \tag{6.16}$$

We distinguish the following cases.   □

Case 1. $k_1 \in I_\alpha$ for some α and

$$\text{dist}(k_1, \partial I_\alpha) \geq \sqrt{\delta}M. \tag{6.17}$$

From the resolvent identity

$$|(R_J(A - E')R_J)^{-1}(k_1, k_2)| \leq$$
$$\sum_{\substack{k_3 \in \partial I_\alpha, k_4 \in J \setminus I_\alpha \\ |k_3 - k_4| = 1}} |(R_{I_\alpha}(A - E')R_{I_\alpha})^{-1}(k_1, k_3)| \, |(R_J(A - E')R_J)^{-1}(k_4, k_2)| \leq$$
$$\sum_{\substack{k_3 \in \partial I_\alpha, k_4 \in J \setminus I_\alpha \\ |k_3 - k_4| = 1}} e^{-c_1|k_1 - k_3|} |(R_J(A - E')R_J)^{-1}(k_4, k_2)| \tag{6.18}$$

from (6.5), (6.16), (6.17). It follows that

$$|(R_J(A - E')R_J)^{-1}(k_1, k_2)| \leq$$
$$\max_{\sqrt{\delta}M \leq |k_1 - k'| \leq M} e^{-c_1|k_1 - k'|} |(R_J(A - E')R_J)^{-1}(k', k_2)|. \tag{6.19}$$

Case 2. Negation of Case 1

Assume $I_\alpha = [a_\alpha, a_\alpha + M], I_\beta = [a_\beta, a_\beta + M]$ consecutive intervals s.t.

$$k_1 \in [a_\alpha + M - \sqrt{\delta}M, a_\beta + \sqrt{\delta}M].$$

Denote

$$\Lambda = [a_\alpha + M - \sqrt{\delta}M, a_\beta + \delta^{1/4}M]. \tag{6.20}$$

Apply the resolvent identity to the decomposition $J = \Lambda \cup (J \setminus \Lambda)$.

Since $|k_1 - k_2| > d_0(k_1, k_2) > 10M, k_2 \notin \Lambda$. Thus

$$|(R_J(A - E')R_J)^{-1}(k_1, k_2)| \leq$$
$$\sum_{\substack{k_3 \in \partial \Lambda, k_4 \in J \setminus \Lambda \\ |k_3 - k_4| = 1}} |(R_\Lambda(A - E')R_\Lambda)^{-1}(k_1, k_3)| \, |(R_J(A - E')R_J)^{-1}(k_4, k_2)|. \tag{6.21}$$

The first factor in (6.21) is estimated by (6.15). Clearly Λ was constructed such that $k_4 \in I_\alpha \cup I_\beta$ satisfies Case 1.

Thus from (6.19)

$$|(R_J(A - E')R_J)^{-1}(k_4, k_2)| \lesssim$$
$$\max_{\text{dist}(k', \partial I_\alpha \cup \partial I_\beta) \leq 1} e^{-c_1|k_4 - k'|} |(R_J(A - E')R_J)^{-1}(k', k_2)|. \tag{6.22}$$

Consider the element $k_3 \in \partial \Lambda$ in (6.21).

Assume first $k_3 \in I_\alpha$, hence $k_3 \leq k_1$. Then $[k_4, k_2] \supset [k_1, k_2]$, hence

$$d_0(k_1, k_2) \leq d_0(k_4, k_2)$$

and

$$d_0(k', k_2) \geq d_0(k_1, k_2) - |k_4 - k'|. \tag{6.22'}$$

For the corresponding contribution in (6.21), we get from the preceding

$$M_1^5 e^{3\delta M} \max_{\substack{\sqrt{\delta} M \leq |k_4 - k'| \leq M \\ d_0(k', k_2) > d_0(k_1, k_2) - |k_4 - k'|}} e^{-c_1|k_4 - k'|} |(R_J(A - E')R_J)^{-1}(k', k_2)|.$$

$$\tag{6.23}$$

Next, assume $k_3 \in I_\beta$, thus $k_3 > k_1$.

The element k' in (6.22) satisfies $\text{dist}(k', \partial I_\beta) \leq 1$.

If $k' < k_4$, then from definition (6.20)

$$|k_4 - k'| = \delta^{1/4} M \tag{6.24}$$

and clearly

$$\begin{aligned} d_0(k', k_2) &\geq d_0(k_1, k_2) - |[k_1, k_2] \cap I_\alpha| \\ &\geq d_0(k_1, k_2) - \delta^{1/2} M \\ &= d_0(k_1, k_2) - \delta^{1/4}|k_4 - k'|. \end{aligned} \tag{6.25}$$

If $k' > k_4$, then

$$|k_4 - k'| > (1 - \delta^{1/4})M \tag{6.26}$$

and

$$\begin{aligned} d_0(k', k_2) &\geq d_0(k_1, k_2) - \delta^{1/2} M - M \\ &> d_0(k_1, k_2) - (1 + 2\delta^{1/4})|k_4 - k'|. \end{aligned} \tag{6.27}$$

The contribution to (6.21) is therefore at most

$$M_1^5 e^{3\delta m} \max_{\substack{\delta^{1/2} M \leq |k_4 - k'| \leq M \\ d_0(k', k_2) > d_0(k_1, k_2) - (1 + 2\delta^{1/4})|k_4 - k'|}} e^{-c_1|k_4 - k'|} |(R_J(A - E')R_J)^{-1}(k', k_2)|$$

$$\tag{6.28}$$

which also covers (6.23).

Clearly

$$(6.28) < e^{-\delta M} \max_{d_0(k'', k_2) \geq d_0(k_1, k_2) - (1 + 2\delta^{1/4})|k' - k''|} e^{-c_1(1 - \delta^{1/3})|k' - k''|} |(R_J(A - E')R_J)^{-1}(k'', k_2)| \tag{6.29}$$

which also covers (6.19).

One may iterate (6.29) as long as $d_0(k'', k_2) > 10M$.

After r steps, we get thus a bound

$$\left|\left(R_J(A - E')R_J\right)^{-1}(k_1, k_2)\right| <$$

$$e^{-r\delta M}e^{-c_1(1-\delta^{1/3})[|k_1'-k_1''|+\cdots+|k_r'-k_r''|]}\left|\left(R_J(A - E')R_J\right)^{-1}(k_r'', k_2)\right|$$

for some $k_1', k_1'', \ldots, k_r', k_r''$ satisfying

$$d_0(k_r'', k_2) \geq d_0(k_1, k_2) - (1 + 2\delta^{1/4})[|k_1' - k_1''| + \cdots + k_r' - k_r''|].$$

Consequently, it follows that

$$\left|\left(R_J(A - E')R_J\right)^{-1}(k_1, k_2)\right| < e^{-c_1(1-3\delta^{1/4})d_0(k_1,k_2)+10M}$$

for each interval $J = J_s$, $s \in S_{E'}'$.

Recall that E' is restricted to $\tau \backslash \tau'$, τ' satisfying (6.12).

Thus to each $E' \in \tau \backslash \tau'$, we may associate a system of intervals $I_\beta^1 \subset [1, N]$ of length $|I_\beta^1| = M_1$, satisfying

$$\sum_\beta \sigma_0(I_\beta^1) > (c_0 - 2\delta)N \tag{6.30}$$

by (6.13), and for each β

$$\left\|\left(R_{I_\beta^1}(A - E')R_{I_\beta^1}\right)^{-1}\right\| < M_1^5 e^{3\delta M} < e^{\delta^{100}M_1} \tag{6.31}$$

by (6.15),

$$\left|\left(R_{I_\beta^1}(A - E')R_{I_\beta^1}\right)^{-1}(k_1, k_2)\right| < e^{-c_1(1-3\delta^{1/4})d_0(k_1,k_2)} \tag{6.32}$$

if $k_1, k_2 \in I_\beta^1, d_0(k_1, k_2) > \delta^{100}M_1 > \frac{M}{\delta}$.

Let next

$$\delta_1 = \delta^{100} > M_1^{-1/2}$$

by (6.2), (6.8). Replacing

$$c_0 \text{ by } c_{0,1} = c_0 - 2\delta \tag{6.33}$$
$$c_1 \text{ by } c_{1,1} = c_1(1 - \delta^{1/5})$$

conditions (6.3), (6.4), (6.5) remain preserved for the system $\{I_\beta^1\}$. Observe that form (6.14), each interval I_β^1 satisfies

$$\sigma_0(I_\beta^1) > \delta M_1. \tag{6.34}$$

To $E' \in \tau \backslash \tau'$, we associate then the measure

$$\sigma_1(\Lambda) = \sum_\beta \sigma_0(I_\beta^1 \cap \Lambda) \tag{6.35}$$

and let for $k_1 < k_2$

$$|k_1 - k_2| \geq d_0(k_1, k_2) \geq d_1(k_1, k_2) = \sigma_1([k_1, k_2]).$$

Thus from (6.34)

$$\sigma_0(I_\beta^1) = \sigma_1(I_\beta^1) > \delta M_1 \tag{6.36}$$

and from (6.30)

$$\sigma_1([1, N]) > (c_0 - 2\delta)N. \tag{6.37}$$

Letting the energy E' range in the set

$$\mathcal{E}_1 = \bigcup_{E \in \mathcal{E}} ((\tau_E \backslash \tau_E') \cap \mathcal{E})$$

it clearly follows from (6.12) that

$$\text{mes}\,(\mathcal{E} \backslash \mathcal{E}_1) < \frac{1}{M_1}. \tag{6.38}$$

For $E \in \mathcal{E}_1$, we repeat the construction (subject to a few modifications) with the associated system $\{I_\alpha^1\}$ satisfying (6.31), (6.32), i.e.

$$\|(R_{I_\alpha^1}(A - E)R_{I_\alpha^1})^{-1}\| < e^{\delta_1 M_1} \tag{6.39}$$

$$|(R_{I_\alpha^1}(A - E)R_{I_\alpha^1})^{-1}(k_1, k_2)| < e^{-c_{11} \cdot d_0(k_1, k_2)} \tag{6.40}$$

for $k_1, k_2 \in I_\alpha^1$ and $d_0(k_1, k_2) > \delta_1 M_1$.

Take then again

$$E' \in \tau = \tau_E = [E - e^{-3\delta_1 M_1}, E + e^{-3\delta_1 M_1}]. \tag{6.41}$$

Replacing E by E' clearly preserves properties (6.39), (6.40).

Define

$$M_2 = M_1^{100}$$

and partition $[1, N]$ in intervals $J_s = [sM_2, (s + 1)M_2]$. Denote

$$S = \left\{ s < \frac{N}{M_2} \,\middle|\, \sigma_1(J_s) > \delta_1 M_2 \right\}$$

where σ_1 is defined by (6.35).

Thus from (6.37)

$$\sum_{s \in S} \sigma_1(J_s) \geq \sigma_1([1, N]) - \frac{N}{M_2} \delta_1 M_2 > (c_{0,1} - \delta_1)N. \tag{6.42}$$

Fixing $s \in S$, set

$$\tau_s = \{E' \in \tau \,|\, \text{dist}(E', \text{Spec}\, R_\Lambda A R_\Lambda)$$
$$< M_2^{-5} e^{-3\delta_1 M_1} \text{ for some interval } \Lambda \subset J_s\}$$

of measure

$$|\tau_s| < M_2^3 M_2^{-5} e^{-3\delta_1 M_1} < M_2^{-2}|\tau|.$$

Hence, there is again a subset $\tau' \subset \tau$

$$|\tau'| < M_2^{-1}|\tau| \qquad (6.43)$$

so that if $E' \in \tau \backslash \tau'$ then, if

$$S' = \{s \in S | E' \notin \tau_s\}$$

we have that

$$|S \backslash S'| < \frac{N}{M_2^2}.$$

Thus, recalling (6.42)

$$\sum_{s \in S'} \sigma_1(J_s) > (c_{0,1} - \delta_1)N - \frac{N}{M_2} > (c_{0,1} - 2\delta_1)N. \qquad (6.44)$$

Fix $E' \in \tau \backslash \tau'$ and $s \in S'_{E'}$. Denote $J = J_s$. Thus

$$\sigma_1(J_s) > \delta_1 M_2 \qquad (6.45)$$

and

$$\|(R_\Lambda(A - E')R_\Lambda)^{-1}\| < M_2^5 e^{3\delta_1 M_1}$$

for all intervals $\Lambda \subset J_s$.

Take $k_1, k_2 \in J, k_1 < k_2$ such that

$$d_1(k_1, k_2) > 10M_1.$$

Case 1. $k_1 \in I_\alpha^1$ for some α and

$$\min_{k \in \partial I_\alpha^1} d_0(k_1, k) \geq \sqrt{\delta_1} M_1.$$

The same argument as in (6.18)-(6.19) gives then that

$$|(R_J(A - E')R_J)^{-1}(k_1, k_2)| \leq$$
$$\max_{\sqrt{\delta_1} M_1 \leq d_0(k_1,k') \leq M_1} e^{-c_{1,1} d_0(k_1,k')} |(R_J(A - E')R_J)^{-1}(k', k_2)|. \qquad (6.46)$$

Case 2. Negation of Case 1.

Let $I_\alpha^1 = [a_\alpha, a_\alpha + M_1], I_\beta^1 = [b_\alpha, b_\alpha + M_1]$ be consecutive intervals such that

$$a_\alpha < k_1 < a_\beta + M_1$$

and one of the following

$$k_1 \notin I_\alpha^1 \cup I_\beta^1$$

or
$$k_1 \in I_\alpha^1, \quad d_0(k_1, a_\alpha + M_1) < \sqrt{\delta_1} M_1$$

or
$$k_1 \in I_\beta^1, \quad d_0(k_1, a_\beta) < \sqrt{\delta_1} M_1.$$

Define $b_\alpha \in I_\alpha^1, b_\beta \in I_\beta^1$ such that

$$d_0(b_\alpha, a_\alpha + M_1) = \sqrt{\delta_1} M_1 \qquad (6.47)$$

$$d_0(a_\beta, b_\beta) = \delta_1^{1/4} M_1 \qquad (6.48)$$

and let
$$\Lambda = [b_\alpha, b_\beta]$$

which is the analogue if (6.20) for the distance d_0.

Observe that (6.47), (6.48) make sense, since by (6.36)

$$d_0(a_\alpha, a_\alpha + M_1) = \sigma_0(I_\alpha^1) > \delta M_1 = \delta_1^{1/100} M_1$$

and similarly for $d_0(a_\beta, a_\beta + M_1)$.

Repeat inequalities (6.21), (6.22).

Inequalities (6.22'), (6.23) need to be replaced by

$$d_1(k', k_2) \geq d_1(k_1, k_2) - d_0(k_4, k')$$

and

$$M_2^5 e^{3\delta_1 M_1} \max_{\substack{\sqrt{\delta_1} M_1 < d_0(k_4,k') < M_1 \\ d_1(k',k_2) \geq d_1(k_1,k_2) - d_0(k_4,k')}} e^{-c_{11} d_0(k_4,k')} \left| (R_J(A - E')R_J)^{-1}(k', k_2) \right|.$$

$$(6.49)$$

Inequalities (6.24), (6.25), (6.26), (6.27) are replaced by

$$d_0(k_4, k') = \delta_1^{1/4} M_1$$
$$d_1(k', k_2) \geq d_1(k_1, k_2) - d_0(b_\alpha, a_\alpha + M_1)$$
$$= d_1(k_1, k_2) - \delta_1^{1/2} M_1$$
$$\geq d_1(k_1, k_2) - \delta_1^{1/4} d_0(k_4, k')$$

$$d_0(k_4, k') = d_0(b_\beta, a_\beta + M_1)$$
$$= \sigma_0(I_\beta^1) - \delta_1^{1/4} M_1 \overset{\text{by (6.36)}}{>} (1 - \delta_1^{1/4 - 1/100}) \sigma_0(I_\beta^1)$$

$$d_1(k', k_2) \geq d_1(k_1, k_2) - d_0(b_\alpha, a_\alpha + M_1) - \sigma_0(I_\beta^1)$$
$$> d_1(k_1, k_2) - \frac{1 + \delta_1^{\frac{1}{2} - \frac{1}{100}}}{1 - \delta_1^{\frac{1}{4} - \frac{1}{100}}} \sigma_0(I_\beta^1)$$
$$> d_1(k_1, k_2) - (1 - 2\delta_1^{\frac{6}{25}}) d_0(k_4, k')$$

Replace (6.28), (6.29) by the bounds

$$M_2^5 e^{3\delta_1 M_1} \max_{\substack{\delta_1^{1/2} M_1 < d_0(k_4, k') < M_1 \\ d_1(k', k_2) \geq d_1(k_1, k_2) - (1 + 2\delta_1^{6/25}) d_0(k_4, k')}} e^{-c_{11} d_0(k_4, k')}$$

$$|(R_J(A - E')R_J)^{-1}(k', k_2)|$$

$$< e^{-\delta_1 M_1} \max_{d_1(k'', k_2) \geq d_1(k_1, k_2) - (1 + 2\delta^{6/25}) d_0(k', k'')} e^{-c_{11}(1 - \delta_1^{1/5}) d_0(k', k'')}$$

$$|(R_J(A - E')R_J)^{-1}(k'', k_2)| \tag{6.50}$$

that also cover (6.49), (6.46).

Iteration of (6.50) implies then again that

$$|(R_J(A - E')R_J)^{-1}(k_1, k_2)| < e^{-c_{11}(1 - \delta_1^{1/5}) d_1(k_1, k_2) + 10 M_1}$$

and

$$< e^{-c_{11}(1 - \delta_1^{1/5}) d_1(k_1, k_2)} \tag{6.51}$$

if

$$d_1(k_1, k_2) > \delta_1^{100} M_2 > \frac{M_1}{\delta_1}.$$

Letting thus

$$\delta_2 = \delta_1^{100}$$

the interval $J = J_s, s \in S'_{E'}$ satisfies thus

$$\|(R_J(A - E')R_J)^{-1}\| < M_2^5 e^{3\delta_1 M_1} < e^{\delta_2 M_2}$$

and (6.51) if $d_1(k_1, k_2) > \delta_2 M_2$.

To $E' \in \tau \setminus \tau'$, we associate then again the system $\{J_s | s \in S'_{E'}\} = \{I_\beta^2\}$ of M_2-intervals, satisfying (6.39), (6.40) with d_0 replaced by d_1, δ_1 by δ_2 and $c_{1,1}$ by

$$c_{1,2} = c_{1,1}(1 - \delta_1^{1/5}).$$

Recall (6.45)

$$\sigma_1(I_\beta^2) > \delta_1 M_2.$$

Letting

$$\sigma_2(\Lambda) = \sum_\beta \sigma_1(\Lambda \cap I_\beta^2)$$

(6.44), (6.33) imply

$$\sigma_2([1, N]) = \sum_\beta \sigma_1(I_\beta^2) > c_{0,2} N$$

where

$$c_{0,2} = c_{0,1} - 2\delta_1 = c_0 - 2\delta - 2\delta_1.$$

The energies E' range in

$$\mathcal{E}_2 = \bigcup_{E \in \mathcal{E}_1} (\tau_E \backslash \tau_{E'}) \cap \mathcal{E}_1$$

with τ_E given by (6.41), τ'_E satisfying (6.43).

Thus

$$\text{mes}\,(\mathcal{E}_2 \backslash \mathcal{E}_1) < \frac{1}{M_2}$$

and recalling (6.38)

$$\text{mes}\,(\mathcal{E} \backslash \mathcal{E}_2) < \frac{1}{M_1} + \frac{1}{M_2}.$$

The continuation of the process is clear.

Eventually, one obtains the conclusion stated in the lemma, with $\mathcal{E}' \subset \mathcal{E}$ satisfying

$$\text{mes}\,(\mathcal{E} \backslash \mathcal{E}') < \frac{1}{M^{10^2}} + \frac{1}{M^{10^4}} + \cdots < \frac{1}{M}$$

and where the measures σ on $[1, N]$ associated to each $E \subset \mathcal{E}'$ will satisfy

$$\sigma([1, N]) > (c_0 - 2\delta - 2\delta_1 - \cdots)N > \frac{c_0}{2}N.$$

The off-diagonal decay exponent for the Green's function $(A - E)^{-1}$ wrt the distance

$$d(k, k') = \sigma([k, k']) \text{ for } k < k'$$

is at least

$$c_1(1 - \delta^{1/5})(1 - \delta_1^{1/5}) \cdots > \frac{1}{2}c_1. \tag{6.52}$$

This proves Lemma 6.1.

Remark. In case the exponent c_1 in Lemma 6.1 is small, the smallness assumption on δ includes

$$\delta < c_1^{10}. \tag{6.53}$$

From (6.6), (6.7), we see that in particular for $E \in \mathcal{E}'$

$$|(A - E)^{-1}(1, N)| < e^{-\frac{1}{4}c_0 c_1 N}.$$

Hence

Corollary 6.54 *Assume A satisfies the assumptions of Lemma 6.1. Then, for $E \in \mathcal{E}'$, $\text{mes}\,(\mathcal{E} \backslash \mathcal{E}') < \frac{1}{M}$, we have that*

$$\frac{1}{N} \log \left\| \prod_N^1 \begin{pmatrix} E - v_n & -1 \\ 1 & 0 \end{pmatrix} \right\| \gtrsim c_0 c_1.$$

Remark. Corollary 6.54 could have been used to derive Proposition 4 from Corollary 3. The argument given in §5 is a bit simpler however since it does not involve the multiscale reasoning.

7   Application to Skew-shift

(7) In this section, we give an easy application of Corollary 3 and Corollary 6.54 to positivity of the Lyapounov exponent for small λ for certain skew shifts.

Recall that for given $\omega \in \mathbb{T}$, the skew-shift T on $\mathbb{T}^2$ is defined by

$$(x,y) \overset{T}{\longmapsto} (x+y, y+\omega)$$

hence

$$T^n(x,y) = \left(x + ny + \frac{n(n-1)}{2}\omega, y + n\omega\right).$$

We will prove

Proposition 5. *Let $\lambda > 0$ and assume $0 < |\omega| < \varepsilon(\lambda)$.*
Then for all $E \in [-2,2]$ except in a set of small complementary measure ($\to 0$ for $\lambda \to 0$), the exponent

$$\gamma_{\lambda,E} \overset{a.e.}{=} \lim_{N \to \infty} \frac{1}{N} \log \left\| \prod_N^1 \begin{pmatrix} E - \lambda\cos(x + ny + \frac{n(n-1)}{2}\omega) & -1 \\ 1 & 0 \end{pmatrix} \right\| > 0.$$

$$(7.1)$$

Proof. We denote

$$H = H_{\omega,\lambda,x,y} = \lambda \cos\left(x + ny + \frac{n(n-1)}{2}\omega\right)\delta_{nn'} + \Delta$$

where ω is assumed small

$$|\omega| < \varepsilon. \qquad (7.2)$$

Let $m \in \mathbb{Z}_+$ satisfy

$$m < \log\frac{1}{\varepsilon}. \qquad (7.3)$$

As in the proof of Proposition 4, we will locally replace H by an almost Mathieu operator with variable angle.   □

Write for $n = n_0 + k, 0 \le k \le m$

$$\cos\left(x + ny + \frac{n(n-1)}{2}\omega\right) =$$

$$\cos\left(x + n_0 y + \frac{n_0(n_0-1)}{2}\omega + \left(y + (n_0 - \frac{1}{2})\omega\right)k\right) + 0(\varepsilon m^2) =$$

$$\cos(\theta + 2\pi\alpha k) + 0(\varepsilon m^2)$$

where

$$\theta = x + n_0 y + \frac{n_0(n_0-1)}{2}\omega$$

and

$$\alpha = y + \left(n_0 - \frac{1}{2}\right)\omega.$$

We repeat the considerations from section 5.

Let thus $\mathcal{E}_\lambda$ be the set of energies from Lemma 2, Corollary 3. We restrict $\mathcal{E}_\lambda$ a bit to ensure that if $E \in \mathcal{E}_\lambda$, then

$$\gamma(\alpha, \lambda, E) > c_1 = c_1(\lambda)$$

for a set of rotation numbers $\alpha \in \mathcal{R}_E \subset \mathbb{T}$ s.t.

$$|\mathcal{R}_E| > c_1(\lambda). \tag{7.4}$$

Moreover, for $m > m_0$ (=fixed integer), the estimate (5.11) for the restricted almost Mathieu Green's function (5.10) holds, i.e.

$$|G_\Lambda(\alpha, \theta)(k_1, k_2)| < e^{-c_1|k_1 - k_2| + 0(m^{1-\sigma/2})} \tag{7.5}$$

for some $\Lambda \in \{[0, m], [0, m - 1], [1, m], [1, m - 1]\}$ and provided

$$\theta \in \Theta_{\alpha, E}$$

where

$$\mathrm{mes}\,(\mathbb{T} \backslash \Theta) < e^{-\frac{1}{2}m^{\sigma/2}}. \tag{7.6}$$

Considering $\cos(x + ny + \frac{n(n-1)}{2}\omega)$, we partition the index set $[1, N]$ in intervals $I = [n_0, n_0 + m]$ of size $m + 1$. Then, from the preceding

$$R_I(H - E)R_I = R_{I-n_0}[H_{\frac{1}{2\pi}(y + (n_0 - \frac{1}{2})\omega), \lambda, x + n_0 y + \frac{n_0(n_0-1)}{2}\omega} - E]R_{I-n_0} + 0(\varepsilon m^2)$$

where $H_{\alpha, \lambda, \theta}$ refers to the almost Mathieu Hamiltonians.

One may write further that

$$R_I(H - E)R_I = R_{I-n_0}[H_{\alpha, \lambda, x + n_0 y + \frac{n_0(n_0-1)}{2}\omega} - E]R_{I-n_0} + 0(\varepsilon m^2)$$

provided

$$\left\| \frac{1}{2\pi}\left(y + \left(n_0 - \frac{1}{2}\right)\omega\right) - \alpha \right\| < \varepsilon. \tag{7.7}$$

By (7.3) and assuming (7.7) holds for some $\alpha \in \mathcal{R}_E$ and $x + n_0 y + \frac{n_0(n_0-1)}{2}\omega \in \Theta_{\alpha, E}$, the estimate (7.5) remains valid for $[R_I(H - E)R_I]^{-1}$, for some

$$I \in \{[n_0, n_0 + m], [n_0, n_0 + m - 1], [n_0 + 1, n_0 + m], [n_0 + 1, n_0 + m - 1]\}. \tag{7.8}$$

Fix $y \in \mathbb{T}$.

Equidistribution considerations of the sequence $\{\frac{1}{2\pi}(y + (n_0 - \frac{1}{2})\omega)\}$, $n_0 \in [1, N]$ and (7.4) easily show that if

$$\mathcal{J}_E = \{n_0 \in [1, n]| \mathrm{dist}_\mathbb{T}\left(\frac{1}{2\pi}(y + (n_0 - \frac{1}{2})\omega), \mathcal{R}_E\right) < \varepsilon\}$$

then

$$|\mathcal{J}_E| > \frac{c_1}{2} N.$$

Proceeding as in §5, we obtain thus for each $n_0 \in \mathcal{J}_E$ a set $\Theta'_{(n_0),E} \subset \mathbb{T}$ satisfying (7.6), i.e.

$$\text{mes}\,(\mathbb{T} \backslash \Theta'_{(n_0),E}) < e^{-\frac{1}{2}m^{\sigma/2}} \tag{7.9}$$

and such that

$$|[R_I(H_{\omega,\lambda,x,y} - E)R_I]^{-1}(k_1, k_2)| < e^{-c_1|k_1 - k_2| + 0(m^{1-\sigma/2})} \tag{7.10}$$

for some I in (7.8), provided

$$x + n_0 y + \frac{n_0(n_0 - 1)}{2}\omega \in \Theta'_{(n_0),E}.$$

The set

$$\left\{ (E, n, x) \in \mathcal{E}_\lambda \times [1, N] \times \mathbb{T} \,|\, n \in \mathcal{J}_E \text{ and } x + ny + \frac{n(n-1)}{2}\omega \notin \Theta'_{(n),E} \right\}$$

is by (7.9) of measure $< N e^{-\frac{1}{2}m^{\sigma/2}}$. By Fubini, there is therefore a subset

$$\Theta' \subset \mathbb{T}$$

$$\text{mes}\,(\mathbb{T}\backslash\Theta') < \frac{1}{10}$$

such that if $x \in \Theta'$, then there is a subset $\mathcal{E}_x \subset \mathcal{E}_\lambda$

$$\text{mes}\,(\mathcal{E}_\lambda\backslash\mathcal{E}_x) = o(1) \tag{7.11}$$

and for $E \in \mathcal{E}_x$

$$|\mathcal{J}_{E,x}| > \frac{c_1}{3} N$$

where

$$\mathcal{J}_{E,x} = \left\{ n_0 \in \mathcal{J}_E \,|\, x + n_0 y + \frac{n_0(n_0 - 1)}{2}\omega \in \Theta'_{(n),E} \right\}.$$

Thus for fixed $x \in \Theta'$ there is for each $E \in \mathcal{E}_x$ a collection of disjoint intervals $\{I_\alpha\} \subset [1, N], |I_\alpha| = m - 1, m, m + 1$, such that

$$\sum |I_\alpha| > \frac{c_1}{4} N$$

and each I_α satisfies (7.10).

Letting $A = R_{[1,N]} H_{\omega,\lambda,x,y} R_{[1,N]}$ in Lemma 6.1, $\mathcal{E} = \mathcal{E}_x$, we get a subset $\mathcal{E}'_x \subset \mathcal{E}_x$,

$$|\mathcal{E}_x\backslash\mathcal{E}'_x| < \frac{1}{m} \tag{7.12}$$

such that by Corollary 6.54

$$\frac{1}{N} \log \left\| \prod_N^1 \left(\begin{array}{cc} E - \lambda \cos(x + ny + \frac{n(n-1)}{2}\omega) & -1 \\ 1 & 0 \end{array} \right) \right\| \gtrsim c_1(\lambda)$$

for all $E \in \mathcal{E}'_x$.

Since

$$\{(x, E) \in \Theta' \times \mathcal{E}_\lambda | E \in \mathcal{E}'_x\}$$

is by (7.11), (7.12) of small complementary measure in $\Theta' \times \mathcal{E}_\lambda$, it follows again form Fubini that

$$\frac{1}{N} \int \log \left\| \prod_N^1 \left(\begin{array}{cc} E - \lambda \cos(x + ny + \frac{n(n-1)}{2}\omega) & -1 \\ 1 & 0 \end{array} \right) \right\| dx \gtrsim c_1(\lambda)$$

for E in a subset $\mathcal{E}' \subset \mathcal{E}_\lambda$ of small complementary measure. This set depends on y, which was fixed in the preceding. Thus, another application of Fubini produces a set $\mathcal{E}' \subset \mathcal{E}_\lambda$ of energies, $\text{mes}(\mathcal{E}') > 2 - o(1)$, such that if $E \in \mathcal{E}'$

$$\frac{1}{N} \int \log \left\| \prod_N^1 \left(\begin{array}{cc} E - \lambda \cos(x + ny + \frac{n(n-1)}{2}\omega) & -1 \\ 1 & 0 \end{array} \right) \right\| dx\,dy \gtrsim c_1(\lambda). \quad (7.13)$$

Here $\mathcal{E}'$ depends on N (and on λ).

Finally, one obtains clearly a set of energies $\mathcal{E}'_\lambda \subset \mathcal{E}_\lambda$, $\text{mes}(\mathcal{E}_\lambda) > 2 - o(1)$, such that if $E \subset \mathcal{E}'_\lambda$, then (7.13) holds for infinitely many $N's$. Therefore

$$\gamma(E, \lambda) = \lim_{N \to \infty} (7.13) \gtrsim c_1(\lambda)$$

for $E \in \mathcal{E}'_\lambda$.

Recalling condition (6.2), (6.53), i.e.

$$m^{-1/2} < \delta < c_1(\lambda)^{10}$$

and (7.2), (7.3), ω is subject here to a bound

$$|\omega| < e^{-c_1(\lambda)^{-20}}.$$

(8) Further remarks on skew shift

(i) Observe that

$$\frac{1}{m} \iint \log \left\| \prod_{n=n_0}^{n_0+m} \left(\begin{array}{cc} E - \lambda \cos(x + ny + \frac{n(n-1)}{2}\omega) & -1 \\ 1 & 0 \end{array} \right) \right\| dx\,dy$$

$$= \frac{1}{m} \iint \log \left\| \prod_{k=0}^{m} \left(\begin{array}{cc} E - \lambda \cos(x + ky + \frac{k(k-1)}{2}\omega) & -1 \\ 1 & 0 \end{array} \right) \right\| dx\,dy. \quad (8.1)$$

Fixing $\lambda > 0$, assume that we established that

$$(8.1) > \delta > 0$$

for $m = m_0(\delta)$ and all E (or for all E in a subset $\mathcal{E} \subset \mathcal{R}$).

(This may possibly be performed numerically).

It follows then from Lemma 6.1 that the Lyapounov exponent $\gamma_{\lambda,E}$ in (7.1) will be strictly positive for all $E \in \mathcal{E}' \subset \mathcal{E}$,

$$\mathrm{mes}\,(\mathcal{E}\backslash\mathcal{E}') < \gamma(m,\delta) \overset{m\to\infty}{\longrightarrow} 0.$$

Thus, in case of absence of absolutely continuous spectrum for

$$H_{\omega,\lambda,x,y} = \lambda \cos(x + ny + \frac{n(n-1)}{2}\omega)\delta_{nn'} + \Delta$$

(this fact is independent of x, y if $\omega \notin \pi\mathbb{Q}$), one may in principle establish any upperbound

$$\mathrm{mes}\,(\sigma_{\omega,\lambda}) < \gamma$$

for any $\gamma > 0$, by numerics.

These considerations apply in fact equally well to any Hamiltonian

$$H_x = v(T^n x)\delta_{nn'} + \Delta$$

with T an ergodic measure preserving transformation of, say, a torus $\mathbb{T}^b$ and v a smooth function of $\mathbb{T}^b$.

(ii) Assume next

$$\omega = 4\pi\frac{p}{q} \in \pi\mathbb{Q} \quad (p, q \in \mathbb{Z}_+). \tag{8.2}$$

Then, from periodicity

$$M_N(x, y, E) = \prod_{N=N_1 q-1}^{0} \begin{pmatrix} E - \lambda \cos(x + ny + \frac{n(n-1)}{2}\omega) & -1 \\ 1 & 0 \end{pmatrix}$$

$$= \prod_{n_1=N_1-1}^{0} \prod_{r=q-1}^{0} \begin{pmatrix} E - \lambda \cos(x + ry + \frac{r(r-1)}{2}\omega + n_1 qy) & -1 \\ 1 & 0 \end{pmatrix} \tag{8.3}$$

Fixing y, one may thus define $A(x) \in SL_2(R)$ by

$$A(x) = \prod_{r=q-1}^{0} \begin{pmatrix} E - \lambda \cos(x + ry + \frac{r(r-1)}{2}\omega) & -1 \\ 1 & 0 \end{pmatrix}$$

and write

$$(8.3) = \prod_{n_1=N_1-1}^{0} A(x + n_1 qy). \tag{8.4}$$

This observation permits us to apply the methods from [B-G] and [G-S] related to the shift (given by qy in (8.4)). In particular, there is the following conclusion

$$\left| \gamma_{E,\lambda} - \frac{1}{N} \iint \log \| M_N(x, y, E) \| dx dy \right| < \varepsilon(q, N, \delta) \qquad (8.5)$$

where we assume $\gamma_{E,\lambda} \equiv \lim_{N \to \infty} \frac{1}{N} \iint \log \| M_N(x, y, E) \| dx dy > \delta > 0$ and $\varepsilon(q, N, \delta) \overset{N \to \infty}{\longrightarrow} 0$ is an explicit upperbound.

From (8.5), positivity of $\gamma_{E,\lambda}$ for arbitrary $\lambda > 0$ (when the property holds) may be established numerically, since (8.5) gives an explicit rate of convergence for $N \to \infty$ (not provided by Kingman's subadditivity theorem).

Then considerations are strongly dependent (at this point) on our assumption (8.2).

References

[A-A]   Aubry S., Andre G. (1980) Analyticity breaking and Anderson localization in commensurate lattices. Ann. Israel Phys. Soc. 3:133–164

[B-G]   Bourgain J., Goldstein M. On nonperturbative localization with quasiperiodic potential. Annals of Math., to appear

[B-F]   Brenner N., Fishman S. (1992) Pseudo-randomness and localization. Nonlinearity 4:211-235

[F-S-W] Fröhlich J., Spencer T., Wittwer P. (1990) Localization for a class of one dimensional quasi-periodic Schrödinger operators. Comm. Math. Physics. 132:5–25

[G-S]   Goldstein M., Schlag W. (1999) Hölder continuity of the integrated density of states for quasi-periodic Schrödinger equations and averages of shifts of subharmonic functions. Preprint, to appear

[G-F]   Griniasty M., Fishman S. (1988) Localization by pseudorandom potentials in one dimension. Phys. Rev. Lett. 60:1334–1337

[H]   Herman M. (1983) Une méthode pour minorer les exposants de Lyapounov et quelques exemples montrant le charactère local d'un theoreme d'Arnold et de Moser sur le tore de dimension 2. Comment. Math. Helv. 58(3):453–502

[Ji]   Jitomirskaya S. (1999) Metal-insulator transition for the almost Mathieu operator. Annals of Math. 150(3):1159–1175

[La]   Last Y. (1995) Almost everything about the Almost Mathieu operator. I XIth International Congress of Math. Physics, Intern. Press Inc, Boston, 366–372

[L-S]   Last Y., Simon B. (1999) Eigenfunctions, transfer matrices, and absolutely continuous spectrum of one-dimensional Schrödinger operators. Inventiones Math. 135:329-367

[S]   Sinai Y.G. (1987) Anderson localization for one-dimensional difference Schrödinger operator with quasi-periodic potential. J. Stat. Phys. 46:861–909

[S-S]   Sorets E., Spencer T. (1991) Positive Lyapounov exponents for Schrödinger operators with quasi-periodic potentials. Comm. Math. Phys. 142(3): 543–566

Anderson Localization for the Band Model

J. Bourgain[1] and S. Jitomirskaya[2]

[1] Institute for Advanced Study, Princeton, NJ 08540, USA
[2] University of California, Irvine, CA 92717, USA

Abstract. In this paper, we show how the methods from [B-G] may be adapted to establish Anderson localization for quasi-periodic lattice Schrödinger operators corresponding to the band model $\mathbb{Z} \times \{1, \ldots, b\}$. Recall that 'Anderson localization' means pure point spectrum with exponentially decaying eigenfunctions. We also discuss the issue of dynamical localization.

1 LDT for Subharmonic Functions

Lemma 1.1 *Let* $u : \mathbb{T} \to \mathbb{R}$ *be periodic with bounded subharmonic extension* $\tilde{u}$ *to* $|\operatorname{Im} z| \leq 1$.
   Assume $\omega \in \mathbb{T}$ *typical.*
   Then, for $\kappa > M^{-1/3}$

$$\operatorname{mes} \left[\theta \in \mathbb{T} \, \Big| \, \Big| \sum_{0 \leq |m| < M} \frac{M - |m|}{M^2} u(\theta + m\omega) - \langle u \rangle \Big| > \kappa \right] < e^{-c\kappa M}. \quad (1.2)$$

Proof. It follows from the assumption and Riesz' theorem that for $z = x + iy$, $|x| \leq 1$, $|y| \leq \frac{1}{2}$

$$\tilde{u}(z) = \int \log |z - w| \mu(dw) + h(z)$$

where h is smooth and μ is a positive measure of bounded mass. Therefore

$$u(\theta) = \langle u \rangle + \sum_{k \in \mathbb{Z} \setminus \{0\}} \hat{u}(k) e^{ik\theta} \quad \text{with} \quad |\hat{u}(k)| < \frac{C}{|k|} \quad (1.3)$$

(cf. [B-G]).
   Let q be an approximant of ω s.t.

$$M < q < M (\log M)^2 \quad (1.4)$$

$$\Rightarrow \quad \|j\omega\| > \frac{1}{q+1} \quad \text{for} \quad j \leq q. \quad (1.5)$$

Write

$$v(\theta) \equiv \sum_{0 \leq |m| < M} \frac{M - |m|}{M^2} u(\theta + m\omega) - \langle u \rangle$$

$$= \sum_{k \neq 0} \hat{u}(k) \left[\sum_{|m| < M} \frac{M - |m|}{M^2} e^{2\pi i m k \omega} \right] e^{2\pi i k \theta} \quad (1.6)$$

with

$$\left| \sum_{|m|<M} \frac{M-|m|}{M^2} e^{2\pi imk\omega} \right| < \frac{1}{1+M^2\|k\omega\|^2}. \tag{1.7}$$

Split the sum (1.6) as

$$\sum_{k\neq 0} = \sum_{0<|k|<M^{1/2}} + \sum_{M^{1/2}\leq|k|<q} + \sum_{q\leq|k|<K} + \sum_{|k|\geq K}$$

$$= (I) + (II) + (III) + (IV)$$

with

$$\log K \sim \kappa M. \tag{1.8}$$

From (1.3), (1.7)

$$|(I)| < \sum_{0<|k|<M^{1/2}} \frac{1}{|k|} \frac{1}{1+M^2\|k\omega\|^2} < \sum_{0\leq|k|<M^{1/2}} \frac{1}{M|k|\,\|k\omega\|} < M^{-\frac{1}{2}+\varepsilon} \tag{1.9}$$

$$|(II)| < \sum_{M^{1/2}\leq|k|<q} \frac{1}{|k|} \frac{1}{1+M^2\|k\omega\|^2} < M^{-1/2} \sum_{|k|<q} \frac{1}{1+M^2\|k\omega\|^2}. \tag{1.10}$$

Observe that if J is an interval in $\mathbb{Z}$ of length q, then

$$\sum_{k\in J} \frac{1}{1+M^2\|k\omega\|^2} < 1 + \sum_{1\leq j\leq q} \frac{1}{1+M^2\frac{j^2}{q^2}} < \frac{q}{M}. \tag{1.11}$$

Thus, from (1.10), (1.11)

$$|(II)| < 2qM^{-3/2} < M^{-\frac{1}{2}+\varepsilon}. \tag{1.12}$$

Similarly, recalling (1.8)

$$|(III)| < \sum_{q<|k|<K} \frac{1}{|k|(1+M^2\|k\omega\|^2)} < \sum_{s=1}^{K/q} \frac{1}{sq} \frac{q}{M} < \frac{\log K}{M} < \frac{\kappa}{2}. \tag{1.13}$$

Finally

$$\|(IV)\|_{L^2_\theta} < \left(\sum_{|k|>K} |k|^{-2} \right)^{1/2} < K^{-1/2}$$

$$\Rightarrow \text{mes } [\theta \in \mathbb{T} |\, |(IV)| > \kappa] < \kappa^{-2} e^{-c\kappa M}.$$

This proves the lemma. $\qquad\square$

Remark. If ω satisfies a weaker Diophantine condition, of the form $\|k\omega\| > c|k|^{-a}$, $k \in \mathbb{Z}\backslash\{0\}$, one can obtain a similar inequality, with the right hand side in (1.2) of the form $e^{-c\kappa M^{\frac{1}{a}-\varepsilon}}$, sufficient for what follows.

2   Band Schrödinger Operators

Define

$$H_{(n,s),(n',s')}(\omega,\theta) = \lambda\delta_{nn'}\delta_{ss'}v_s(\theta+n\omega) + \Delta \qquad (n \in \mathbb{Z}, s = 1,\dots,b) \tag{2.1}$$

where

$$\Delta((n,s),(n',s')) = \begin{cases} 1 & \text{if } |n-n'|+|s-s'|=1 \\ 0 & \text{otherwise} \end{cases} \tag{2.2}$$

and $v_s(1 \le s \le b)$ are non-constant periodic and real analytic (assume bounded analytic extension to $|\operatorname{Im} z| < 2$). We will also write $H(\theta)$ or H for $H(\omega,\theta)$.

For $\Lambda \subset \mathbb{Z} \times \{1,\dots,b\}$, denote R_Λ the restriction operator and $R_{[1,N]} = R_{[1,N]\times\{1,\dots,b\}}$. C and $O(1)$ will stand for various constants, independent of $\lambda, \theta, \omega, E$.

The next lemma is an analogue of the Soretz-Spencer estimate for the Lyapounov exponent.

Lemma 2.3 *Fix $\delta > 0$. Then there is λ_0 s.t. for $\lambda > \lambda_0$, $|E| < C\lambda$, any ω, we have*

$$\int_{\mathbb{T}} \left[\frac{1}{bN} \log|\det(R_{[1,N]}(H-E)R_{[1,N]})|\right] d\theta > (1-\delta)\log\lambda. \tag{2.4}$$

Proof. Since the analytic extensions

$$|v_s(z)| < C \quad \text{for } |\operatorname{Im} z| \le 1 \tag{2.5}$$

we get an analytic function

$$F(z) = \det(R_{[1,N]}(H(z)-E)R_{[1,N]}) \tag{2.6}$$

bounded by

$$|F(z)| < (C|\lambda| + |E| + 4)^{bN}. \tag{2.7}$$

Hence

$$u = \frac{1}{bN}\log|F| \tag{2.8}$$

has subharmonic extension to $|\operatorname{Im} z| \le 1$ bounded from above by

$$u(z) < \log|\lambda| + O(1) \tag{2.9}$$

if $|E| < C|\lambda|$.

The main point, depending on analyticity, is that there is $\varepsilon > 0$ such that for all $|E_1| < C$, there exists

$$\frac{\delta}{2} < y < \delta \tag{2.10}$$

for which

$$\inf_{x \in \mathbb{R}, 1 \le s \le b} |v_s(x + iy) - E_1| > \varepsilon \tag{2.11}$$

Fix $|E| < C\lambda$. Thus there is y satisfying (2.10) and

$$|\lambda v_s(\theta + n\omega + iy) - E| > \lambda\varepsilon \quad \text{for all } \theta \in \mathbb{T}, n \in \mathbb{Z}, s = 1, \dots, b. \tag{2.12}$$

Take

$$\lambda > \lambda_0 \gtrsim \frac{1}{\varepsilon}. \tag{2.13}$$

One easily sees that

$$|F(\theta + iy)| > (\lambda\varepsilon)^{bN} \left(1 - \frac{C}{\lambda\varepsilon}\right)^{bN}, \tag{2.14}$$

hence

$$u(\theta + iy) > \log(\lambda\varepsilon - C) > \log\varepsilon\lambda - o(1). \tag{2.15}$$

To get (2.14), we factor out the diagonal part of the determinant, use (2.12) as a lower bound on the diagonal elements and Hadamard's bound for the inverse of the second factor.

Denote μ_{iy} the harmonic measure of iy in the strip $0 \le \operatorname{Im} z \le 1$. It follows from subharmonicity, (2.9), (2.15)

$$\log\varepsilon\lambda - O(1) < u(iy)$$

$$\le \int_{\operatorname{Im} z=0} u(\theta)\mu_{iy}(d\theta) + \int_{\operatorname{Im} z=1} u(\theta + i)\mu_{iy}(d\theta)$$

$$< \int_{\operatorname{Im} z=0} u(\theta)\mu_{iy}(d\theta) + \delta(\log\lambda + O(1)) \tag{2.16}$$

since

$$\mu_{iy}(\operatorname{Im} z = 1) \le \delta. \tag{2.17}$$

Consequently

$$\int_{\operatorname{Im} z=0} u(\theta)\mu_{iy}(d\theta) > \log\lambda - \log\frac{1}{\varepsilon} - O(1) - \delta(\log\lambda)$$

$$> (1 - 2\delta)\log\lambda. \tag{2.18}$$

Replace $0 + iy$ by $\theta_0 + iy$ for arbitrary $\theta_0 \in R$. Estimate (2.18) remains valid for μ_{θ_0+iy} instead of μ_{iy}. Averaging over $\theta_0 \in \mathbb{T}$, we obtain

$$\int_{\mathbb{T}} u(\theta)d\theta > (1 - 2\delta)\log\lambda. \qquad (2.19)$$

This proves the Lemma. $\qquad\qquad\qquad\qquad\qquad\qquad\qquad\qquad\qquad\qquad\qquad$ □

Remark. If the v_s are trigonometric polynomials, one may simply apply Herman's subharmonicity argument to get a lower bound $\log\lambda - C$ in (2.4).

The next Lemma provides an upperbound on the minors $\mu_{(n,s);(n',s')}$ of $R_{[1,N]}(H - E)R_{[1,N]}$.

Lemma 2.20 *There is a uniform estimate*

$$\frac{1}{bN}\log|\mu_{(n,s);(n',s')}| < \left(1 - \frac{|n - n'|}{bN}\right)\log\lambda + O(1). \qquad (2.21)$$

Proof. We have that

$$|\mu_{(n,s);(n',s')}| \le \sum_{\gamma}\left|\det\left[R_{([1,N]\times\{1,...,b\})\backslash\gamma}(H - E)R_{([1,N]\times\{1,...,b\})\backslash\gamma}\right]\right| \qquad (2.22)$$

where the sum extends to all paths γ joining (n, s) and (n', s'). We write $R_{\gamma c}$ for $R_{([1,N]\times\{1,...,b\})\backslash\gamma}$

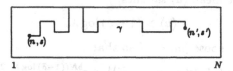

Obviously

$$|\det R_{\gamma c}(H - E)R_{\gamma c}| \le (C|\lambda| + |E| + 4)^{bN-\ell(\gamma)} \qquad (2.23)$$

with

$$\ell(\gamma) = \text{ length of } \gamma \ge |n - n'|. \qquad (2.24)$$

Thus

$$|\mu_{(n,s);(n',s')}| < \sum_{\ell \ge |n-n'|} 4^{\ell}(C|\lambda|)^{bN-\ell} < (C|\lambda|)^{bN}\left(\frac{4}{C|\lambda|}\right)^{|n-n'|} \qquad (2.25)$$

$$\log|\mu_{(n,s),(n,s)}| < (bN - |n - n'|)\left(\log|\lambda| + O(1)\right) \qquad (2.26)$$

which is (2.21). $\qquad\qquad\qquad\qquad\qquad\qquad\qquad\qquad\qquad\qquad\qquad\qquad\qquad$ □

3 Green's Function Estimates

Proposition 3.1 *Denote*

$$G_N(E + oi; \theta) = \left(R_{[1,N]}(H - E)R_{[1,N]} \right)^{-1}$$

the Green's function. We have chosen λ large enough (see below) and E is fixed ($|E| < C\lambda$).

Then for all θ outside a set of measure $< e^{-cM}$, there is $|m| < M$ s.t. the Green's function $G_N = G_N(E + oi; \theta + m\omega)$ satisfies the bound

$$|G_N\big((n, s); (n', s')\big)| < e^{-(|n - n'| - \frac{N}{50})\log\lambda}. \tag{3.2}$$

Proof. Denote

$$u(\theta) = \frac{1}{2bN} \log(|\det R_{[1,N]}(H(\theta) - E)R_{[1,N]}|^2 + 1) \tag{3.2'}$$

which has a bounded (by $\log\lambda + 0(1)$) subharmonic extension to $|\operatorname{Im} z| \leq 1$.

From (1.2) in Lemma 1.1 and M large enough

$$\sum_{0 \leq |m| < M} \frac{M - |m|}{M^2} u(\theta + m\omega) > \langle u \rangle - 1 \tag{3.3}$$

for all θ outside a set $\Omega \subset \mathbb{T}$ of measure

$$\operatorname{mes} \Omega < e^{-cM} \qquad c = c(\lambda). \tag{3.4}$$

Fixing $\delta > 0$ (to be specified) and letting λ be large enough, it follows from Lemma 2.3 that the mean $\langle u \rangle$ satisfies

$$\langle u \rangle > (1 - \delta)\log\lambda. \tag{3.5}$$

Therefore, we may choose $|m| < M$ so that

$$|\det \big(H_N(\theta + m\omega) - E \big)| > e^{bN((1-\delta)\log\lambda - 2)}, \tag{3.6}$$

where we write H_N for $R_{[1,N]}H R_{[1,N]}$. Combined with the upperbound (2.21) on the minors, we obtain thus

$$|G_N(\theta + m\omega, E + oi)\big((n, s); (n', s')\big)| =$$

$$\frac{|\mu_{(n,s);(n',s')}|}{|\det \big(H_N(\theta + m\omega) - E \big)|} <$$

$$\exp bN\left[\left(1 - \frac{|n - n'|}{bN} \right) \log\lambda + O(1) - (1 - \delta)\log\lambda \right] <$$

$$\exp \big[(\delta bN - |n - n'|)\log\lambda + CbN \big]. \tag{3.7}$$

For $\delta = \frac{1}{100b}$ and λ sufficiently large (depending on b), it follows

$$|G_N\big((n, s); (n', s')\big)| < e^{(\frac{N}{50} - |n - n'|)\log\lambda} \tag{3.8}$$

as claimed. $\qquad\square$

Proposition 3.9 *There is a constant C s.t. for all θ, there is $0 \le j < N^C$ for which the Green's function $G_N(E + oi; \theta + j\omega)$ satisfies the estimate (3.2).*

Proof. Returning to the proof of Proposition 3.1 and denoting Ω the exceptional set where

$$\sum_{0 \le |m| < M} \frac{M - |m|}{M^2} u(\theta' + m\omega) \le \langle u \rangle - 1 \tag{3.10}$$

(u defined by (3.2′); M to be specified), it clearly suffices to show that $\theta + j\omega \notin \Omega$ for some $0 \le j < N^C \equiv N_1$.

Condition (3.10) may be rewritten as

$$\prod_{0 \le |m| < M} \left[1 + (\det R_{[1,N]}(H(\theta' + m\omega) - E) R_{[1,N]})^2 \right]^{(M - |m|)} \le e^{2bM^2 N(\langle u \rangle - 1)} \tag{3.11}$$

where the left side is of degree $< bM^2 N$ in the $\{v(\theta' + m\omega)\}$. If v is given by a trigonometric polynomial, (3.11) is of the form

$$P(\cos \theta', \sin\theta') \le 0 \tag{3.12}$$

with

$$\deg P < CbM^2 N. \tag{3.13}$$

In the general real analytic case, property (3.2) clearly admits perturbing v by $e^{-CN \log \lambda}$ and hence replacement by a trigonometric polynomial of degree $\sim N$. Hence

$$\deg P < 0(M^2 N^2) \tag{3.14}$$

in (3.12) and $\Omega \subset \mathbb{T}$ may be taken to be union of at most $(MN)^{10}$ intervals I_α. Recalling further (3.4)

$$|I_\alpha| \le \text{mes } \Omega < e^{-cM}. \tag{3.15}$$

Take

$$M = (\log N)^2. \tag{3.16}$$

The points in the orbit $\{\theta + j\omega \mid |j| < N_1\}$ may be assumed at least $N_1^{-2} \gg e^{-cM}$ separated in $\mathbb{T}$, so that each interval I_α contains at most one of its points. Letting $N_1 > (MN)^{10}$, it follows thus that

$$\{\theta + j\omega \mid j = 0, 1, \dots, N_1\} \not\subset \Omega \tag{3.17}$$

for any θ. This proves Proposition 3.9. □

Remark. Although the proof above requires a Diophantine condition on ω, the Lemma can be formulated without it. Namely, one can show (cf. [J-L]) that for any ω and θ, there is $|j| < g_n$ s.t. the Green's function $G_N(E + oi, \theta + j\omega)$ satisfies the estimate (3.2), where g_n is an approximant of ω s.t. $g_n > CN^2$.

4 Anderson Localization

Theorem 4.1 *Consider the band Schrödinger operator* (2.1). *Fix* $\theta \in \mathbb{T}$. *Then for λ large enough, A.L. holds for almost all ω.*

Remark. The result and the proof extends with obvious changes to operators (2.1) on $\mathbb{Z} \times \{1, \ldots, b\}^d$, or, more generally, on $\mathbb{Z} \times G$, where G is any finite graph.

Proof. The basic scheme of the argument is analogous to the 1D quasi-periodic case (i.e. $b = 1$) treated in [B-G]. We briefly recall some main points.

Fix N. Consider the set $\mathfrak{S} \subset \mathbb{T}^2 \times R$ of elements (ω, θ', E) satisfying

$$\|G_{N_1}(E)\| \geq C^N \tag{4.2}$$

and

$$\sum_{0 \leq |m| < M} \frac{M - |m|}{M^2} u(\theta' + m\omega) \leq (1 - \delta) \log \lambda \tag{4.3}$$

where

$$u = \frac{1}{2bN} \log \left(|\det R_N\big(H(\omega, \theta') - E\big) R_N|^2 + 1 \right) \tag{4.4}$$

and

$$M = \frac{1}{100} N. \tag{4.5}$$

Notice that $G_{N_1}(E)$ in (4.2) also depends on ω. The constant C in (4.2) is taken sufficiently large, $N_1 < N^{100}$. The frequency vector ω is subject to a diophantine condition $\|k\omega\| > c|k|^{-3}$ say, restricted at this stage to $|k| < 2N$.

Using the Hilbert-Schmidt norm, condition (4.2) may be replaced by a polynomial condition

$$P_0(E, \omega) \geq 0 \tag{4.6}$$

with

$$\deg P_0 \leq N^C. \tag{4.7}$$

Considering (4.3) as a condition in ω, θ', E, we get, cf. (3.11)

$$\prod_{0 \leq |m| < M} \left[1 + (\det R_N\big(H(\omega, \theta') - E\big) R_N)^2 \right]^{(M - |m|)} \leq \lambda^{2(1-\delta)bM^2 N}. \tag{4.8}$$

This condition may again be substituted for a polynomial inequality

$$P_1(E, \omega, \cos\theta', \sin\theta') > 0 \tag{4.9}$$

with

$$\deg P_1 < N^C. \tag{4.10}$$

Thus $\mathfrak{S}$ is defined by (4.6), (4.9) and ω satisfies the restricted DC.

The next point is that the projection

$$\text{mes Proj}_{\omega,\theta'}(\mathfrak{S}) < e^{-cN} \tag{4.11}$$

since, using (3.4), one may verify that

$$\text{mes Proj}_{\theta'}(\mathfrak{S}_\omega) < e^{-cM} \tag{4.12}$$

for fixed diophantine ω.

Our aim (as in [B-G]) is to ensure that

$$\{(\omega, \theta + \ell\omega)|\ell \sim e^{(\log N)^2}\} \cap \text{Proj}_{\omega,\theta'}(\mathfrak{S}) = \phi. \tag{4.13}$$

The argument is the same as in [B-G].

The main point is the fact that for fixed $\theta' \in \mathbb{T} \backslash \Omega'$

$$\text{Proj}_\omega(\mathfrak{S}_{\theta'}) = \bigcup_{\alpha < N^C} I_\alpha \tag{4.14}$$

with $I_\alpha \subset [0, 1]$ intervals of size $< e^{-\frac{c}{2}N}$ and where

$$\text{mes } \Omega' < e^{-\frac{c}{2}N}. \tag{4.15}$$

This is a consequence of previous semi-algebraic description of $\mathfrak{S}$ and the measure estimate (4.11). $\qquad\qquad\square$

5   Remark on Dynamical Localization

The result stated in Theorem 4.1, holds also for the dynamical localization. Thus denoting e^{itH} the Schrödinger group with Hamiltonian H considered above, we have that

$$\sup_{t \in \mathbb{R}} \sum_{\substack{x \in \mathbb{Z} \\ s=1,\dots,b}} x^2 |(e^{itH}\psi)(x, s)|^2 < \infty \tag{5.1}$$

for any initial data ψ of sufficiently rapid decay (the weight x^2 in (5.1) may be replaced by an arbitrary power x^{2r}).

The argument indicated below applies to any of the situations considered in [B-G] or [B-G-S] where AL is established. As for the previous section, we will follow the exposition from [B-G] (see §7 in particular). Since the bandwidth b (assumed fixed) plays no role in what follows, take $b = 1$. Denote

(φ_α) the ℓ^2-normalized eigenfunctions of H (for which AL holds) and E_α corresponding eigenvalues. Then

$$(e^{itH}\psi)(x) = \sum_\alpha e^{itE_\alpha}\langle \psi, \varphi_\alpha\rangle\varphi_\alpha(x), \qquad (5.2)$$

hence

$$|(e^{itH}\psi)(x)| \le \sum_\alpha |\langle \psi, \varphi_\alpha\rangle|\,|\varphi_\alpha(x)| \qquad (5.3)$$

and

$$\left[\sum_{x\in\mathbb{Z}} x^2 |(e^{itH}\psi)(x)|^2\right]^{1/2} \le \sum_\alpha |\langle \psi, \varphi_\alpha\rangle|\left(\sum x^2 |\varphi_\alpha(x)|^2\right)^{1/2}. \qquad (5.4)$$

Take first $\psi = e_0 \equiv \delta_{0,x}$, so that $\langle \psi, \varphi_\alpha\rangle = \varphi_\alpha(0)$.
   Fix $\varepsilon > 0$ and α such that

$$|\varphi_\alpha(0)| > \varepsilon. \qquad (5.5)$$

Let

$$n \gtrsim \log \frac{1}{\varepsilon}. \qquad (5.6)$$

One may then obtain some n_0

$$n_0 \sim n^C \qquad (5.7)$$

(C a suitable constant related to Proposition 3.9) such that

$$|\varphi_\alpha(-n_0 - 1)| + |\varphi_\alpha(n_0 + 1)| < e^{-n} \qquad (5.8)$$

(cf. [B-G]).
   Denote

$$G_\Lambda(E) = \left[R_\Lambda(H - E)R_\Lambda\right]^{-1}.$$

Since

$$R_{[-n_0,n_0]}\varphi_\alpha = -G_{[-n_0,n_0]}(E_\alpha)R_{[-n_0,n_0]}(H - E_\alpha)R_{\mathbb{Z}\setminus[-n_0,n_0]}\varphi_\alpha, \quad (5.9)$$

it follows from (5.5), (5.8) that

$$\varepsilon < \|R_{[-n_0,n_0]}\varphi_\alpha\| \le \|G_{[-n_0,n_0]}(E_\alpha)\|(|\varphi_\alpha(n_0 + 1)| + |\varphi_\alpha(-n_0 - 1)|) \quad (5.10)$$

$$\|G_{[-n_0,n_0]}(E_\alpha)\| > \varepsilon.e^n > e^{n/2}. \qquad (5.11)$$

Following [B-G] and using the suitable restriction of ω, one concludes then from (5.11) that

$$|\varphi_\alpha(N)| < e^{-c|N|} \qquad (5.12)$$

for all

$$2^{(\log n)^2} < |N| < 2^{(\log n)^3}. \qquad (5.13)$$

In the present context, we may take $c = 1$ in (5.12). In the [B-G] context, the exponent c depends on a lower bound for the Lyapounov exponents

$$L(E_\alpha) > \delta_0. \qquad (5.14)$$

Recall also that ensuring the preceding at stage n required removal of a subset of measure $< e^{-\frac{1}{4}(\log n)^2}$ from the ω-parameter set.

Thus, taking into account (5.6), (5.12), (5.13), we conclude that

$$|\varphi_\alpha(j)| < e^{-|j|} \text{ if } |j| > 4^{(\log\log\frac{1}{\epsilon})^2}. \qquad (5.15)$$

Denote $\mathcal{F}_\epsilon$ the set of α's for which (5.5) holds and $\Lambda = [-J_\epsilon, J_\epsilon], J_\epsilon = 4^{(\log\log\frac{1}{\epsilon})^2}$. By orthonormality of the eigenvectors $\{\varphi_\alpha\}$ and (5.15), for sufficiently small ϵ

$$\begin{aligned}
|\mathcal{F}_\epsilon| = \sum_{\alpha \in \mathcal{F}_\epsilon} \|\varphi_\alpha\|_2^2 &< 2 \sum_{\alpha \in \mathcal{F}_\epsilon} \|R_\Lambda \varphi_\alpha\|_2^2 \\
&\leq 2 \sum_\alpha \|R_\Lambda \varphi_\alpha\|_2^2 \\
&\leq 2 \text{ rank } R_\Lambda < 5 J_\epsilon. \qquad (5.16)
\end{aligned}$$

Also, for $\alpha \in \mathcal{F}_\epsilon$

$$\begin{aligned}
\sum_x x^2 |\varphi_\alpha(x)|^2 &\leq J_\epsilon^2 + \sum_{|x|>J_\epsilon} x^2 e^{-c|x|} \\
&< J_\epsilon^2 + 1. \qquad (5.17)
\end{aligned}$$

Hence, returning to (5.4), we obtain the bound

$$\sum_{\substack{0<\epsilon<1 \\ \epsilon \text{ dyadic}}} |\mathcal{F}_\epsilon|.\epsilon.(J_\epsilon + 1) \lesssim$$

$$\sum_\epsilon \epsilon J_\epsilon^2 = \sum_\epsilon \epsilon 4^{(\log\log\frac{1}{\epsilon})^2} < C. \qquad (5.18)$$

This proves (5.1) for $\psi = e_0$.

Replacing e_0 by a unit vector e_k, the preceding may be repeated providing we let moreover $n_0 > |k|$ in (5.10), hence also $\log n \gtrsim \log k$. Hence J_ϵ is replaced by $4^{(\log\log\frac{1}{\epsilon})^2 + (\log k)^2}$, leading to the bound $4^{(\log k)^2}$ in (5.18). Writing

$$\psi = \sum \psi_k e_k \qquad (5.19)$$

this permits us to obtain (5.1) provided

$$|\psi_k| < e^{-C(\log k)^2}. \tag{5.20}$$

The condition may be weakened to

$$|\psi_k| < |k|^C \tag{5.21}$$

for some constant C, by more adequate restriction of ω. More precisely, fix $k \in \mathbb{Z}$ and replace H by $H^{(k)}$ obtained by k-shift

$$H^{(k)}(x, x') = H(x + k, x' + k). \tag{5.22}$$

If one ensures the same property for $H^{(k)}$ as exploited above for H, but considering now only scales

$$n > n_k \sim \frac{1}{\kappa} e^{C\sqrt{\log(|k|+2)}} \tag{5.23}$$

removal of a subset of measure at most

$$\sum_{\substack{n \text{ dyadic} \\ n > n_k}} e^{-\frac{1}{4}(\log n)^2} < 2e^{-\frac{1}{4}(\log n_k)^2} < \kappa k^{-2} \tag{5.24}$$

in the ω-parameter set is required. Thus summing (5.24) in k, the total contribution is at most $\kappa = o(1)$.

Fixing k and ε, condition (5.6) is replaced by

$$n > n_k + C \log \frac{1}{\varepsilon}, \tag{5.25}$$

hence, from (5.23), (5.25)

$$J_\varepsilon^{(k)} = e^{[\log(n_k + C \log \frac{1}{\varepsilon})]^2}$$
$$< C_\kappa e^{2(\log\log \frac{1}{\varepsilon})^2} \cdot |k|^C. \tag{5.26}$$

We have that

$$|\varphi_\alpha(j)| < e^{-c|j-k|} < e^{-\frac{c}{2}|j|} \text{ if } |j| > J_\varepsilon^{(k)} \gg |k| \tag{5.27}$$

whenever φ_α is an eigenfunction of H satisfying

$$|\varphi_\alpha(k)| > \varepsilon. \tag{5.28}$$

From (5.27), the resulting bound in (5.1) is thus

$$\sup_{t \in \mathbb{R}} \sum_{x \in \mathbb{Z}} x^2 |(e^{itH}\psi)(x)|^2 < C_\kappa |k|^C \text{ for } \psi = e_k. \tag{5.29}$$

A condition on ψ of the form (5.21) for suitable power C therefore suffices to ensure (5.1).

References

[B-G]   Bourgain J., Goldstein M. (1999) On non-perturbative localization with
          quasiperiodic potential. Preprint. Annals of Math., to appear
[B-G-S]   Bourgain J., Goldstein M., Schlag W. (2000) Anderson localization for
          Schrödinger operators on $\mathbb{Z}$ with potentials given by the skew-shift.
          Preprint
[J-L]   Jitomirskaya S., Last Y. (1999) Power Law subordinary and singular spec-
          tra II, Line operators. CMP, to appear

Convex Bodies with Minimal Mean Width

A.A. Giannopoulos[1], V.D. Milman[2] and M. Rudelson[3]

[1] Department of Mathematics, University of Crete, Iraklion, Greece
[2] School of Mathematical Sciences, Tel Aviv University, Tel Aviv 69978, Israel
[3] Department of Mathematics, University of Missouri, Columbia, MO 65211, USA

1 Introduction

Let K be a convex body in $\mathbb{R}^n$, and $\{TK \mid T \in SL(n)\}$ be the family of its *positions*. In [GM] it was shown that for many natural functionals of the form

$$T \mapsto f(TK), \qquad T \in SL(n),$$

the solution T_0 of the problem

$$\min\{f(TK) \mid T \in SL(n)\}$$

is *isotropic* with respect to an appropriate measure depending on f. The purpose of this note is to provide applications of this point of view in the case of the *mean width functional* $T \mapsto w(TK)$ under various constraints.

Recall that the *width* of K in the direction of $u \in S^{n-1}$ is defined by $w(K, u) = h_K(u) + h_K(-u)$, where $h_K(y) = \max_{x \in K}\langle x, y \rangle$ is the *support function* of K. The width function $w(K, \cdot)$ is translation invariant, therefore we may assume that $o \in \operatorname{int}(K)$. The mean width of K is given by

$$w(K) = \int_{S^{n-1}} w(K, u)\sigma(du) = 2\int_{S^{n-1}} h_K(u)\sigma(du),$$

where σ is the rotationally invariant probability measure on the unit sphere S^{n-1}.

We say that K has *minimal mean width* if $w(TK) \geq w(K)$ for every $T \in SL(n)$. The following isotropic characterization of the minimal mean width position was proved in [GM]:

Fact. *A convex body K in $\mathbb{R}^n$ has minimal mean width if and only if*

$$\int_{S^{n-1}} h_K(u)\langle u, \theta \rangle^2 \sigma(du) = \frac{w(K)}{2n}$$

for every $\theta \in S^{n-1}$. Moreover, if $U \in SL(n)$ and UK has minimal mean width, we must have $U \in O(n)$. □

Research of the second named author partially supported by the Israel Science Foundation founded by the Academy of Sciences and Humanities. Research of the third named author was supported in part by NSF Grant DMS-9706835.

Our first result is an application of this fact to a "reverse Urysohn inequality" problem: The classical Urysohn inequality states that $w(K) \geq (|K|/\omega_n)^{1/n}$ where ω_n is the volume of the Euclidean unit ball D_n, with equality if and only if K is a ball. A natural question is to ask for which bodies K

$$a_n := \max_{|K|=1} \min_{T \in SL(n)} w(TK)$$

is attained, and what is the precise order of growth of a_n as $n \to \infty$. Examples such as the regular simplex or the cross-polytope show that $a_n \geq c\sqrt{n}\sqrt{\log(n+1)}$. On the other hand, it is known that every symmetric convex body K in $\mathbb{R}^n$ has an image TK with $|TK| = 1$ for which

$$w(TK) \leq c_1\sqrt{n}\log[d(X_K, \ell_2^n) + 1],$$

where $X_K = (\mathbb{R}^n, \|\cdot\|_K)$ and d denotes the Banach-Mazur distance. This statement follows from an inequality of Pisier [Pi], combined with work of Lewis [L], Figiel and Tomczak-Jaegermann [FT]. John's theorem [J] implies that

$$\min_{T \in SL(n)} w(TK) \leq c_2\sqrt{n}\log(n+1),$$

for every symmetric convex body K with $|K| = 1$, and a simple argument based on the difference body and the Rogers-Shephard inequality [RS] shows that the same holds true without the symmetry assumption. Therefore,

$$c\sqrt{n}\sqrt{\log(n+1)} \leq a_n \leq c_3\sqrt{n}\log(n+1).$$

Here, we shall give a precise estimate for the minimal mean width of zonoids (this is the class of symmetric convex bodies which can be approximated by Minkowski sums of line segments in the Hausdorff sense):

Theorem A. *Let Z be a zonoid in $\mathbb{R}^n$ with volume $|Z| = 1$. Then,*

$$\min_{T \in SL(n)} w(TZ) \leq w(Q_n) = \frac{2\omega_{n-1}}{\omega_n},$$

where $Q_n = [-1/2, 1/2]^n$.

For our second application, we consider the class of origin symmetric convex bodies in $\mathbb{R}^n$. Every symmetric body K induces a norm $\|x\|_K = \min\{\lambda \geq 0 : x \in \lambda K\}$ on $\mathbb{R}^n$, and we write X_K for the normed space $(\mathbb{R}^n, \|\cdot\|_K)$. The polar body of K is defined by $\|x\|_{K^\circ} = \max_{y \in K} |\langle x, y \rangle| = h_K(x)$, and will be denoted by K°. Whenever we write $(1/a)|x| \leq \|x\|_K \leq b|x|$, we assume that a, b are the smallest positive numbers for which this inequality holds true for every $x \in \mathbb{R}^n$.

We consider the average

$$M(K) = \int_{S^{n-1}} \|x\|_K \sigma(dx)$$

of the norm $\|\cdot\|_K$ on S^{n-1}, and define $M^*(K) = M(K^\circ)$. Thus, $M^*(K)$ is half the mean width of K. We will say that K has *minimal* M if $M(K) \leq M(TK)$ for every $T \in SL(n)$. Equivalently, if K° has minimal mean width.

Our purpose is to show that if K has minimal M, then the volume radius of K is bounded by a function of b and M. Actually, it is of the order of b/M. The precise formulation is as follows:

Theorem B. *Let K be a symmetric convex body in $\mathbb{R}^n$ with minimal M, such that $(1/a)|x| \leq \|x\|_K \leq b|x|$, $x \in \mathbb{R}^n$. Then,*

$$\frac{b}{M} \leq \left(\frac{|K|}{|(1/b)D_n|} \right)^{1/n} \leq c \frac{b}{M} \log \left(\frac{2b}{M} \right),$$

where $c > 0$ is an absolute constant.

Our last result concerns optimization of the width functional under a different condition. We say that an n-dimensional symmetric convex body K is in the *Gauss-John position* if the minimum of the functional

$$\mathbb{E}\|g\|_{TK}$$

under the constraint $TK \subseteq D_n$ is attained for $T = I$. That is, K° has minimal mean width under the condition $TK \subseteq D_n$ (it minimizes M under the condition $a(TK) \leq 1$).

We can consider this optimization problem only for positive self-adjoint operators T. Since the norm of T should be bounded to guarantee that $TK \subseteq D_n$ and the norm of T^{-1} should be bounded as well, there exists T for which the minimum is attained. Denote by γ the standard Gaussian measure in $\mathbb{R}^n$. Then, we have the following decomposition.

Theorem C. *Let K be in the Gauss-John position. Then there exist: $m \leq n(n+1)/2$, contact points $x_1, \ldots, x_m \in \partial K \cap S^{n-1}$ and numbers $c_1, \ldots, c_m > 0$ such that $\sum_{i=1}^m c_i = 1$ and*

$$\int_{\mathbb{R}^n} (x \otimes x - I)\|x\|_K d\gamma(x) = \int_{\mathbb{R}^n} \|x\|_K d\gamma(x) \cdot \left(\sum_{i=1}^m c_i x_i \otimes x_i \right).$$

The Gauss-John position is not equivalent to the classical John position. Examples show that, when K is in the Gauss-John position, the distance between D_n and the John ellipsoid may be of order $\sqrt{n/\log n}$.

2   Reverse Urysohn Inequality for Zonoids

The proof of Theorem A will make use of a characterization of the minimal surface position, which was given by Petty [Pe] (see also [GP]): Recall that the area measure σ_K of a convex body K is defined on S^{n-1} and corresponds to the usual surface measure on K via the Gauss map: For every Borel $V \subseteq S^{n-1}$, we have

$$\sigma_K(V) = \nu(\{x \in \mathrm{bd}(K) : \text{the outer normal to } K \text{ at } x \text{ is in } V\}),$$

where ν is the $(n-1)$-dimensional surface measure on K. If $A(K)$ is the surface area of K, we obviously have $A(K) = \sigma_K(S^{n-1})$. We say that K has *minimal surface area* if $A(K) \le A(TK)$ for every $T \in SL(n)$. With these definitions, we have:

2.1 Theorem. *A convex body K in $\mathbb{R}^n$ has minimal surface area if and only if*

$$\int_{S^{n-1}} \langle u, \theta \rangle^2 \sigma_K(du) = \frac{A(K)}{n}$$

for every $\theta \in S^{n-1}$. Moreover, if $U \in SL(n)$ and UK has minimal surface area, we must have $U \in O(n)$. $\square$

Recall also the definition of the projection body ΠK of K: it is the symmetric convex body whose support function is defined by $h_{\Pi K}(\theta) = |P_\theta(K)|$ where $P_\theta(K)$ is the orthogonal projection of K onto $\theta^\perp$, $\theta \in S^{n-1}$. It is known that Z is a zonoid in $\mathbb{R}^n$ if and only if there exists a convex body K in $\mathbb{R}^n$ such that $Z = \Pi K$. By the formula for the area of projections, this can be written in the form

$$h_Z(x) = \frac{1}{2} \int_{S^{n-1}} |\langle x, u \rangle| \sigma_K(du).$$

Then, the characterization of the minimal mean width position and Theorem 2.1 imply the following:

2.2 Lemma. *Let $Z = \Pi K$ be a zonoid. Then, Z has minimal mean width if and only if K has minimal surface area.*

Proof. The proof (modulo the characterization of the minimal mean width position) may be found in [Pe]: By Cauchy's surface area formula,

$$A(K) = \frac{n\omega_n}{\omega_{n-1}} \int_{S^{n-1}} h_Z(\theta) \sigma(d\theta).$$

If f_2 is a spherical harmonic of degree 2, the Funk-Hecke formula shows that

$$\int_{S^{n-1}} f_2(u) |\langle u, \tau \rangle| \sigma(du) = c_n f_2(\tau)$$

for all $u, \tau \in S^{n-1}$, where c_n is a constant depending only on the dimension. Therefore,

$$\int_{S^{n-1}} f_2(u) h_Z(u) \sigma(du) = \frac{1}{2} \int_{S^{n-1}} \int_{S^{n-1}} f_2(u) |\langle u, \tau \rangle| \sigma(du) \sigma_K(d\tau)$$

$$= \frac{c_n}{2} \int_{S^{n-1}} f_2(\tau) \sigma_K(d\tau).$$

Since $u \mapsto \langle u, \theta \rangle^2$ is homogeneous of degree 2, this implies

$$\int_{S^{n-1}} h_Z(u) \langle u, \theta \rangle^2 \sigma(du) = \frac{c_n}{2} \int_{S^{n-1}} \langle u, \theta \rangle^2 \sigma_K(du)$$

for every $\theta \in S^{n-1}$. The characterizations of the minimal mean width and the minimal surface area positions make it clear that Z has minimal mean width if and only if K has minimal surface area. $\qquad\square$

Our next lemma is a well-known fact, proved by K. Ball [B]:

2.3 Lemma *Let $\{u_j\}_{j \leq m}$ be unit vectors in $\mathbb{R}^n$ and $\{c_j\}_{j \leq m}$ be positive numbers satisfying*

$$I = \sum_{j=1}^{m} c_j u_j \otimes u_j.$$

If $Z = \sum_{j=1}^{m} \alpha_j [-u_j, u_j]$ for some $\alpha_j > 0$, then

$$|Z| \geq 2^n \prod_{j=1}^{m} \left(\frac{\alpha_j}{c_j} \right)^{c_j}. \qquad\square$$

We apply this result to the projection body of a convex body with minimal surface area.

2.4 Lemma *If K has minimal surface area, then*

$$A(K) \leq n |\Pi K|^{1/n}.$$

Proof. We may assume that K is a polytope with facets F_j and normals u_j, $j = 1, \ldots, m$. Then, Theorem 2.1 is equivalent to the statement

$$I = \sum_{j=1}^{m} c_j u_j \otimes u_j$$

where $c_j = n|F_j|/A(K)$ (see [GP]). On the other hand,

$$\Pi K = \frac{A(K)}{2n} \sum_{j=1}^{m} c_j [-u_j, u_j].$$

We now apply Lemma 2.3 for $Z = \Pi K$, with $\alpha_j = \frac{A(K)}{2n} c_j$:

$$|\Pi K| \geq 2^n \left(\frac{A(K)}{2n} \right)^n. \qquad\square$$

Remark. An alternative proof of Lemma 2.4 may be given through Barthe's reverse Brascamp-Lieb inequality (see [Ba]). In the previous argument, equality can hold only if $(u_j)_{j \leq m}$ is an orthonormal basis of $\mathbb{R}^n$. This means that if K is a polytope then equality in Lemma 2.4 can hold only if K is a cube.

Proof of Theorem A. Let Z be a zonoid with minimal mean width and volume $|Z| = 1$. By Lemma 2.2, Z is the projection body ΠK of some convex body K with minimal surface area. We have

$$w(Z) = 2 \int_{S^{n-1}} h_Z(u)\sigma(du) = 2 \int_{S^{n-1}} |P_u(K)|\sigma(du) = \frac{2\omega_{n-1}}{n\omega_n} A(K).$$

By Lemma 2.4, the area of K is bounded by $n|Z|^{1/n} = n$. We have equality when K is a cube, and this corresponds to the case $Z = Q_n$. Therefore,

$$w(Z) \leq w(Q_n) = \frac{2\omega_{n-1}}{\omega_n}. \qquad \square$$

Remark. Urysohn's inequality and Theorem A show that if Z is a zonoid with $|Z| = 1$, then

$$\alpha_n \sqrt{\frac{2}{\pi e}} \sqrt{n} \leq \min_{T \in SL(n)} w(TZ) \leq \beta_n \sqrt{\frac{2}{\pi}} \sqrt{n},$$

where $\alpha_n, \beta_n \to 1$ as $n \to \infty$.

3 Volume Ratio of Symmetric Convex Bodies with Minimal M

For the proof of Theorem B we will need the following fact which was proved in [GM]:

3.1 Theorem. *Let K be a symmetric convex body in $\mathbb{R}^n$ with minimal M. Then, for every $\lambda \in (0,1)$ there exists a $[(1-\lambda)n]$-dimensional subspace E of $\mathbb{R}^n$ such that*

$$\frac{b}{r(\lambda)}|x| \leq \|x\|_K \leq b|x| \quad , \quad x \in E, \tag{3.1}$$

where $r(\lambda) \leq c\frac{b}{M\lambda^{1/2}}\log(\frac{2b}{M\lambda})$, and $c > 0$ is an absolute constant. $\square$

Actually, the proof of Theorem 3.1 shows that the statement holds true for a random $[(1-\lambda)n]$-dimensional subspace E of $\mathbb{R}^n$. One can assume that for every $k \leq n - \frac{n}{c\log^2 n}$ we have the result with probability greater than $1 - \frac{1}{n}$ (this formulation is correct when $n \geq n_0$, where $n_0 \in \mathbb{N}$ is absolute). This assumption on the measure of subspaces satisfying (3.1) implies that there is an increasing sequence of subspaces $E_1 \subset E_2 \subset \ldots \subset E_{k_0}$, where $k_0 = [n - \frac{n}{c\log^2 n}]$ and $\dim E_k = k$, so that (3.1) holds for each E_k with $r = r(k/n)$.

We will also need the following

3.2 Lemma. *Let K be a symmetric convex body in $\mathbb{R}^n$, such that $(1/a)|x| \leq \|x\|_K \leq b|x|$. If E is a k-dimensional subspace of $\mathbb{R}^n$, then*

$$\frac{|K|}{|D_n|} \leq \left(Ca(\frac{n}{n-k})^{1/2}\right)^{n-k} \frac{|K \cap E|}{|D_k|},$$

where $C > 0$ is an absolute constant.

Proof. Let E be a k-dimensional subspace of $\mathbb{R}^n$. Replacing K by $(1/a)K$, we may assume that $a = 1$, so $K \subset D_n$. Using Brunn's theorem we see that

$$|K| = \int_{P_{E^\perp}(K)} |K \cap (E + y)| dy \leq |P_{E^\perp}(K)||K \cap E|$$

$$\leq |K \cap E||D_{n-k}|.$$

This shows that

$$\frac{|K|}{|D_n|} \leq \frac{|D_k||D_{n-k}|}{|D_n|} \frac{|K \cap E|}{|D_k|} \leq \left(C\frac{n}{n-k}\right)^{(n-k)/2} \frac{|K \cap E|}{|D_k|}. \qquad \square$$

Proof of Theorem B. We first observe that

$$ab \leq Cn \log n.$$

Indeed, let $e \in S^{n-1}$ be such that $\|e\|_K = b$ and let γ be a standard normal variable. Then

$$cb = E\|\gamma e\| \leq E\|g\|_K.$$

Similarly,

$$ca \leq E\|g\|_{K^\circ}.$$

Multiplying these inequalities, we obtain

$$ab \leq CE\|g\|_K E\|g\|_{K^\circ} \leq Cn \log n.$$

The last inequality follows from the fact that M is minimal for K, and Pisier's inequality [Pi].

Assume now that $b = 1$. Let $t = [\log n]$. For $s = 1, 2, \ldots, t$ put $k_s = [(1 - 1/s)n]$ and let $E_s = E_{k_s}$ be a subspace from our flag. Then by Theorem 3.1 we have $(1/a_s)|x| \leq \|x\|_K \leq |x|$ on E_s, where

$$a_s \leq r((n - k_s)/n) \leq \frac{c}{M}s^{1/2} \log\left(\frac{2s}{M}\right) \leq s \cdot \frac{c}{M} \log\left(\frac{2}{M}\right) =: s \cdot c(M).$$

Now, Lemma 3.2 shows that

$$\frac{|K|}{|D_n|} \leq \left(Ca\sqrt{t}\right)^{n-k_t} \frac{|K \cap E_t|}{|D_{k_t}|} \leq (Cn \log^2 n)^{n/\log n} \frac{|K \cap E_t|}{|D_{k_t}|}$$

$$\leq C^n \frac{|K \cap E_t|}{|D_{k_t}|}$$

and

$$\frac{|K \cap E_{s+1}|}{|D_{k_{s+1}}|} \leq (Ca_s s)^{k_{s+1}-k_s} \frac{|K \cap E_s|}{|D_{k_s}|}$$

$$\leq (Cc(M)s^2)^{k_{s+1}-k_s} \frac{|K \cap E_s|}{|D_{k_s}|},$$

for all $s = 1, 2, \ldots, t$. Since $a_1 \leq c(M)$, we have $|K \cap E_1|/|D_{k_1}| \leq c(M)^{k_1}$. Hence, multiplying the inequalities above, we get

$$\frac{|K|}{|D_n|} \leq C^n (Cc(M))^{k_t - k_1} c(M)^{k_1} \prod_{s=2}^{t} s^{2(k_{s+1}-k_s)}.$$

By the definition of k_s,

$$\prod_{s=2}^{t} s^{2(k_{s+1}-k_s)} \leq \exp\left(cn \cdot \sum_{s=2}^{t} \frac{\log s}{s^2}\right) \leq e^{cn},$$

therefore

$$\left(\frac{|K|}{|(1/b)D_n|}\right)^{1/n} \leq C_1 c(M).$$

The left hand side inequality is an immediate consequence of Hölder's inequality:

$$\left(\frac{|K|}{|(1/b)D_n|}\right)^{1/n} = b\left(\int_{S^{n-1}} \|x\|_K^{-n} \sigma(dx)\right)^{1/n}$$

$$\geq b\left(\int_{S^{n-1}} \|x\|_K\right)^{-1} = \frac{b}{M}. \qquad \square$$

4   Gauss-John Position

We prove Theorem C. Consider the following optimization problem:

$$F(T) = \int_{\mathbb{R}^n} \|T^{-1}x\|_K \, d\gamma(x) \to \min \qquad (4.1)$$

under the constraint

$$H_x(T) = |Tx|^2 - 1 \leq 0 \quad \text{for} \quad x \in K.$$

Assume that the body K is in the Gauss-John position, namely the minimum in (4.1) is attained for $T = I$. Let W be the set of the contact points of K: $W = \partial K \cap \partial S^{n-1}$. First we apply an argument of John [J] to show that we can consider only finitely many constraints. Since the paper [J] is not easily available, we shall sketch the argument. Let T be a self-adjoint operator and let $T_s = I + sT$. We shall prove that if

$$\frac{d}{ds}H_x(T_s)|_{s=0} < 0$$

for every $x \in W$, then

$$\frac{d}{ds}F(T_s)|_{s=0} \geq 0.$$

Indeed, assume that

$$a = \sup_{x \in W} \frac{d}{ds}H_x(T_s)|_{s=0} < 0.$$

Let W_ε be an ε-neighborhood of W: $W_\varepsilon = \{x \in K | \text{dist}(x, W) < \varepsilon\}$. There exists an $\varepsilon > 0$ such that

$$\frac{d}{ds}H_x(T_s)|_{s=0} < \frac{a}{2}$$

for every $x \in W_\varepsilon$. So, there exists $s_0 > 0$ such that for any $0 < s < s_0$ and any $x \in W_\varepsilon$

$$H_x(T_s) < H_x(I) \leq 0.$$

On the other hand, if $x \in K \setminus W_\varepsilon$ then

$$H_x(T_s) \leq |H_x(T_s) - H_x(I)| + H_x(I).$$

Here,

$$|H_x(T_s) - H_x(I)| = \left||T_s x|^2 - |x|^2\right| \leq \|T\|(1 + \|T\|)s$$

and

$$\sup_{x \in K \setminus W_\varepsilon} H_x(I) = \sup_{x \in K \setminus W_\varepsilon} |x|^2 - 1 < 0$$

since $K \setminus W_\varepsilon$ is compact.

Thus for a sufficiently small s we have $H_x(T_s) < 0$ for all $x \in K$. So, since I is the solution of the minimization problem (4.1),

$$\frac{d}{ds}F(T_s)|_{s=0} \geq 0.$$

Since $\frac{d}{ds}H_x(T_s)|_{s=0} = \langle \nabla H_x(I), T \rangle$ and $\frac{d}{ds}F(T_s)|_{s=0} = \langle \nabla F(I), T \rangle$, this means that the vector $-\nabla F(I)$ cannot be separated from the set $\{\nabla H_x(I) | x \in W\}$ by a hyperplane. By Carathéodory's theorem, there exist $M \leq n(n+1)/2$ contact points $x_1 \ldots x_M \in W$ and numbers $\lambda_1 \ldots \lambda_M > 0$ such that

$$-\nabla F(I) = \sum_{i=1}^{M} \lambda_i \nabla H_{x_i}(I) = \sum_{i=1}^{M} \lambda_i x_i \otimes x_i. \tag{4.2}$$

Now we have to calculate $\nabla F(I)$.

We have

$$F(T) = (2\pi)^{-n/2} \int_{\mathbb{R}^n} \|T^{-1}x\|_K e^{-|x|^2/2} dx$$

$$= (2\pi)^{-n/2} \det T \cdot \int_{\mathbb{R}^n} \|x\|_K e^{-|Tx|^2/2} dx,$$

so,

$$\nabla F(I) = \left((2\pi)^{-n/2} \int_{\mathbb{R}^n} \|x\|_K e^{-|x|^2/2} dx \right) I$$

$$- (2\pi)^{-n/2} \int_{\mathbb{R}^n} \|x\|_K e^{-|x|^2/2} x \otimes x dx$$

$$= \int_{\mathbb{R}^n} (I - x \otimes x) \cdot \|x\|_K d\gamma(x).$$

Combining it with (4.2) we obtain

$$\int_{\mathbb{R}^n} (I - x \otimes x) \cdot \|x\|_K d\gamma(x) + \sum_{i=1}^{M} \lambda_i x_i \otimes x_i = 0.$$

Taking the trace, we get

$$\mathrm{Tr} \left(\int_{\mathbb{R}^n} (I - x \otimes x) \cdot \|x\|_K d\gamma(x) \right)$$

$$= n \int_{\mathbb{R}^n} \|x\|_K d\gamma(x) - \int_{\mathbb{R}^n} |x|^2 \cdot \|x\|_K d\gamma(x)$$

$$= n \int_0^\infty r^n e^{-n^2/2} dr \int_{S^{n-1}} \|\omega\|_K dm(\omega) - \int_0^\infty r^{n+2} e^{-n^2/2} dr \int_{S^{n-1}} \|\omega\|_K dm(\omega)$$

$$= - \int_{\mathbb{R}^n} \|x\|_K d\gamma(x).$$

Finally, putting $\lambda_i = c_i \int_{\mathbb{R}^n} \|x\|_K d\gamma(x)$, we obtain the decomposition

$$\int_{\mathbb{R}^n} (I - x \otimes x) \|x\|_K d\gamma(x) = \int_{\mathbb{R}^n} \|x\|_K d\gamma(x) \left(\sum_{i=1}^{M} c_i x_i \otimes x_i \right),$$

where $\sum_{i=1}^{M} c_i = 1$. This completes the proof of Theorem C.   □

We proceed to compare D_n with the John ellipsoid in the Gauss-John position:

Proposition. *Let K be a symmetric convex body in $\mathbb{R}^n$ which is in the Gauss-John position. Then,*

(i) $(2/\pi)^{1/2} n^{-1} D_n \subset K \subset D_n$;

(ii) *It may happen that $\frac{c\sqrt{\log n}}{n} D_n$ is not contained in K.*

Proof. (i) Let T_0 be an operator which puts K into the maximal volume position. Then $\min F(T) \leq F(T_0) \leq n$. From the other side, if there exists $y \in S^{n-1}$ such that $\|y\|_{K^\circ} < (2/\pi)^{1/2} n^{-1}$ then

$$\int_{\mathbb{R}^n} \|x\|_K d\gamma(x) \geq \int_{\mathbb{R}^n} |\langle x, y/\|y\|_{K^\circ}\rangle| d\gamma(x) \geq (2/\pi)^{1/2} \cdot 1/\|y\|_{K^\circ} > n.$$

(ii) Let $K = B_1^{n-1} + [-e_n, e_n]$. Let T be a positive self-adjoint operator such that TK is in the Gauss-John position. We first prove that T is a diagonal operator. Let $G \subset O(n)$ be the group generated by the operators $U_i = I - 2e_i \otimes e_i$, $i = 1, \ldots, n$ and let m be the uniform measure on G. Notice that $U_i K = K$ for every i. Then,

$$\int_{\mathbb{R}^n} \|x\|_{TK} d\gamma(x) = \int_G \int_{\mathbb{R}^n} \|Ux\|_{TU(K)} d\gamma(x) dm(U)$$

$$\geq \int_{\mathbb{R}^n} \left\|\left(\int_G U^{-1}T^{-1}U \, dm(U)\right) x\right\| d\gamma(x).$$

Put

$$W = \int_G U^{-1}T^{-1}U dm(U) = \text{diag}(T^{-1}).$$

We claim that

$$W \geq \left(\text{diag}(T)\right)^{-1}.$$

Indeed, since for any i

$$\langle e_i, \text{diag}(T^{-1})e_i\rangle = \langle e_i, (T^{-1})e_i\rangle$$

and

$$\langle e_i, \left(\text{diag}(T)\right)^{-1}e_i\rangle = \langle e_i, Te_i\rangle^{-1},$$

the claim follows from the fact that for any $\theta \in S^{n-1}$

$$\langle \theta, T^{-1}\theta\rangle \cdot \langle \theta, T\theta\rangle \geq 1.$$

Let $S = \text{diag}(T)$. Since $W \geq S^{-1}$, we have

$$F(T) = \int_{\mathbb{R}^n} \|x\|_{TK} d\gamma(x) \geq \int_{\mathbb{R}^n} \|Wx\|_K d\gamma(x) \geq \int_{\mathbb{R}^n} \|S^{-1}x\|_K d\gamma(x) = F(S).$$

[Notice that since $TK \subset D_n$,

$$SK = \left(\int_G U^{-1}TU \, dm(U)\right)(K) \subset D_n,$$

so the restrictions of the optimization problem (4.1) are satisfied.]

Let now $G' \subset O(n)$ be the group generated by the operators $U_{ij} = I - e_i \otimes e_i - e_j \otimes e_j + e_i \otimes e_j + e_j \otimes e_i$ for $i, j = 1, \ldots, n-1$, $i \neq j$. Arguing the same way we can show that there exist $a, b > 0$ such that $F(S) \geq F(T_0)$, where

$$T_0 = a \left(\sum_{i=1}^{n-1} e_i \otimes e_i \right) + b e_n \otimes e_n .$$

Since the vertices of $T_0 K$ are contact points,

$$a^2 + b^2 = 1.$$

We have

$$\|x\|_{T_0 K} = \max \left(a^{-1} \sum_{i=1}^{n-1} |x_i|, \, b^{-1} |x_n| \right).$$

Denote $\|x\|_1 = \sum_{i=1}^{n-1} |x_i|$ and let $t = t(x) = (b/a) \cdot \|x\|_1$. Then,

$$\psi(b) = \int_{\mathbb{R}^n} \|x\|_{T_0 K} d\gamma(x)$$

$$= \int_{\mathbb{R}^{n-1}} \left(\frac{1}{\sqrt{2\pi}} \int_{-t}^{t} a^{-1} \|x\|_1 e^{-x_n^2/2} dx_n + \frac{2}{\sqrt{2\pi}} \int_t^\infty b^{-1} x_n e^{-x_n^2/2} dx_n \right) d\gamma(x)$$

$$= \int_{\mathbb{R}^{n-1}} \left(\frac{2}{a} \|x\|_1 \Phi(t) + \frac{2}{\sqrt{2\pi}} b^{-1} e^{-t^2/2} \right) d\gamma(x),$$

where $\Phi(t) = (1/\sqrt{2\pi}) \int_0^t e^{-u^2/2} du$. We have to show that $b \leq \frac{c\sqrt{\log n}}{n}$. We may assume that $b \geq c/n$. Putting $a = (1 - b^2)^{1/2}$ and differentiating, we get after some calculations

$$\frac{d}{db} \psi(b) = \int_{\mathbb{R}^{n-1}} \left(2a^{-3} b \|x\|_1 \Phi(t) - \frac{2}{\sqrt{2\pi}} b^{-2} e^{-t^2/2} \right) d\gamma(x) .$$

Since $b \geq c/n$ and $\|x\|_1 \geq Cn$ with probability at least $1/2$, we have $\Phi(t) > c$ with probability $1/2$, for some absolute constant $c > 0$. So,

$$\frac{d}{db} \psi(b) \geq \bar{c} - Cb^{-2} \exp(-cn^2 b^2),$$

which is positive when $b \geq c\sqrt{\log n}/n$. $\qquad \square$

Remark. The dual problem

$$f(T) = \int_{\mathbb{R}^n} \sup_{y \in TK} \langle x, y \rangle d\gamma(x) \to \max$$

under the constraint

$$h_x(T) = |Tx|^2 - 1 \leq 0 \quad \text{for} \quad x \in K$$

is very different. The examples suggest that the matrix T for which the maximum is attained may be singular.

References

[B]   Ball K.M. (1991) Shadows of convex bodies. Trans. Amer. Math. Soc.
        327:891–901

[Ba]   Barthe F. (1998) On a reverse form of the Brascamp-Lieb inequality. Invent.
        Math. 134:335–361

[BM]   Bourgain J., Milman V.D. (1987) New volume ratio properties for convex
        symmetric bodies in $\mathbb{R}^n$. Invent. Math. 88:319–340

[FT]   Figiel T., Tomczak-Jaegermann N. (1979) Projections onto Hilbertian sub-
        spaces of Banach spaces. Israel J. Math. 33:155–171

[GM]   Giannopoulos A.A., Milman V.D. Extremal problems and isotropic posi-
        tions of convex bodies. Israel J. Math., to appear

[GP]   Giannopoulos A.A., Papadimitrakis M. (1999) Isotropic surface area mea-
        sures. Mathematika 46:1–13

[J]   John F. (1948) Extremum problems with inequalities as subsidiary condi-
        tions. Courant Anniversary Volume, Interscience, New York, 187–204

[L]   Lewis D.R. (1979) Ellipsoids defined by Banach ideal norms. Mathematika
        26:18–29

[MS]   Milman V.D., Schechtman G. (1986) Asymptotic theory of finite dimen-
        sional normed spaces. Lecture Notes in Mathematics, 1200, Springer, Berlin

[Pe]   Petty C.M. (1961) Surface area of a convex body under affine transforma-
        tions. Proc. Amer. Math. Soc. 12:824–828

[Pi]   Pisier G. (1982) Holomorphic semi-groups and the geometry of Banach
        spaces. Ann. of Math. 115:375–392

[RS]   Rogers C.A., Shephard G. (1957) The difference body of a convex body.
        Arch. Math. 8:220–233

Euclidean Projections of a p-convex Body

O. Guédon[1] and A.E. Litvak[2]

[1] Equipe d'Analyse, Université Paris 6, 4 Place Jussieu, Case 186, 75005 Paris, France
[2] Department of Mathematics, Technion, Haifa 32000, Israel

Abstract. In this paper we study Euclidean projections of a p-convex body in $\mathbb{R}^n$. Precisely, we prove that for any integer k satisfying $\ln n \leq k \leq n/2$, there exists a projection of rank k with the distance to the Euclidean ball not exceeding $C_p(k/\ln(1+\frac{n}{k}))^{1/p-1/2}$, where C_p is an absolute positive constant depending only on p. Moreover, we obtain precise estimates of entropy numbers of identity operator acting between ℓ_p and ℓ_r spaces for the case $0 < p < r \leq \infty$. This allows us to get a good approximation for the volume of p-convex hull of n points in $\mathbb{R}^k$, $p < 1$, which shows the sharpness of the announced result.

1 Introduction

Our work is motivated by so-called "Isomorphic Dvoretzky Theorem" proved recently by Milman and Schechtman [M-S1], [M-S2] (see also [G1]). The theorem states that given $1 \leq k \leq n/2$ and a centrally symmetric body K in $\mathbb{R}^n$ there exists a k-dimensional subspace $E \subset \mathbb{R}^n$ such that Banach-Mazur distance between $K \cap E$ and the Euclidean ball is bounded by $C \max\{1, \sqrt{k/\ln(1+\frac{n}{k})}\}$, where C is an absolute constant (see below the precise definitions). In the dual setting it means that there exists a projection P of rank k such that Banach-Mazur distance between PK and the Euclidean ball has the same upper bound. Since the condition of the symmetry is natural for functional analysis but not so natural for convex geometry, the work of Milman and Schechtman leads to the investigation of similar question about general convex bodies and even quasi-convex bodies. It was shown by Gordon, Guédon and Meyer [G-G-M] that the same estimate holds for non-symmetric convex bodies as well (for $k \leq cn/(\ln^2 n)$). The authors have used a different approach, based on Rudelson's result ([R1], [R2]) about John decomposition. Moreover, very recently Litvak and Tomczak-Jaegermann [L-T] was able to show that the proof of Milman and Schechtman can be extended to the non-symmetric case for all $1 \leq k \leq n/2$. In fact, the authors extended to the non-symmetric case so-called "The proportional Dvoretzky-Rogers factorization" ([B-S], [S-T]), the only place, which needs symmetry in the proof.

In this paper we investigate the behavior of projections of quasi-convex (not necessarily symmetric) bodies. The study quasi-convex bodies has been of interest in the last years, since it turns out that many crucial results of

The second named author holds the Lady Davis Fellowship.

the asymptotic theory hold for non-convex (but quasi-convex) bodies as well. It is rather surprising, because convexity was essentially used in the first proofs of most results. Let us note here that contrary to the convex case the statements about projections and sections of the quasi-convex bodies are completely different and do not follow one from the another, since one cannot use duality in the quasi-convex setting.

We prove that given p-convex body K, there exists a projection P of rank k such that Banach-Mazur distance from PK to the Euclidean space is bounded by $C_p \max\{1, (k/\ln(1 + \frac{n}{k}))^{1/p-1/2}\}$ and Banach-Mazur distance from PK to its convex hull is bounded by $C_p' \max\{1, (k/\ln(1 + \frac{n}{k}))^{1/p-1}\}$, where $C_p, C_p' > 0$ depend on p only. Of course, we use crucially the corresponding "convex" result. However, the straightforward extension can not be done. Moreover, recall that proofs in the convex case deals with sections and the result for projections follows by duality. We do not know any reasonable estimate for sections of p-convex bodies. The reason is that a projection, as any linear operator, preserves the convex hull of the set, while the convex hull of a section of a set can be very far from the section of the convex hull of the set, as was shown by Kalton [K1]. We expect that estimates for sections are much better for small p.

To show the sharpness of estimates above, we study entropy numbers of identity operators acting between ℓ_p and ℓ_r spaces, when $0 < p < r \leq \infty$. Such investigation was already done by Schütt [Sc] (see also [Pi] and [H]) for $p \geq 1$ and by Edmunds and Triebel [E-T] for $p < 1$. We give here a different proof which leads to a better dependence of the constants on p, when p tends to 0. Our proof also allows to estimate the corresponding Gelfand numbers in the case $r \leq 2$. As a corollary we obtain estimates for the volume of p-convex hull of a set of points in $\mathbb{R}^k$.

2   Definitions and Notation

By a body we always mean a compact set in R^n containing the origin as an interior point and star shaped with respect to the origin. Let K be an arbitrary body in $\mathbb{R}^n$, the gauge functional of K is defined by $\|x\|_K = \inf\{t \geq 0 \mid x \in tK\}$. By ellipsoid we always mean a linear image of the canonical Euclidean ball (thus all ellipsoids below are centered at origin). Given bodies K, B in $\mathbb{R}^n$ we define the Banach–Mazur distance by

$$d(K, B) = \inf\{\lambda > 0 \mid K - z \subset u(B - x) \subset \lambda(K - z)\},$$

where infimum is taken over all linear operators $u : \mathbb{R}^n \to \mathbb{R}^n$, and all $x, z \in \mathbb{R}^n$. We also define the following distance

$$d_0(K, B) = \inf\{\lambda > 0 \mid K \subset uB \subset \lambda K\},$$

where infimum is taken over all linear operators $u : \mathbb{R}^n \to \mathbb{R}^n$. Clearly, if K and B are centrally symmetric bodies, then $d(K, B) = d_0(K, B)$ and it is the standard Banach–Mazur distance. For $q \in (1,2]$, and a body K, we define the constant $T_q(K)$ as the smallest possible constant C such that for every m, every $x_1, ..., x_m \in K$ the following inequality holds

$$\inf_{\varepsilon_i = \pm 1} \left\{ \left\| \sum_{i=1}^{m} \varepsilon_i x_i \right\|_K \right\} \leq C m^{1/q}.$$

The constant $T_q(K)$ is closely connected to the equal-norms type constant (see e.g. [G-K]).

Let $p \in (0,1]$. A body K is called p-convex if for any $x, y \in K$, and any $\lambda, \mu \in [0,1]$, $\lambda^p + \mu^p = 1$, the point $\lambda x + \mu y$ belongs to K. Correspondingly, the non-negative homogeneous functional $\| \cdot \|_K$ on $\mathbb{R}^n$ is called p-norm if for every $x, y \in \mathbb{R}^n$ we have $\|x+y\|_K^p \leq \|x\|_K^p + \|y\|_K^p$. Let us note that we do not require the symmetry in our definition. Similarly, a body K is called quasi-convex if there is a constant C such that $K + K \subset CK$ and the non-negative homogeneous functional $\| \cdot \|$ on $\mathbb{R}^n$ is called C-quasi-norm (or just quasi-norm) if for every $x, y \in \mathbb{R}^n$ we have $\|x+y\|_K \leq C \max\{\|x\|_K, \|y\|_K\}$. By Aoki-Rolewicz theorem (see e.g. [K-P-R], or [Kö] p.47), for every C-quasi-norm on $\mathbb{R}^n$ there is a p-norm $\| \cdot \|_0$, where $2^{1/p} = 2C$, such that $\|x\| \leq \|x\|_0 \leq 2C\|x\|$ for every $x \in \mathbb{R}^n$. Because of it we restrict ourselves to the study of p-convex bodies only, however all results can be equally well stated for quasi-convex bodies.

Given $p \in (0,1]$ and a set A, its p-convex hull is defined as

$$p\text{-conv } A = \left\{ \sum_{i=1}^{m} \lambda_i x_i \mid m \in \mathbb{N}, x_i \in A, \lambda_i \geq 0, \sum_{i=1}^{m} \lambda_i^p = 1 \right\}.$$

If $p = 1$ we write conv A. The p-absolute convex hull is p-conv $(A \cup -A)$ and we denote it by p-absconv$A$. It was shown in [B-B-P] that for $p \in (0,1)$

$$p\text{-conv } A = \left\{ \sum_{i=1}^{m} \lambda_i x_i \mid m \in \mathbb{N}, x_i \in A, \lambda_i \geq 0, 0 < \sum_{i=1}^{m} \lambda_i^p \leq 1 \right\}.$$

Thus for any $p \in (0,1)$ and for every set A, the origin belongs to closure of p-conv A.

Given p-convex body K, we define

$$\alpha_m = \alpha_m(K) = \sup\left\{ m^{-1} \left\| \sum_{i=1}^{m} x_i \right\|_K \mid x_i \in K, i \leq m \right\}$$

and if δ_K denotes $d_0(K, \text{conv } K)$ then it is known by a result of Peck [P] that $\delta_K = \inf\{\lambda > 0 \mid \text{conv } K \subset \lambda K\} = \sup \alpha_m$. Note also that by property of p-norm, $\alpha_m \leq m^{-1+1/p}$.

Given set $A \subset \mathbb{R}^n$, by vol(A) we denote the volume of A.

Recall that given bodies K, B in $\mathbb{R}^n$ the covering number $N(K, B)$ is the smallest number of translation of B needed to cover K. For a positive integer k the entropy number $e_k(K, B)$ is the smallest ε such that $N(K, \varepsilon B) \leq 2^{k-1}$. The following properties of covering and entropy numbers are well-known (see e.g. [Pis2], pp.56-63, with obvious modifications in the quasi-convex case). The sequence $\{e_k(K, B)\}_k$ is non-increasing. If B is a body and K is a p-convex body then for every positive ε

$$\varepsilon^{-n}\mathrm{vol}(B)/\mathrm{vol}(K) \leq N(B, \varepsilon K) \leq \mathrm{vol}(B + \varepsilon K/2^{1/p})/\mathrm{vol}(\varepsilon K/2^{1/p}). \quad (1)$$

For every sets K_1, K_2, K_3 in $\mathbb{R}^n$ and every positive integers k and m we have $e_{k+m-1}(K_1, K_2) \leq e_k(K_1, K_3)e_m(K_3, K_2)$. Given an operator $u : E \to F$ we denote $e_k(u) = e_k(uB_E, B_F)$, where B_E is the unit ball of E and B_F is the unit ball of F and the Gelfand numbers $c_k(u)$ are defined for $k = 1, \ldots, n$ by

$$c_k(u) = \inf\{\|u : E_k \to F\|, \; E_k \subset E \text{ with codim } E_k < k\}.$$

For $0 < p, q \leq \infty$, $x \in \mathbb{R}^n$ we denote by $\mathrm{id}_{p,q}^n$ the identity operator from ℓ_p^n to ℓ_q^n and $|x|_p = (\sum_{i=1}^{n} |x_i|^p)^{1/p}$. By B_p^n we denote the unit ball of ℓ_p^n, i.e. $\{x \in \mathbb{R}^n, |x|_p \leq 1\}$. As usual $e_1, e_2, \ldots, e_n$ denotes the canonical basis of $\mathbb{R}^n$.

3   Large Rank Projections of p-convex Bodies

In [G-K] (Lemmas 2, 3 with remark after it) the authors proved that if p-normed space has equal-norms type $q > 1$ then the distance from the space to the corresponding normed space is bounded by a constant depending on p, q, and type constant only. We start with the following non-symmetric analog of their result.

Lemma 1. *Let $p \in (0, 1)$, $q \in (1, 2]$, K be a p-convex body and B be a symmetric body with respect to the origin. Define ϕ as $\phi = (1/p - 1/q)/(1 - 1/q)$ then*

(i) $\delta_K \leq \left(\dfrac{c}{p(q-1)}\right)^{1/p\,(1-1/\phi)} (T_q(B)\, d_0(K, B))^{1-1/\phi}$,

(ii) $\delta_K \leq \left(\dfrac{c}{p(q-1)}\right)^{(\phi-1)/p} (T_q(B)\, d_0(\mathrm{conv}\, K, B))^{\phi-1}$,

(iii) $d_0(K, B) \leq \left(\dfrac{c}{p(q-1)}\right)^{(\phi-1)/p} T_q(B)^{\phi-1} d_0(\mathrm{conv}\, K, B)^\phi$,

where $c > 0$ is an absolute constant.

The proof is essentially the same as in the symmetric case but for completeness we outline it here.

Proof. Let $d = d_0(K, B)$ and $T = T_q(B)$. Without loss of generality we can assume that $(1/d)B \subset K \subset B$. Let m be a positive integer and x_i, $i = 1, \ldots, 2^m$ be a family of points in K, then $x_i \in B$, $i \leq 2^m$ and by definition there is a choice of signs ε_i, $i \leq 2^m$, such that $\|\sum \varepsilon_i x_i\|_B \leq T2^{m/q}$. Since the body B is symmetric we can assume that $A = \{i \mid \varepsilon_i = 1\}$ has cardinality larger than 2^{m-1}. Thus

$$\left\| \sum_{i=1}^{2^m} x_i \right\|_K^p = \left\| \sum_{i=1}^{2^m} \varepsilon_i x_i + 2 \sum_{i \notin A} x_i \right\|_K^p$$

$$\leq d^p \left\| \sum_{i=1}^{2^m} \varepsilon_i x_i \right\|_B^p + 2^p \left\| \sum_{i \notin A} x_i \right\|_K^p \leq d^p T^p 2^{mp/q} + 2^{mp} \alpha_{2^{m-1}}.$$

Thus for any $k \leq m$

$$\alpha_{2^m}^p \leq \alpha_{2^k}^p + d^p T^p \sum_{i=k+1}^{\infty} 2^{-ip(1-1/q)} \leq 2^{k(1-p)} + d^p T^p \frac{2^{-kp(1-1/q)}}{p(1-1/q)\ln 2}.$$

Choosing k from $2^{k(1-p)}(p(1-1/q)\ln 2) = d^p T^p 2^{-kp(1-1/q)}$ we get the first estimate. The second and third estimates follow from the inequality $d_0(K, B) \leq \delta_K d_0(\text{conv } K, B)$. □

This lemma allows us to extend the "Isomorphic Dvoretzky Theorem" to the p-convex setting.

Theorem 2. *There exists an absolute positive constant c such that for every p-convex body K in $\mathbb{R}^n$, $0 < p < 1$, for all integer $1 \leq k \leq n/2$ there exists a projection P of rank k such that*

$$d_0(PK, B_2^k) \leq C_p \max \left\{ 1, \left(\frac{k}{\ln(1 + \frac{n}{k})} \right)^{\frac{1}{p} - \frac{1}{2}} \right\},$$

where $C_p \leq (c/p)^{2/p^2}$.

Proof. It is known [L-T] (see also [M-S1], [M-S2], [G1], [G-G-M]) that for all integer $k = 1, \ldots, [n/2]$ there exists a projection P of rank $k + 1$ such that

$$d(P(\text{conv } K), B_2^{k+1}) \leq A := c \max \left\{ 1, \sqrt{\frac{k}{\ln(1 + \frac{n}{k})}} \right\}.$$

In other words there are an ellipsoid centered at the origin $\mathcal{E}$ and a vector $a \in \mathbb{R}^n$ such that $P\mathcal{E} \subset P(\text{conv } K) - a \subset A(P\mathcal{E})$. Let Q be an orthogonal projection of rank k with $\text{Ker } Q \subset \text{span } \{\text{Ker } P, a\}$. Then $QP = Q$ which gives $Q\mathcal{E} \subset QK \subset AQ\mathcal{E}$. The result follows now by Lemma 1, since $T_2(Q\mathcal{E}) = 1$,

$\phi = 2(1/p - 1/2)$ and $(\text{conv } QK) = Q(\text{conv } K)$. $\square$

Remark 1. If δ_K is not large then the trivial estimate $c_p \delta_K \sqrt{\frac{k}{\ln(1+\frac{n}{k})}}$ can be better than the one given in the theorem. Thus the theorem is of interest for "essentially" non-convex bodies only.

Remark 2. The theorem with the same proof holds without restriction "origin is an interior point of K" (assuming that interior of K is not empty).

Remark 3. The theorem holds for $k > n/2$ as well. Indeed, let $\varepsilon \in (0, 1/2)$ and $k = [(1 - \varepsilon)n]$. Recently, the first name author ([G2], Theorem 2.3) has shown that for any convex body K with baricenter at origin there is a k-dimensional section E such that $d_0(K \cap E, B_2^k) \leq c\sqrt{k}/\varepsilon^{3/2}$, where c is an absolute constant. The proof is based on corresponding result from [L-M-P] and estimate for the parameter $\tilde{M} = \int_K |x|_2 dx / \text{vol}(K)$ ([G2], Theorem 2.2). In dual formulation it means that for any convex body K there are a projection of rank k and a point $a \in K$ such that

$$d(P(K - a), B_2^k) \leq c\sqrt{k}/\varepsilon^{3/2}. \tag{2}$$

Thus repeating the proof above, we get that the theorem holds for any $k = [(1 - \varepsilon)n] > n/2$ with an additional factor $\varepsilon^{3(1/2 - 1/p)}$. Note also that (2) with slightly worse dependence on ε follows immediately from so-called "The proportional Dvoretzky-Rogers factorization" [L-T] (see also [B-S] and [S-T] for the symmetric case).

It is interesting to note that dependence on n and k in the theorem is optimal as simple example of the B_p^n shows. To prove it we need the following volume estimate.

Lemma 3. *There exists an absolute positive constant c such that for every set of points $x_1, \ldots, x_n$ in the k-dimensional space $\mathbb{R}^k$, the following estimate holds*

$$\left(\frac{\text{vol}(p\text{-absconv}\{x_1, \ldots, x_n\})}{\text{vol}(\text{absconv}\{x_1, \ldots, x_n\})} \right)^{1/k} \leq C_p \min \left\{ 1, \left(\frac{\ln(1 + \frac{n}{k})}{k} \right)^{\frac{1}{p} - 1} \right\},$$

where $C_p = \left(\frac{c}{p} \ln(2/p) \right)^{1/p}$.

We obtain this lemma as a corollary of entropy estimates at the end of our paper.

Remark. In particular the lemma implies that for every set of points $x_1, \ldots, x_n$ in the k-dimensional Euclidean ball B_2^k we have

$$\left(\frac{\text{vol}(p\text{-absconv}\{x_1, \ldots, x_n\})}{\text{vol } B_2^k} \right)^{1/k} \leq C_p \min \left\{ 1, \left(\frac{\ln(1 + \frac{n}{k})}{k} \right)^{\frac{1}{p} - \frac{1}{2}} \right\},$$

where $C_p = \left(\frac{c}{p}\ln(2/p)\right)^{1/p}$. Indeed, it follows by well-known estimate of vol(absconv$\{x_1,\ldots,x_n\}$) obtained independently by Bárány-Füredy [B-F], Carl-Pajor [C-P], and Gluskin [G].

Proposition 4. *For every* $p \in (0,1]$, *for all integer* $k = 1,\ldots,n$, *and all projections* P *of rank* k *we have*

$$d(PB_p^n, B_2^k) \geq \frac{1}{C_p}\max\left\{1, \left(\frac{k}{\ln(1+\frac{n}{k})}\right)^{\frac{1}{p}-\frac{1}{2}}\right\},$$

where C_p *is the same constant as in the lemma above.*

Proof. Let $\mathcal{E}$ be an ellipsoid satisfying $PB_p^n \subset \mathcal{E}$ and d be the best constant such that

$$1/d\ \mathcal{E} \subset PB_p^n \subset \mathcal{E}.$$

Denote by v the isomorphism on $\mathbb{R}^k$ such that $v(\mathcal{E}) = B_2^k$ and define for all $i = 1,\ldots,n$, $x_i = vPe_i$. It is clear that for all $i = 1,\ldots,n$, $|x_i|_2 \leq 1$ and that

$$1/d \leq \left(\frac{\text{vol}(v(PB_p^n))}{\text{vol}B_2^k}\right)^{1/k} = \left(\frac{\text{vol}(PB_p^n)}{\text{vol}\mathcal{E}}\right)^{1/k}.$$

As $v(PB_p^n) = p\text{-absconv}\{x_1,\ldots,x_n\}$, we conclude applying Lemma 3 and remark after it, that

$$d \geq \frac{1}{C_p}\max\left\{1, \left(\frac{k}{\ln(1+\frac{n}{k})}\right)^{\frac{1}{p}-\frac{1}{2}}\right\}.$$

It is now enough to choose an ellipsoid $\mathcal{E}$ which realizes the distance from PB_p^n to the Euclidean ball.   □

We would like to end this section with the version of the theorem dealing with the estimates for distance from projection of p-convex body to its convex hull. Recall that by strong form of Carathéodory theorem if K is a k-dimensional p-convex body then $\delta_K \leq k^{1/p-1}$. Repeating the previous argument we obtain the following theorem.

Theorem 5. *There exists an absolute constant c such that for every p-convex body K in $\mathbb{R}^n$, $0 < p < 1$, for all integer $k \leq n/2$, there exists a projection P of rank k such that*

$$\delta_{PK} \leq c_p\max\left\{1, \left(\frac{k}{\ln(1+\frac{n}{k})}\right)^{\frac{1}{p}-1}\right\},$$

where $c_p \leq (c/p)^{2/p^2}$.

Moreover, if $K = B_p^n$ then for all projection P of rank k, we have

$$\delta_{PK} \geq \frac{1}{C_p} \max\left\{1, \left(\frac{k}{\ln(1 + \frac{n}{k})}\right)^{\frac{1}{p}-1}\right\},$$

where $C_p \leq \left(\frac{c}{p}\ln(2/p)\right)^{1/p}$.

Remark 1. As above we do not need the restriction "origin is an interior point of K". Also, as it follows from the proof, there exists one projection which satisfies both estimates of Theorem 2 and 5.

Remark 2. The last part of Theorem 5 shows that one can not expect in general that there exists a projection of p-convex body of sufficiently large rank which is almost convex. It was known ([K2]) that to get "convex" projection (i.e. projection such that $\delta_{PK} \leq C_p$) the rank of the projection can not exceed $c_p \log n$. Thus the formula above is quantification of this observation. The similar question about sections of p-convex body remains open.

4   Entropy Numbers in the p-convex Case

In this section we start by extending a few known convex results to the quasi-convex case. First, we extend result of Schütt [Sc] about entropy numbers of identity operator acting between ℓ_p and ℓ_r spaces (see also [Pi], [H] for the upper estimates), to the case $0 < p < r \leq \infty$. When $p < 1$, the upper estimates are already proved by Edmunds and Triebel [E-T]. The proof we give here is new and gives better dependence in p when p goes to 0.

Theorem 6. *Let $0 < p < r \leq \infty$ and n be an integer. Let $p' = \min\{1, p\}$ and $r' = \min\{1, r\}$. There are absolute positive constants c_0, c_1, c_2 such that for all integer k one has*

(i) *if $k < \lceil \log_2 n \rceil$ then $e_k(\mathrm{id}_{p,r}^n) \in [1/2, 1]$.*

(ii) *if $\lceil \log_2 n \rceil \leq k \leq n$ then*

$$c_0^{1/p'} f_1(n, k) \leq e_k(\mathrm{id}_{p,r}^n) \leq 2 \cdot 2^{1/r} C_p^{1/p - 1/r} f_1(n, k),$$

where $f_1(n, k) = \left\{\dfrac{\ln(1 + \frac{n}{k})}{k}\right\}^{1/p - 1/r}$ and $C_p = \dfrac{c_1}{p'} \ln\left(\dfrac{2}{p'}\right)$.

(iii) *if $k > n$ then*

$$c_2 \sqrt{r'/p'}\, f_2(n, k) \leq e_k(\mathrm{id}_{p,r}^n) \leq 2 \cdot 6^{1/r'} C_p^{1/p - 1/r} f_2(n, k),$$

where $f_2(n, k) = 2^{-k/n} n^{1/r - 1/p}$ and C_p as above.

The proofs of (i) and (iii) are standard, but we show it for completeness. In the case $k > c_3 n/p'$ one can replace the constant in front of $f_2(n, k)$ with $c_4 8^{1/p'}$. The case $p \geq 1$ holds by Theorem 1 of [Sc], so we restrict ourselves to the case $p < 1$. We will need two auxiliary lemmas which are essentially 12.1.11 and 12.1.12 of [Pi], adapted to the quasi-convex case. The proofs are identical and we only need to substitute the triangle inequality by the quasi-triangle one. All the spaces considered are $\mathbb{R}^n$ equipped with some quasi-norm. When there is some quasi-normed indexed by a parameter i, E_i denotes the space $(\mathbb{R}^n, \|\cdot\|_i)$.

Lemma 7. *For $i = 0, 1$, let $\|\cdot\|_i$ be a symmetric C_i-quasi-norm on $\mathbb{R}^n$ and for $\theta \in [0, 1]$ assume that a quasi-norm $\|\cdot\|_\theta$ satisfies $\|x\|_\theta \leq \|x\|_0^\theta \|x\|_1^{1-\theta}$ for all $x \in \mathbb{R}^n$. Then for every quasi-normed space F, for every linear operator $T : F \longrightarrow \mathbb{R}^n$, for every integer k, m one has*

$$e_{m+k-1}(T : F \longrightarrow E_\theta) \leq (C_0 e_m(T : F \longrightarrow E_0))^\theta (C_1 e_k(T : F \longrightarrow E_1))^{1-\theta}.$$

Remark. This lemma holds true (even without the constants C_0 and C_1 in the last formula) for Gelfand numbers instead of entropy numbers.

Lemma 8. *Let $A > 0$, $\theta \in [0, 1]$, and assume that quasi-norms $\|\cdot\|_0$, $\|\cdot\|_1$, $\|\cdot\|_\theta$ satisfy*

$$\inf \{a\|y\|_0 + b\|z\|_1 \mid y + z = x\} \leq A a^\theta b^{1-\theta} \|x\|_\theta \tag{3}$$

for every $a \geq 0$, $b \geq 0$ and $x \in \mathbb{R}^n$. If F is a C-quasi-normed space then for every linear operator $T : \mathbb{R}^n \longrightarrow F$, for every integer k and m one has

$$e_{m+k-1}(T : E_\theta \longrightarrow F) \leq AC (e_m (T : E_0 \longrightarrow F))^\theta (e_k (T : E_1 \longrightarrow F))^{1-\theta}.$$

Remark. We will use this lemma with $\theta = p < 1$, $E_0 = \ell_p^n$, $E_1 = F = \ell_\infty^n$, $E_\theta = \ell_1^n$. In this case, we can take $A = C = 2$. Indeed, let $x \in \mathbb{R}^n$, $a > 0$, $b > 0$ and assume, without loss of generality, that $x_i \geq 0$ for every i and $\sum x_i = 1$. Now, let

$$v = \min \left\{ 1, \left(\frac{a}{b} \frac{1-p}{p} \right)^p \right\}$$

and define y and z as follows, $z_i = \min \{v, x_i\}$ for every i, $y = x - z$. Then

$$a|y|_p + b|z|_\infty \leq a N^{-1+1/p} + bv,$$

where $N = |\{i \mid x_i > v\}|$. Since $Nv \leq 1$ and $(1-p)^{p-1} p^{-p} \leq 2$, inequality (3) follows with $A = 2$.

Proof of Theorem 6.

Case 1. $k < [\log_2 n]$.

One trivially has $e_k \left(\mathrm{id}^n_{p,r} \right) \leq 1$. Assume that $B^n_p \subset \cup^N_1 x_i + \varepsilon B^n_r$ with $N \leq 2^{k-1}$, then there exist $l \in \{1, \ldots, N\}$ and two different vectors of the canonical basis e_i and e_j such that $e_i \in x_l + \varepsilon B^n_r$ and $e_j \in x_l + \varepsilon B^n_r$. Thus

$$2^{1/r} = |e_i - e_j|_r \leq \varepsilon \, 2^{\max\{1, 1/r\}}.$$

That proves the case (i).

Case 2. $[\log_2 n] \leq k \leq n$.

Upper estimate. We start to estimate $e_k \left(\mathrm{id}^n_{p,\infty} \right)$. Let us denote by $A = A(n,p)$ the smallest constant which satisfies for all $1 \leq k \leq n$

$$e_k \left(\mathrm{id}^n_{p,\infty} \right) \leq A \left\{ \frac{\ln(1 + \frac{n}{k})}{k} \right\}^{1/p}.$$

By Lemma 7 used with $F = E_0 = \ell^n_p$, $E_1 = \ell^n_\infty$, $\theta = p$, $E_\theta = \ell^n_1$ and $m = 1$, we get

$$e_k \left(\mathrm{id}^n_{p,1} \right) \leq 4 e_k \left(\mathrm{id}^n_{p,\infty} \right)^{1-p}. \tag{4}$$

Factorizing identity from ℓ^n_p to ℓ^n_∞ through ℓ^n_1, we obtain by properties of entropy numbers and by (4)

$$e_k \left(\mathrm{id}^n_{p,\infty} \right) \leq e_{[(1-p)k]} \left(\mathrm{id}^n_{p,1} \right) e_{[pk]} \left(\mathrm{id}^n_{1,\infty} \right)$$
$$\leq 4 \left(e_{[(1-p)k]} \left(\mathrm{id}^n_{p,\infty} \right) \right)^{1-p} e_{[pk]} \left(\mathrm{id}^n_{1,\infty} \right).$$

It is well known that for all $k = 1, \ldots, n$,

$$e_k \left(\mathrm{id}^n_{1,\infty} \right) \leq c \min \left\{ 1, \frac{\ln(1 + \frac{n}{k})}{k} \right\}.$$

Since for $a \in (0,1)$

$$\frac{\ln(1 + \frac{n}{ak})}{ak} \leq \min \left\{ \frac{1}{a^2}, \frac{2}{a} \ln \left(\frac{2}{a} \right) \right\} \frac{\ln(1 + \frac{n}{k})}{k},$$

we obtain

$$e_k \left(\mathrm{id}^n_{p,\infty} \right) \leq 4c \left(\frac{1}{1-p} \right)^{2\frac{1-p}{p}} \frac{2}{p} \ln \left(\frac{2}{p} \right) A^{1-p} \left\{ \frac{\ln(1 + \frac{n}{k})}{k} \right\}^{1/p}$$

for every $[\log_2 n] \leq k \leq n$. Hence, by definition of A,

$$A \leq 4c \left(\frac{1}{1-p} \right)^{2\frac{1-p}{p}} \frac{2}{p} \ln \left(\frac{2}{p} \right) A^{1-p} \leq \frac{c_1}{p} \ln \left(\frac{2}{p} \right) A^{1-p},$$

which implies the desired estimate for $r = \infty$ and $k \leq n$.

The proof of the upper estimate for every r and $k \leq n$ is based on the interpolation inequality: if $0 < p < q < r \leq \infty$ and $\theta \in [0, 1]$ be such that $\frac{\theta}{p} + \frac{1-\theta}{r} = \frac{1}{q}$ then for all $x \in \mathbb{R}^n$, $|x|_q \leq |x|_p^\theta |x|_r^{1-\theta}$. Using Lemma 7 with $F = E_0 = \ell_p^n$, $E_1 = \ell_\infty^n$, $\theta = p/r$, $E_\theta = \ell_r^n$ and $m = 1$ we obtain

$$e_k(\mathrm{id}_{p,r}^n) \leq 2^{1+1/r}(e_k(\mathrm{id}_{p,\infty}^n))^{1-p/r}$$

which is the announced result.

Lower estimate. It follows from Lemma 8 (and remark after it) that

$$e_{2k}(\mathrm{id}_{1,\infty}^n) \leq 4(e_{2k}(\mathrm{id}_{p,\infty}^n))^p(e_1(\mathrm{id}_{\infty,\infty}^n))^{1-p}.$$

Applying properties of entropy numbers we obtain

$$e_{2k}(\mathrm{id}_{1,\infty}^n)^{1/p} \leq 4^{1/p}e_{2k}(\mathrm{id}_{p,\infty}^n) \leq 4^{1/p}e_k(\mathrm{id}_{p,r}^n)e_k(\mathrm{id}_{r,\infty}^n).$$

We deduce the lower bound for $e_k(\mathrm{id}_{p,r}^n)$ combining the upper bound obtained in the preceding part for $e_k(\mathrm{id}_{r,\infty}^n)$ and the well known estimate: $e_k(\mathrm{id}_{1,\infty}^n) \geq c\sqrt{k/\ln(1 + n/k)}$ for these values of k.

Case 3. $k > n$.

This case follows from standard volume consideration. By (1) we have $e_{m+1}(\mathrm{id}_{r,r}^n) \leq 3^{1/r'}2^{-m/n}$ for every m. Since

$$e_k(\mathrm{id}_{p,r}^n) \leq e_n(\mathrm{id}_{p,r}^n)e_{k-n+1}(\mathrm{id}_{r,r}^n)$$

by the preceding result we obtain

$$e_k(\mathrm{id}_{p,r}^n) \leq 2 \cdot 6^{1/r'} \left\{ \frac{c_1}{p} \ln\left(\frac{2}{p}\right) \right\}^{1/p-1/r} 2^{-k/n}n^{1/r-1/p}.$$

On the other hand if $N(B_p^n, \varepsilon B_r^n) \leq 2^{k-1}$ then by (1)

$$\varepsilon \geq \left(\frac{\mathrm{vol}\,(B_p^n)}{N\mathrm{vol}\,(B_r^n)} \right)^{1/n} \geq c\sqrt{\frac{\min\{1, r\}}{\min\{1, p\}}}\, 2^{-k/n}\, n^{1/r-1/p},$$

where c is an absolute positive constant. This proves the theorem.   □

Corollary 9. *Let $X = (\mathbb{R}^n, \|\cdot\|)$ be a quasi normed space and u be a linear operator from $\mathbb{R}^n$ to X. For all integers $k \leq n$ and $m \geq 1$ one has*

$$e_{k+m-1}(u : \ell_p^n \to X) \leq C_p \min\left\{ 1, \left(\frac{\ln(1 + \frac{n}{k})}{k} \right)^{\frac{1}{p}-1} \right\} e_m(u : \ell_1^n \to X),$$

where $C_p \leq \left(\frac{c}{p} \ln\left(\frac{2}{p}\right) \right)^{1/p}$ for an absolute constant $c > 0$.

Remark. In some cases the estimate of $e_m(u : \ell_1 \to X)$ is well known. For example, the case of Banach space X with non-trivial type was studied in [Pis1] and in [C].

Proof. Let $u : \ell_p^m \to X$ and factorize it as $u = v \, \mathrm{id}_{p,1}^n$ where $v : \ell_1^n \to X$. By property of entropy numbers we obtain for all integers k and m

$$e_{k+m-1}(u : \ell_p^n \to X) \le e_k(\mathrm{id}_{p,1}^n) e_m(v).$$

Now Corollary 9 follows from Theorem 6. □

Remark. Repeating the argument of Theorem 6 one can get the same upper estimate of Gelfand numbers $c_k(\mathrm{id}_{p,r}^n)$ for the case $0 < p < r \le 2$. In particular, Corollary 9 remains true for Gelfand numbers instead of entropy numbers.

From this entropy estimate we shall deduce a good approximation from above for the volume of the p-convex hull of n points $x_1, \ldots, x_n$ in $\mathbb{R}^k$ as was stated in Lemma 3.

Proof of Lemma 3. Consider the operator $u : \mathbb{R}^n \to \mathbb{R}^k$ defined by $u(e_i) = x_i$ for all integer $i = 1, \ldots, n$. Let X be $(\mathbb{R}^k, \| \cdot \|)$, where unit ball of $\| \cdot \|$ is $K = \mathrm{absconv}\{x_1, \ldots, x_n\}$. By the previous corollary applied with $m = 1$, we have

$$e_k(u : \ell_p^n \to X) \le C_p \min\left\{1, \left(\frac{\ln(1 + \frac{n}{k})}{k}\right)^{\frac{1}{p}-1}\right\} \|u : \ell_1^n \to X\|$$

$$= C_p \min\left\{1, \left(\frac{\ln(1 + \frac{n}{k})}{k}\right)^{\frac{1}{p}-1}\right\}.$$

Clearly,

$$2\, e_k(u : \ell_p^n \to X) \ge \left(\frac{\mathrm{vol}\, u(B_p^n)}{\mathrm{vol}\, K}\right)^{1/k}.$$

Since $u(B_p^n) = p\text{-}\mathrm{absconv}\{x_1, \ldots, x_n\}$, the result follows. □

Acknowledgments

The work on this paper was started during Workshop on Geometric Functional Analysis at the Pacific Institute of the Mathematical Sciences. The authors wish to thank the Institute and organizers of the Workshop for their hospitality. The second named author thanks E. Gluskin for a discussion concerning entropy numbers.

References

[B-F]   Bárány I., Füredy Z. (1988) Approximation of the sphere by polytopes having few vertices. Proc. Amer. Math. Soc. 102(3):651–659

[B-L]   Bergh J., Löfström J. (1976) Interpolation Spaces. An Introduction. Grundlehren der Mathematischen Wissenschaften, No. 223. Springer-Verlag, Berlin-New York

[B-S]   Bourgain J., Szarek S.J. (1988) The Banach-Mazur distance to the cube and the Dvoretzky-Rogers factorization. Israel J. Math. 62(2):169–180

[B-B-P]   Bastero J., Bernués J., Peña A. (1995) The theorems of Carathéodory and Gluskin for $0 < p < 1$. Proc. Amer. Math. Soc. 123(1):141–144

[C]   Carl B. (1985) Inequalities of Bernstein-Jackson type and the degree of compactness of operators in Banach spaces. Ann. Inst. Fourier 35:79–118

[C-P]   Carl B., Pajor A. (1988) Gelfand numbers of operators with values in a Hilbert space. Invent. Math. 94:479–504

[E-T]   Edmunds D.E., Triebel H. (1989) Entropy numbers and approximation numbers in function spaces. Proc. London Math. Soc. (3) 58(1):137–152

[G]   Gluskin E.D. (1988) Extremal properties of orthogonal parallelepipeds and their applications to the geometry of Banach spaces. (Russian) Mat. Sb. (N.S.) 136(178), no. 1:85–96

[G-G-M]   Gordon Y., Guédon O., Meyer M. (1998) An isomorphic Dvoretzky's theorem for convex bodies. Studia Math. 127(2):191–200

[G-K]   Gordon Y., Kalton N.J. (1994) Local structure theory for quasi-normed spaces. Bull. Sci. Math. 118:441–453

[G1]   Guédon O. (1997) Gaussian version of a theorem of Milman and Schechtman. Positivity 1:1–5

[G2]   Guédon O. (1998) Sections euclidiennes des corps convexes et inégalités de concentration volumique. Thèse de doctorat de mathématiques, Université de Marne-la-Vallée

[H]   Höllig K. (1980) Diameters of classes of smooth functions. Quantitative Approximation (Proc. Internat. Sympos., Bonn 1979), Academic Press, New York-London, 163–175

[K1]   Kalton N.J. (1986) Banach envelopes of nonlocally convex spaces. Canad. J. Math. 38(1):65–86

[K2]   Kalton N.J. Private communication

[K-P-R]   Kalton N.J., Peck N.T., Roberts J.W. (1984) An F-space sampler. London Mathematical Society Lecture Note Series, 89, Cambridge University Press, Cambridge-New-York

[Kö]   König H. (1986) Eigenvalue distribution of compact operators. Operator Theory: Advances and Applications, 16, Birkhäuser Verlag, Basel-Boston, Mass.

[L-M-P]   Litvak A.E., Milman V.D., Pajor A. (1999) The covering numbers and "low M^*-estimate" for quasi-convex bodies. Proc. Amer. Math. Soc. 127:1499–1507

[L-T]   Litvak A.E., Tomczak-Jaegermann N. Random aspects of high-dimensional convex bodies. GAFA, Lecture Notes in Math., Springer, Berlin-New York. This volume

[M-S1]   Milman V.D., Schechtman G. (1995) An "isomorphic" version of Dvoretzky's theorem. C.R. Acad. Sci. Paris t. 321 Série I, 541–544

[M-S2]   Milman V.D., Schechtman G. (1999) An "isomorphic" version of Dvoretzky's theorem II. Convex Geometric Analysis (Berkeley, CA, 1996), Math. Sci., Res. Inst. Publ., 34, Cambridge Univ. Press, Cambridge, 159–164

[P]   Peck N.T. (1981) Banach-Mazur distances and projections on p-convex spaces. Math. Z. 177(1):131–142

[Pi]   Pietsch A. (1978) Operator ideals. Mathematische Monographien [Mathematical Monographs], 16, VEB Deutscher Verlag der Wissenschaften, Berlin

[Pis1]   Pisier G. (1980-1981) Remarques sur un résultat non publié de B. Maurey. Séminaire d'Analyse Fonctionnelle, École Polytechnique-Palaiseau, exposé 5

[Pis2]   Pisier G. (1989) The Volume of Convex Bodies and Banach Space Geometry. Cambridge University Press, Cambridge

[R1]   Rudelson M. (1997) Contact points of convex bodies. Israel J. Math. 101:93–124

[R2]   Rudelson M. (1999) Random vectors in the isotropic position. J. Funct. Anal. 164(1):60–72

[Sc]   Schütt C. (1984) Entropy numbers of diagonal operators between symmetric Banach spaces. J. Approx. Theory 40(2):121–128

[S-T]   Szarek S.J., Talagrand M. (1989) An "isomorphic" version of the Sauer-Shelah lemma and the Banach-Mazur distance to the cube. In: Geometric Aspects of Functional Analysis (1987–88), Lecture Notes in Math., 1376, Springer, Berlin-New York, 105–112

Remarks on Minkowski Symmetrizations

B. Klartag

School of Mathematical Sciences, Tel Aviv University, Tel Aviv 69978, Israel

Abstract. Here we extend a result by J. Bourgain, J. Lindenstrauss, V.D. Milman on the number of random Minkowski symmetrizations needed to obtain an approximated ball, if we start from an arbitrary convex body in $\mathbb{R}^n$. We also show that the number of "deterministic" symmetrizations needed to approximate an Euclidean ball may be significantly smaller than the number of "random" ones.

1 Background and Notation

Let K be a compact convex (symmetric) set in $\mathbb{R}^n$ and let u be any vector in $S^{n-1} = \{u; |u| = 1\}$ where $|\cdot|$ denotes the standard Euclidean norm on $\mathbb{R}^n$. We denote by $\pi_u \in O(n)$ the reflection with respect to the hyperplane through the origin orthogonal to u, i.e. $\pi_u x = x - 2\langle x, u \rangle u$.

The Minkowski symmetrization (sometimes called Blaschke symmetrization) of K with respect to u is defined to be the convex set $\frac{1}{2}(\pi_u K + K)$. Denote by $\|\cdot\|$ the norm whose unit ball is K, and by $M^*(K)$ the half mean width of K: $M^*(K) := \int_{S^{n-1}} \|x\|^* d\sigma(x)$, where σ is the normalized rotation invariant measure on S^{n-1}, and $\|\cdot\|^*$ is the dual norm.

It is easy to verify that $M^*(K) = M^*(\frac{1}{2}(\pi_u K + K))$, so the mean width is preserved under Minkowski symmetrizations. Since successive Minkowski symmetrizations make the body symmetric with respect to more and more hyperplanes, one might expect convergence to a ball. However, surprisingly, very few symmetrizations are needed for that convergence, as stated and proved in [BLM]:

Proposition 1.1 *If we start with arbitrary body K, and perform $cn \log n + c(\epsilon)n$ "random" Minkowski symmetrizations, with high probability we obtain a body $\hat{K}$ such that*

$$(1 - \epsilon)M^*D \subset \hat{K} \subset (1 + \epsilon)M^*D$$

where $D = \{u; |u| \leq 1\}$.

"Random" means that the $N = cn \log n + c(\epsilon)n$ symmetrizations are performed with respect to $u_1, ..., u_N \in S^{n-1}$, and the u_i's are chosen independently and uniformly (i.e. according to the probability rotation invariant measure on the sphere).

Supported by the Israel Science Foundation founded by the Academy of Sciences and Humanities.

Now we give a formal description of the body $\hat{K}$. Take the body K, symmetrize with respect to $u_1, ..., u_T$ (in that order). The body achieved is $\hat{K} = \frac{1}{2^T} \sum_{D \subset \{1,..,T\}} \prod_{i \in D} \pi_{u_i}(K)$.

The technique we use forces us to work in the dual space, so the next formula would be needed:

$$|||x||| = \frac{1}{2^T} \sum_{D \subset \{1,..,T\}} \left\| \prod_{i \in D} \pi_{u_i} x \right\|^*$$

where $\| \cdot \|^*$ is the dual norm to K, and $||| \cdot |||$ is the dual norm to $\hat{K}$.

Note that our notation doesn't specify the exact order of multiplications in expressions like $\prod_{i \in D} \pi_{u_i} x$. One may think that this might be a source for problems, since the reflections do not generally commute, and when passing to the dual, the order is reversed. However, we do not rely anywhere on the order of multiplications.

In the following sections, the number of random (and non random) Minkowski symmetrizations needed in order to obtain approximately a ball is investigated. Tight estimates (from below and above) for the random case are given, and actually there exists a simple formula to describe this quantity. The behavior in the deterministic case is not as clear to us as in the random case, but few results about that behavior are presented.

By c, C we define universal constants, which are not the same at different appearances.

2   Random Case

Denote by $a(K)$ the half-diameter of K. i.e. $a(K) = \sup_{x \in S^{n-1}} \|x\|^*$.

This section relies heavily on statements and proofs from [BLM]. One of the main lemmas from [BLM] needed is the following:

Lemma 2.1 *Assume $a(K) = 1$, and $M^*(K) \leq \frac{1}{8}$, then with $N_1 < c_1 n$ we have*

$$\forall x \in \mathbb{R}^n \quad |||x||| = 2^{-N_1} \sum_D \left\| \prod_{i \in D} \pi_{u_i} x \right\|^* \leq \frac{1}{2}|x|$$

with probability at least $1 - \exp(-c_2 n)$.

This means that the diameter of the new body is at most half the diameter of the original body, while its mean width remains the same. This happens after random $c_1 n$ symmetrizations.

If the new body $\hat{K}$ satisfies $a(\hat{K})/M^*(\hat{K}) \geq 8$ we can repeat the same process, and reduce the diameter by an additional factor of 2. After $\log \frac{a(K)}{M^*(K)}$ iterations, each involving less than $c_1 n$ symmetrizations, we achieve a body that satisfies $a(\hat{K})/M^*(\hat{K}) \leq 8$, with exponential probability (exponential in the dimension n).

When we have reached this stage, we can use the next lemma from [BLM]:

Lemma 2.2 *Assume $a(K) \leq 1$, and $M^*(K) \geq \frac{1}{8}$, then for all $\epsilon > 0$ and $n > n(\epsilon)$ and with $N_2 = c(\epsilon)n$*

$$\forall x \in \mathbb{R}^n \quad (1 - \epsilon)M^*|x| \leq 2^{-N_2} \sum_D \left\| \prod_{i \in D} \pi_{u_i} x \right\|^* \leq (1 + \epsilon)M^*|x|$$

with probability of at least $1 - \exp(-c_1(\epsilon)n)$.

In this paper, we are interested only in the "isomorphic" symmetrization procedure, meaning we want our body to be close to a ball up to a factor, say 4: $\frac{1}{2}D \subset K \subset 2D$ (as opposed to "almost-isometric" symmetrizations, where dependence in ϵ is added). Our analysis here leads us to the following formulation of Proposition 1.1:

Proposition 2.3 *K is a convex symmetric body. Perform $c_1 n \log \frac{a}{M^*}$ "random" Minkowski symmetrizations. With probability greater than $1 - \exp(-c_2 n)$ we obtain a body $\hat{K}$ such that*

$$\frac{1}{2}M^*D \subset \hat{K} \subset 2M^*D$$

where c_1, c_2 are numerical constants, and "random" means that the symmetrizations are chosen independently and uniformly on the sphere.

Since for every body $K \subset \mathbb{R}^n$, the ratio $a(K)/M^*(K)$ is bounded by $\sqrt{n}$, the worst body we can find demands $cn \log n$ symmetrizations. There exist bodies - such as the n dimensional cube - that satisfy $a(K)/M^*(K) < Const$ independent of n. Those bodies become close to an Euclidean ball only after cn random symmetrizations. Therefore, Proposition 2.3 is slightly more informative than Proposition 1.1.

We will analyze now the process of performing random Minkowski symmetrizations, starting with a specific body. This specific body I is just a simple segment. Let $v \in S^{n-1}$, and $I = [-v, v]$.

The dual norm of the segment is $\|x\|^* = |\langle x, v \rangle|$. We will denote the dual norm after T symmetrizations with respect to the vectors $u_1, \ldots u_T$ by

$$\|x\|_T = \frac{1}{2^T} \sum_{D \subset \{1,\ldots,T\}} |\langle \prod_{i \in D} \pi_{u_i} x, v \rangle|$$

We will denote by I_T the body after T symmetrizations, the body that $\| \cdot \|_T$ is its dual norm. Note that the diameter of I is 1, while its mean width is $\approx 1/\sqrt{n}$. Therefore, also $M^*(I_T) \approx 1/\sqrt{n}$, and if I_T is close to a ball, then we must have $a(I_T) \leq 2/\sqrt{n}$.

In the next few lines we will show a lower bound on the diameter of I_T, and we will use this result to get that Proposition 2.3 is tight for the case of a segment.

Now, a straightforward calculation (by induction on k) yields the following equality: for any $x \in \mathbb{R}^n$,

$$\int_{S^{n-1}} \cdots \int_{S^{n-1}} \langle x, \prod_{i=1}^{k} \pi_{u_i} x \rangle d\sigma(u_1) \ldots d\sigma(u_k) = \left(1 - \frac{2}{n}\right)^k |x|^2$$

Roughly, this equality states that if we take a point in S^{n-1}, reflect it randomly k times, and calculate the scalar-product of the point we got with the original point - then the result has expected value of $(1 - \frac{2}{n})^k$.

The expectation of $\|v\|_T$ (where the expectation is over all the possible symmetrizations) is

$$\mathbb{E}\|v\|_T = \mathbb{E}\left[\frac{1}{2^T} \sum_{D \subset \{1,..,T\}} |\langle \prod_{i \in D} \pi_{u_i} v, v \rangle| \geq \left(1 - \frac{2}{n}\right)^T\right]$$

Since the diameter of I_T is $\sup_{x \in S^{n-1}} \|x\|_T$, the expectation of the diameter of I_T is surely greater than $\mathbb{E}\|v\|_T$, and we conclude that $\mathbb{E}[a(I_T)] > \exp(-\frac{2T}{n})$.

Now we can move to a general body $K \subset \mathbb{R}^n$. This body contains a maximal interval: There exists $v \in \mathbb{R}^n$ with $|v| = a(K)$ such that $I = [-v, v] \subset K$. If we apply the same set of symmetrizations to I and to K, then $I_T \subset K_T$, and $\mathbb{E}a(I_T) \leq \mathbb{E}a(K_T)$.

If the body K_T is close to a ball in exponential (in the dimension) probability, then with large probability $a(K_T) \leq 2M^*$, and clearly $\mathbb{E}a(K_T) \leq 2M^*$. But

$$\mathbb{E}a(I_T) \leq 2M^* \implies a(K)\left(1 - \frac{2}{n}\right)^T \leq 2M^*$$

which means that

$$T > \frac{1}{2}n \log \frac{a}{2M^*}$$

In other words, it takes at least $\frac{1}{2}n \log \frac{a}{2M^*}$ random symmetrizations just to reduce the diameter of K to size of $2M^*$.

The above discussion was actually a proof of the following theorem. Fix a body K. For $u_1, .., u_T \in S^{n-1}$, define $\chi_{u_1,..,u_T}^K$ to be indicator of the following event:

$$\frac{1}{2}M^* D \subset \frac{1}{2^T} \sum_{D \subset \{1,..,T\}} \prod_{i \in D} \pi_{u_i}(K) \subset 2M^* D$$

i.e. $\chi_{u_1,..,u_T}^K$ equals 1 if the event occurs, and equals 0 otherwise.

Now, for a body K, define $T(K)$ as the minimal T such that

$$measure\{(u_1, .. u_T) \in (S^{n-1})^T : \chi_{u_1,..,u_T}^K = 1\} > 1 - e^{-n}$$

Theorem 2.4 *There exist numerical constants C_1, C_2 such that:*

$$C_1 n \log \frac{a}{M^*} \leq T(K) \leq C_2 n \log \frac{a}{M^*}$$

(or in a shorter form: $T(K) \approx n \log \frac{a}{M^*}$).

We know that after typical $T(K)$ symmetrizations, our body becomes close to a ball, and further symmetrizations would keep it in such a shape. If we apply less than $T(K)$ symmetrizations, with high probability we are far away from a ball.

There is a parameter of the body - namely, the diameter - that decays regularly during the process of symmetrizations. If we look at the change of the diameter during the process of symmetrizations, we observe a phase-transition: By Lemma 2.1 and the discussion above we can see that there are numerical constants such that for $T \leq cT(K)$:

$$\exp\left(-C_1 \frac{T}{n}\right) \leq \frac{E[a(K_T)]}{a(K)} \leq \exp\left(-C_2 \frac{T}{n}\right)$$

But for $T \geq CT(K)$, the diameter stabilized, and with high probability it is very close to $M^*(K)$.

3 Deterministic Examples

Until now, we were interested only in the question: what happen to a body going through "typical" or "random" symmetrizations. A question that naturally arises is whether there exists a special choice of symmetrizations such that an approximated ball is achieved much faster than in a "typical" choice of symmetrizations.

In the spirit of the results in [MS], where symmetrizations of another kind are discussed, one might expect the answer to be "no". In the case of Milman and Schechtman investigating, the random and the non-random behavior essentially coincide. One cannot significantly improve the convergence by choosing specific symmetrizations. "Random" symmetrizations are almost as good as the best ones.

However, in our question it is not like that. We will see a few examples where $cn \log n$ random symmetrizations are needed, while only cn specific symmetrizations suffice.

3.1 Example 1

Again, a segment. $v = \frac{1}{\sqrt{n}}(1, 1, ..., 1) \in S^{n-1}$ and $I = [-v, v]$. According to formula 2.4 we need at least $cn \log n$ random symmetrizations.

Choose the directions to be $e_1, .., e_n$ - the standard unit vectors. Denote by I_n the interval I after going through symmetrizations with respect to $e_1, .., e_n$, and by $||| \cdot |||$ the dual norm to I_n. For every $x = (x_1, .., x_n) \in \mathbb{R}^n$:

$$|||x||| = \frac{1}{2^n} \sum_{D \subset \{1,..,n\}} |\langle \prod_{i \in D} \pi_{e_i} x, v \rangle| = \frac{1}{\sqrt{n}} Ave_{\epsilon \in \{-1,1\}^n} |\sum_i \epsilon_i x_i|$$

By Khinchine's inequality we obtain:

$$\frac{1}{\sqrt{2n}}|x| \le |||x||| \le \frac{1}{\sqrt{n}}|x|$$

or

$$\frac{1}{\sqrt{2n}}D \subset I_n \subset \frac{1}{\sqrt{n}}D$$

Also we proved that for the segment, there exists a choice of directions such that n symmetrizations are sufficient to achieve approximately a ball (actually $n-1$ suffice. The last symmetrization is unnecessary).

3.2   Example 2

Define the cube Q as $Q = \{x = (x_1, .., x_n) \in \mathbb{R}^n : \forall i \ -1 \le x_i \le 1\}$, the L_∞^n ball. Here we will prove the next lemma, which will be very useful in obtaining another example of bodies with an essentially different number of random and deterministic symmetrizations needed to approximate the Euclidean ball.

Lemma 3.1 *If $K \subset \alpha Q$, and $K = convS$, where $|S| = N$, then by symmetrizing with respect to the standard unit vectors $e_1, .., e_n$, a body $\hat{K}$ is obtained, and it satisfies:*

$$diam(\hat{K}) \le c\alpha\sqrt{\log N}$$

Proof. Denote $S = \{v_1, .., v_N\}$. K is a convex hull of S, so the dual norm of K is $||x||^* = \max_i |\langle x, v_i \rangle|$. The dual norm of the symmetrized body will be $||| \cdot |||$. For every $x \in \mathbb{R}^n$:

$$|||x||| = \frac{1}{2^n} \sum_{D \subset \{1,..,n\}} ||\prod_{i \in D} \pi_{e_i} x|| = \frac{1}{2^n} \sum \max_j |\langle \prod_{i \in D} \pi_{e_i} x, v_j \rangle|$$

For $1 \le j \le N$ and $\epsilon \in \{\pm 1\}^n$ define

$$f_x^j(\epsilon) = |\langle \sum_i \epsilon_i x_i e_i, v_j \rangle|$$

Then:

$$|||x||| = \mathbb{E}_\epsilon [\max_j f_x^j(\epsilon)]$$

$|||x|||$ is the expectation of maximum of N random variables (expectation by variable $\epsilon \in \{\pm 1\}^n$). We would like to bound from above $|||x|||$, and for that purpose we will estimate the ψ_2 norm (see, for example, in [BLM]) of those variables. The coordinates of v_j are $v_j = (v_{j1}, .., v_{jn}) \in \mathbb{R}^n$.

For every $j \in \{1, .., N\}$ we have

$$\|f_x^j\|_p = \left(Ave_{\epsilon \in \pm 1^n} |\sum_i v_{ji} x_i \epsilon_i|^p \right)^{\frac{1}{p}} \le c\sqrt{p} \sqrt{\sum_i (x_i v_{ji})^2} \le c\sqrt{p}|x|||v_j||_\infty$$

The first inequality follows by Khinchine's inequality. Since $K \subset \alpha Q$, for every $x \in S^{n-1}$ we get that $\|f_x^j\|_p \le c\sqrt{p}\alpha$. Since the pth moment of f_x^j is bounded by $c\alpha\sqrt{p}$, we get that $\|f_x^j\|_{\psi_2} \le c\alpha$.

Moreover since $|||x||| = \mathbb{E}_\epsilon [\max_j f_x^j(\epsilon)]$, we can use the well known estimate for the expectation of maximum of ψ_2 variables (e.g. [TL]):

$$\forall x \in S^{n-1} \quad |||x||| \le c\alpha\sqrt{\log N}$$

Thus the lemma is proved.   □

Application of that lemma to the case of $B(l_1^n) = \{x \in \mathbb{R}^n : \sum_i |x_i| \le 1\}$ is easy. $B(l_1^n)$ has diameter 1, and $M^* B(l_1^n) \approx \sqrt{\log n/n}$. Again, $cn \log n$ random symmetrizations are needed to make this body close to an Euclidean ball.

For simplicity, assume $\exists k$ with $n = 2^k$ (otherwise, embed $B(l_1^n)$ in such a space). Let $w_1, .., w_n \in S^{n-1}$ be the normalized Walsh vectors. Since $w_1, .., w_n$ is an orthonormal system, we can write as well $K = B(l_1^n) = conv_{i=1,..n}\{\pm w_i\}$. But since $K \subset \frac{1}{\sqrt{n}} Q$ we are in a position to use Lemma 3.1.

Our process of symmetrizations here consists of 2 steps. In the first step, we symmetrize with respect to the standard unit vectors $e_1, .., e_n$. By Lemma 3.1 we obtain a body K_n which has diameter $diam(K_n) \le c\sqrt{\log n}/\sqrt{n}$ or $K_n \subset cM^* D$ (because $M^*(K_n) = M^* B(l_1^n) \approx \sqrt{\log n}/\sqrt{n}$). For the second step, we will choose an additional cn symmetrizations "randomly", and according to Proposition 2.3 we will get an approximated ball.

To summarize, we used specific n symmetrizations, and after that additional cn symmetrizations - and achieved a body close to a ball.

3.3 General Convex Body

Let $e_1, .., e_n$ be the standard orthonormal basis in $\mathbb{R}^n$. Call a body K a 1-unconditional body if it satisfies $\forall 1 \le i \le n, \pi_{e_i} K = K$. The norm that K is its unit ball is a 1-unconditional norm.

The following lemma is quite known:

Lemma 3.2 *Let $K \subset \mathbb{R}^n$ be a convex 1-unconditional body with $M^*(K) = 1$. Then there exist a numerical constant c such that*

$$K \subset c\sqrt{n}B(l_1^n)$$

Proof. Take $x = (x_1, .., x_n) \in K$. Since K is a 1-unconditional body, all the vectors of the form $(\pm x_1, .., \pm x_n)$ are also inside K, and so is their convex hull. Denote by Q_x the convex hull of all such vectors. Clearly, Q_x is a rectangular parallelepiped. Q_x is equal to a Minkowski sum of n segments - those are the segments $[-x_1 e_1, x_1 e_1], .., [-x_n e_n, x_n e_n]$. Since mean width is additive with respect to Minkowski sum, and the mean width of a unit segment is $\approx \frac{1}{\sqrt{n}}$, we can compute the mean width of Q_x, that is: $M^*(Q_x) \approx \frac{1}{\sqrt{n}} \sum_i |x_i|$. On the other hand, the mean width of Q_x is less than 1, since $M^*(K) = 1$ and $Q_x \subset K$. We conclude that there exist c such that $\frac{1}{\sqrt{n}} \sum_i |x_i| \leq c$, and $x \in c\sqrt{n}B(l_1^n)$. This is true for any $x \in K$, hence the lemma is proved. $\square$

Assume $K \subset \mathbb{R}^n$ is a convex body, $M^*(K) = 1$. Apply n symmetrizations to K with respect to $e_1, .., e_n$. Then the resulting body is a 1-unconditional body with mean width 1, and therefore is contained in $c\sqrt{n}B(l_1^n)$. We have proved that using cn symmetrization, $\sqrt{n}B(l_1^n)$ can be transformed to a body which is very close to an Euclidean ball of radius $c\sqrt{\log n}$ (this is just a renormalization of Example 2). In particular, this body has a diameter of $c\sqrt{\log n}$ at most. Since our body was inscribed by $c\sqrt{n}B(l_1^n)$, if we apply to it the same set of cn symmetrizations, we achieve a body with diameter less than $c\sqrt{\log n}$.

Note that we start with a general body K, use only cn deterministic symmetrizations - and reduce the diameter of the body to be just $c\sqrt{\log n}$. At this stage of symmetrizations, the symmetrized body satisfies $\frac{a}{M^*} \leq c\sqrt{\log n}$. Now we can turn to "random" symmetrizations. By theorem 2.4, we need additional $cn \log \frac{a}{M^*} \leq cn \log \log n$ "random" symmetrizations to achieve approximately an Euclidean ball. Therefore we proved the following theorem:

Theorem 3.3 *Let K be a convex body in $\mathbb{R}^n$. For $N = cn \log \log n$, there exist N vectors in S^{n-1} such that if we symmetrize K with respect to those vectors, we obtain a body $\hat{K}$ such that*

$$\frac{1}{2}M^*D \subset \hat{K} \subset 2M^*D$$

where c is a numerical constant, M^* is the mean width of K.

Thus, for a wide class of bodies (bodies with large diameter), there is an essential difference between the number of "random" and "deterministic" Minkowski symmetrizations needed in order to approximate an Euclidean ball.

3.4   Remarks

1. Note that in the case of Example 1, it is clear that at least $n - 1$ symmetrizations are needed just to get an n dimensional body, so one cannot expect convergence to a ball in less than $n - 1$ symmetrizations.
2. We still don't know whether cn deterministic symmetrizations are sufficient for every convex body in $\mathbb{R}^n$.

I would like to thank Vitali Milman for presenting me to the subject, for helping me in fruitful discussions, and for encouraging me to write this paper.

References

[BLM]   Bourgain J., Lindenstrauss J., Milman V.D. (1988) Minkowski sums and symmetrizations. In: Lindenstrauss J., Milman V.D. (Eds.) Geometric Aspects of Functional Analysis (1986–87), Lecture Notes in Math., 1317, Springer-Verlag, 44–66

[MS]   Milman V.D., Schechtman G. (1997) Global vs. local asymptotic theories of finite dimensional normed spaces. Duke J. 90:73–93

[TL]   Talagrand M., Ledoux M. (1991) Probability in Banach spaces. A Series of Modern Surveys in Mathematics 23, Springer-Verlag

Average Volume of Sections of Star Bodies

A. Koldobsky[1] and M. Lifshits[2]

[1] Department of Mathematics, University of Missouri, Columbia, MO 65211, USA
[2] St-Petersburg State University, Russia and Université Lille-I, France

Abstract. We study the asymptotic behavior, as the dimension goes to infinity, of the volume of sections of the unit balls of the spaces ℓ_q^n, $0 < q \le \infty$. We compute the precise asymptotics of the average volume of central sections and then prove a concentration inequality of exponential type. For the case of non-central hyperplane sections of the cube, we prove a local limit theorem confirming the conjecture on the asymptotically Gaussian dependence of the volume of sections on the distance from the hyperplane to the origin. Note that a weak limit theorem was established very recently in [ABP] for a larger class of bodies. Our calculations are based on connections between volume and the Fourier transform.

1 Introduction

For a star body K in $\mathbb{R}^n$ and an integer $0 < p < n$, consider the average volume of p-dimensional sections of K :

$$AV_p(K) = \int_{Gr(n,p)} \mathrm{vol}_p(K \cap H)\, dH,$$

where $Gr(n,p)$ is the Grassman manifold of p-dimensional subspaces of $\mathbb{R}^n$ equipped with the probability Haar measure. Our study of the quantities $AV_p(K)$ and, in particular, of their behavior as the dimension goes to infinity is motivated by connections with the local theory of Banach spaces.

In this paper we consider the case where $K = B_q^n$ is the unit ball of the space ℓ_q^n, $0 < q \le \infty$. In Sections 4-5 we compute precise asymptotics of the averages $AV_p(B_q^n)$, as $n \to \infty$ and $p = p(n)$, using the Fourier transform formula introduced in Section 3. The results are different for the cases where $p = n - d$ with d fixed, where $p \sim \alpha n$, $0 < \alpha < 1$, and where p is fixed and $n \to \infty$.

A problem that logically follows is to estimate the concentration properties of $\mathrm{vol}_p(K \cap H)$ as a function of H on the Grassman manifold. In Section 6, we prove an exponential concentration inequality for the volume of hyperplane sections of the bodies B_q^n, $0 < q \le \infty$: there exist n_0 and c depending on q only so that for all $n > n_0$ and all $\epsilon \in (0,1)$,

$$mes\{\xi \in S^{n-1} : \ |V(\xi) - m_V| > \epsilon\} \le 4\exp\{-c\,\epsilon^2(n-1)\}, \quad (1.1)$$

The first named author was supported in part by the NSF Grant DMS-9996431. The second named author was supported in parts by RFBR and INTAS Grant 99-01-00112.

where $V(\xi)$ is the volume of the central hyperplane section of B_q^n orthogonal to ξ, and $mes(\cdot)$ is the probability uniform measure on S^{n-1}. Our calculations are based on the connections between volumes and the Fourier transform and on Lévy's isoperimetric inequality.

In Section 7, we consider another problem from the local theory that was communicated to us by V. Milman. The problem is to show that the volume of sections of origin-symmetric convex bodies by hyperplanes located at distance r from the origin converges (in some sense) to the Gaussian density function of r, as the dimension goes to infinity. Note that both Laplace [La] and Polya [P] proved this for the sections of the n-cubes perpendicular to the main diagonal, and that a simple calculation gives an affirmative answer for the Euclidean balls. We confirm this conjecture for the hyperplane sections of the cubes by proving first that

$$\lim_{n\to\infty} \frac{1}{\mathrm{vol}_n(B_\infty^n)|S^{n-1}|} \int_{S^{n-1}} \mathrm{vol}_{n-1}(B_\infty^n \cap (r\xi + \xi^\perp))d\xi = \sqrt{3/2\pi}\exp\{-3r^2/2\},$$

and then noting that exactly the same concentration argument, as in Section 6, works for non-central hyperplane sections and leads to a local limit theorem similar to (1.1). After this work was completed, we learned about an excellent earlier paper [ABP], where an exponential concentration inequality (for the distribution function instead of the density) was established for a class of bodies including B_q^n, $1 \leq q \leq \infty$. The result of our Theorem 6.2 can be proved by methods from [ABP] (except for the case $0 < p < 1$, where one can not use Busemann's theorem). However, in the case of non-central sections, we do not immediately see how can one deduce the local limit result (for the density) of our Section 7 from the weak limit theorem (for the distribution function) of [ABP]. In both Sections 6 and 7, our methods are completely different from those of [ABP]. Our argument is based on the approximation of spheric averages by Gaussian ones and on a Fourier transform representation for the volumes, while the proofs in [ABP] use the tools of convexity. We have also learned after this work was completed that important earlier papers [BV] and [V] contain limit theorems for the density (in individual directions) with convergence in L_1 and L_∞ norms. These results, however, do not imply the exponential concentration. The methods there are also different from ours. For several related probabilistic results, see [DF], [R], [S], [W].

In the sequel, $f(n) \sim g(n)$ means that $\lim_{n\to\infty} f(n)/g(n) = 1$.

2   An Upper Bound for the Average Volume of Sections

Let K be a body that is star-shaped with respect to the origin. We call K a *star body* if the origin is an interior point of K and the Minkowski functional of K (defined by $\|x\|_K = \min\{a > 0 : x \in aK\}$) is continuous on $\mathbb{R}^n$.

Let $Gr(n, p)$ be the Grassman manifold of p-dimensional subspaces of R^n. In the sequel, we consider the Grassman manifolds equipped with their normalized Haar measures, while the Haar measures on the sphere S^{n-1} and its sections are not normalized. For every continuous function f on S^{n-1},

$$\int_{S^{n-1}} f(x)\, dx = \frac{|S^{n-1}|}{|S^{p-1}|} \int_{Gr(n,p)} \left(\int_{S^{n-1} \cap H} f(x)\, dx \right) dH, \qquad (2.1)$$

where $|S^{p-1}| = 2\pi^{p/2}/\Gamma(p/2)$ is the surface area of the unit sphere S^{p-1} in $\mathbb{R}^p$.

We use an elementary formula for the p-dimensional volume of the section of K by a subspace $H \in Gr(n,p)$:

$$\mathrm{vol}_p(K \cap H) = \int_{S^{n-1} \cap H} \int_0^{1/\|\xi\|} r^{p-1} dr\, d\xi = p^{-1} \int_{S^{n-1} \cap H} \|\xi\|^{-p}\, d\xi. \quad (2.2)$$

We need the following simple fact.

Lemma 2.1 *For every pair of integers* $0 \le d < n$,

$$1 \le \frac{n^{(n-d)/n}|S^{n-d-1}|}{(n-d)|S^{n-1}|^{(n-d)/n}} = \frac{\left(\Gamma(n/2+1)\right)^{(n-d)/n}}{\Gamma(\frac{n-d}{2}+1)} \le e^{d/2}. \qquad (2.3)$$

Proof. To prove the lower bound, use the well-known fact that the function $\log(\Gamma(x))$ is convex. We have

$$\frac{\log(\Gamma(\frac{n}{2}+1)) - \log(\Gamma(1))}{n/2} \ge \frac{\log(\Gamma((n-d)/2+1)) - \log(\Gamma(1))}{(n-d)/2},$$

which implies the result.

To prove the upper bound, let us write the inequality (2.3) in the form

$$\frac{\Gamma(n/2+1)}{\Gamma((n-d)/2+1)} \frac{1}{(\Gamma(n/2+1))^{d/n} e^{d/2}} \le 1. \qquad (2.4)$$

Using again the log-convexity of the Γ-function, we get

$$\Gamma^2(n/2+1) \le \Gamma(n/2+3/2)\Gamma(n/2+1/2) = (n/2+1/2)\Gamma^2(n/2+1/2).$$

Therefore,

$$\frac{\Gamma(n/2+1)}{\Gamma(n/2+1/2)} \le (n/2+1)^{1/2}. \qquad (2.5)$$

It immediately follows that for bigger d

$$\frac{\Gamma(n/2+1)}{\Gamma((n-d)/2+1)} \le (n/2)^{d/2}, \quad d > 1. \qquad (2.6)$$

For the second fraction in (2.4), we use Stirling's formula in the form of [A, p.24]: for every $z \geq 1/2$,

$$\Gamma(z+1) \geq \sqrt{2\pi}(z+1)^{z+1/2}e^{-z-1} = \frac{\sqrt{2\pi}}{e}(z/e)^z(z+1/z)^z\sqrt{z+1}$$

$$\geq (z/e)^z\sqrt{z+1}.$$

Letting in (2.7) $z = n/2$ we obtain for the second fraction in (2.4) the upper bound

$$(n/2)^{-d/2}(n/2+1)^{-d/2n}. \qquad (2.7)$$

Combining this bound with (2.6) immediately proves (2.4) for $d > 1$. For $d = 1$ after combining (2.7) and (2.5) we still have to check

$$(1+2/n)^{1/2}(n/2+1)^{-1/2n} < 1.$$

This reduces to

$$(1+2/n)^{n/2} < e < (n/2+1)^{1/2}$$

and is true for $n \geq 16$. In the case $d = 1, n < 16$ the inequality can be checked directly. $\qquad\square$

The following inequality follows from a more general result of Lutwak [Lu]. We give here a simple proof.

Proposition 2.2 *Let K be a star body in R^n. Then for every positive integer $p < n$*

$$AV_p(K) \leq R_{n,p} = \frac{|S^{p-1}|}{p\,|S^{n-1}|^{p/n}}\,(n\,\mathrm{vol}_n(K))^{p/n}. \qquad (2.8)$$

In particular, for $p = n - d$ we have

$$AV_{n-d}(K) \leq e^{d/2}(\mathrm{vol}_n(K))^{(n-d)/n}, \qquad (2.9)$$

and if $p = n/\alpha$ and $K = K_n \subset \mathbb{R}^n$ so that $\mathrm{vol}_n(K_n) = c$ then

$$R_{n,p} \sim (\pi n)^{1/2\alpha - 1/2}\alpha^{n/2\alpha + 1/2}\mathrm{vol}_n(K)^{1/\alpha}. \qquad (2.10)$$

Proof. Denote by $c = |S^{p-1}|/(p|S^{n-1}|)$. By (2.1), (2.2) and Hölder's inequality,

$$\int_{Gr(n,p)} \mathrm{vol}_p(K \cap H)\,dH = p^{-1}\int_{Gr(n,p)}\int_{S^{n-1}\cap H} \|\xi\|^{-p}\,d\xi$$

$$= c\int_{S^{n-1}} \|\xi\|^{-p}\,d\xi \leq c\Big(\int_{S^{n-1}} \|\xi\|^{-n}\,d\xi\Big)^{p/n}\Big(\int_{S^{n-1}} 1\,d\xi\Big)^{1-p/n}$$

$$= c\,(n\,\mathrm{vol}_n(K))^{p/n}\,|S^{n-1}|^{1-p/n},$$

which gives (2.8). The estimates (2.9) and (2.10) follow from Lemma 2.1 and

$$|S^{p-1}| \sim \frac{\pi^{p/2-1/2} 2^{p/2} e^{p/2}}{p^{p/2-1/2}}, \qquad |S^{n-1}|^{p/n} \sim \frac{\pi^{p/2-p/2n} 2^{p/2} e^{p/2}}{n^{p/2-p/2n}}. \qquad \square$$

The estimate (2.8) turns into equality if $K = B_2^n$ is the Euclidean ball. Also the estimates (2.9) and (2.10) are asymptotically sharp in this case. The results of Section 5.1 of this paper, together with the formula $\mathrm{vol}_n(B_q^n) = (2\Gamma(1+1/q))^n / \Gamma(1+n/q)$, imply that

$$\lim_{n \to \infty} \frac{AV_{n-d}(B_q^n)}{(\mathrm{vol}_n(B_q^n))^{(n-d)/n}} = \left(\frac{4\Gamma(1/q)(\Gamma(1+1/q))^2 e^{2/q}}{2\pi\Gamma(3/q)} \right)^{d/2}.$$

In particular, if $q = \infty$ the limit is equal to $(6/\pi)^{d/2}$. Note that the minimal volume of a d-codimensional section of the n-cube is 1 (see [Va]), and the maximal volume is $2^{d/2}$ (see [Ba3]).

The situation is different for proportional sections. For example, if $K_n = B_\infty^n$ is the cube with side 2 and $p = n/2$, then the main term of the asymptotic upper bound (2.10) is

$$\alpha^{n/2\alpha} \mathrm{vol}_n(K_n)^{1/\alpha} = 2^{n/4+n/2} = (2.828\ldots)^{n/2},$$

while, as it will be shown in Section 4.3.2, the leading term of the true asymptotics of $AV_{n/2}(K_n)$ is $(2.516\ldots)^{n/2}$.

3   Volumes of Sections and the Fourier Transform

The Fourier transform formulas for the volume of sections have already been applied to several geometric problems. The first formula of this kind was known to Laplace [La], who proved that the volume of the section of $\frac{1}{2}B_\infty^n$ perpendicular to the main diagonal is equal to

$$\frac{2\sqrt{n}}{\pi} \int_0^\infty \left(\frac{\sin(t)}{t} \right)^n dt,$$

and applied the law of large numbers to show that the limit of this expression, as $n \to \infty$, is equal to $\sqrt{6/\pi}$. This formula was generalized in [P], [H], [Va], [Ba1], [Ba3], [MeP], [K1] to the case of arbitrary central sections of the bodies B_q^n, $0 < q \le \infty$ and applied to find the maximal and minimal sections of some of these bodies. It was noticed in [K1] that, in the case of hyperplane central sections, these formulas represent particular cases of the following relation: for every origin-symmetric star body K in $\mathbb{R}^n$ and every $\xi \in S^{n-1}$,

$$\mathrm{vol}_{n-1}(K \cap \xi^\perp) = \frac{1}{\pi(n-1)} (\|x\|_K^{-n+1})^\wedge(\xi), \qquad (3.1)$$

where $\hat{f}$ stands for the Fourier transform of f in the sense of distributions.

The latter formula was generalized in [K3, Lemma 7] to central sections of arbitrary dimension. We say that a star body K in $\mathbb{R}^n$ is k-*smooth* if the restriction of the Minkowski functional of K to the sphere S^{n-1} belongs to the space $C^{(k)}(S^{n-1})$ of continuously differentiable up to order k functions. If K is a $(k-1)$-smooth symmetric star body in $\mathbb{R}^n$, $1 \leq k < n$, then for every $(n-k)$-dimensional subspace H of $\mathbb{R}^n$,

$$\text{vol}_{n-k}(K \cap H) = \frac{1}{n-k} \int_{S^{n-1} \cap H} \|x\|_K^{-n+k} \, dx$$

$$= \frac{1}{n-k} \frac{1}{(2\pi)^k} \int_{S^{n-1} \cap H^\perp} (\|x\|_K^{-n+k})^\wedge(\theta) \, d\theta. \quad (3.2)$$

The formulas (3.1) and (3.2), in conjunction with Propositions 3.2 and 3.3 below, reproduce the formulas for central sections of the bodies B_q^n that appeared in [H], [Ba1], [Ba3], [MeP], [K1].

The formulas (3.1) and (3.2) can be used to get an expression for the average volume of sections in terms of the Fourier transform. However, applying (3.2) requires a certain approximation argument, because of the smoothness condition. Instead, we present here a simple direct proof. If $0 < p < n$ then the function $\|x\|_K^{-p}$ is locally integrable on $\mathbb{R}^n$. The Fourier transform $(\|\cdot\|_K^{-p})^\wedge$ is defined as the distribution satisfying $\langle(\|x\|_K^{-p})^\wedge, \phi\rangle = \langle\|x\|_K^{-p}, \hat{\phi}\rangle$ for every test function $\phi \in \mathcal{S}(\mathbb{R}^n)$. Note that the distribution $(\|\cdot\|_K^{-p})^\wedge$ is homogeneous of degree $-n+p$.

Lemma 3.1 *Let K be a star body in R^n so that $(\|\cdot\|_K^{-p})^\wedge$ is a locally integrable function on R^n, and p is an integer, $0 < p < n$. Then*

$$AV_p(K) = \frac{|S^{n-p-1}|}{p(2\pi)^{n-p}|S^{n-1}|} \int_{S^{n-1}} (\|x\|_K^{-p})^\wedge(\theta) \, d\theta.$$

Proof. We have

$$\langle(\|x\|_K^{-p})^\wedge, \exp(-\|x\|_2^2/2)\rangle = (2\pi)^{n/2}\langle\|x\|_K^{-p}, \exp(-\|x\|_2^2/2)\rangle$$

$$= (2\pi)^{n/2} \int_{R^n} \|x\|_K^{-p} \exp(-\|x\|_2^2/2) \, dx$$

$$= (2\pi)^{n/2} \int_{S^{n-1}} \|\theta\|_K^{-p} \, d\theta \int_0^\infty t^{n-p-1} \exp(-t^2/2) \, dt$$

$$= 2^{n-p/2-1}\pi^{n/2}\Gamma((n-p)/2) \int_{S^{n-1}} \|\theta\|_K^{-p} \, d\theta. \quad (3.3)$$

On the other hand, since $(\|x\|_K^{-p})^\wedge$ is a locally integrable homogeneous function of degree $-n+p$ on R^n, we have

$$\langle(\|x\|_K^{-p})^\wedge, \exp(-\|x\|_2^2/2)\rangle = \int_{S^{n-1}} (\|x\|_K^{-p})^\wedge(\theta) \, d\theta \int_0^\infty t^{p-1} \exp(-t^2/2) \, dt$$

$$= 2^{p/2-1}\Gamma(p/2) \int_{S^{n-1}} (\|x\|_K^{-p})^\wedge(\theta) \, d\theta. \quad (3.4)$$

Finally, using (2.1) and (2.2), we get

$$\int_{Gr(n,p)} \text{vol}_p(K \cap H) \, dH = \frac{1}{p} \int_{Gr(n,p)} \int_{S^{n-1} \cap H} \|\theta\|_K^{-p} \, d\theta$$

$$= \frac{|S^{p-1}|}{p|S^{n-1}|} \int_{S^{n-1}} \|\theta\|_K^{-p} \, d\theta. \qquad (3.5)$$

The desired result now follows from (3.3), (3.4) and (3.5). □

It is known (see [K3]) that $(\|\cdot\|_K^{-p})^\wedge$ is a continuous function on $\mathbb{R}^n \setminus \{0\}$ if K is a $(n-p-1)$-smooth star body. This immediately implies that $(\|\cdot\|_K^{-p})^\wedge$ is locally integrable because this function is also homogeneous of degree $-n+p > -n$. However, Lemma 3.1 can also be applied to the bodies B_q^n that are not necessarily smooth. In order to apply Lemma 3.1 to the bodies B_q^n we use simple direct computations of the Fourier transform from [K2, Lemma 3 and Lemma 8]:

Proposition 3.2 *If $p \in (0, n)$ then the Fourier transform of the function $\|x\|_\infty^{-p}$ is equal to a locally integrable on $\mathbb{R}^n$ function*

$$\xi \mapsto 2^n p \int_0^\infty t^{-p-1} \prod_{k=1}^n \frac{\sin(t\xi_k)}{\xi_k} \, dt.$$

Denote by γ_q the Fourier transform of the function $z \mapsto \exp(-|z|^q)$, $z \in \mathbb{R}$.

Proposition 3.3 *Let $q > 0$, $n \in \mathbb{N}$, $0 < p < n$. Then the Fourier transform of the function $\|x\|_q^{-p}$ is equal to a locally integrable on $\mathbb{R}^n$ function*

$$\xi \mapsto \frac{q}{\Gamma(p/q)} \int_0^\infty t^{n-p-1} \prod_{k=1}^n \gamma_q(t\xi_k) \, dt.$$

4   Sections of the Cube

Combining Lemma 3.1 with Proposition 3.2 we get

$$AV_p(B_\infty^n) = 2^p \pi^{p-n} |S^{n-p-1}| \, I_n, \qquad (4.1)$$

where

$$I_n = \frac{1}{|S^{n-1}|} \int_{S^{n-1}} \int_0^\infty t^{n-p-1} \prod_{k=1}^n \sin(t\xi_k)/(t\xi_k) \, dt d\xi.$$

In this section we study the asymptotics of these integrals and the corresponding average volumes as $n \to \infty$. We consider three typical cases. First, we consider "hypersections", where $p = n - d$ and d is fixed, while $n \to \infty$.

The second is the case of proportional sections, where $p \sim \alpha n$, $0 < \alpha < 1$. Finally, in the case of low-dimensional sections we have p fixed and $n \to \infty$. In fact, the methods exposed below are applicable to any reasonable dependence $p = p(n)$.

The Fourier transform of the function $\|x\|_\infty^{-p}$ is a sign-changing function if $p < n - 3$ which shows that the cube in $\mathbb{R}^n$ is a p-intersection body only if $p \geq n - 3$ (see [K2, Th.1], [K4] for details). Therefore it would be interesting to see how large is the set of those points $\xi \in S^{n-1}$ where $(\|x\|_\infty^{-p})^\wedge(\xi) < 0$. For this reason, along with the integrals I_n we study the asymptotics of the integrals

$$A_n = \frac{1}{|S^{n-1}|} \int_{S^{n-1}} \int_0^\infty t^{n-p-1} \prod_{k=1}^n |\sin(t\xi_k)/(t\xi_k)| dt d\xi.$$

The results of Ball [Ba1, Th.4], [Ba3, Th.6 and Proposition 4] suggest that the behaviour of the integrals A_n must be similar to that of I_n if $d = n - p$ is fixed. We show below that this is, indeed, the case. Moreover, these integrals still have the same main asymptotic term $n^{(1-\alpha)n/2}$ but differ at most by c^n if $p \sim \alpha n$, $0 < \alpha < 1$. However, if p is fixed, the integrals A_n grow much faster than I_n.

To compute the asymptotics of our integrals, we first link the integral over the sphere to the integral over the Gaussian distribution G with zero mean and covariance $\frac{1}{n}U$, where U is the unit matrix. Then the correspondent Gaussian integrals take the form

$$I_n^G = \int_0^\infty t^{n-p-1}[E_G \sin(t\xi_1)/(t\xi_1)]^n dt = \int_0^\infty t^{n-p-1} g\left(\frac{t}{\sqrt{n}}\right)^n dt;$$

$$A_n^G = \int_0^\infty t^{n-p-1}[E_G|\sin(t\xi_1)/(t\xi_1)|]^n dt = \int_0^\infty t^{n-p-1} h\left(\frac{t}{\sqrt{n}}\right)^n dt$$

where

$$g(\tau) = E\sin(\tau X)/(\tau X), \qquad h(\tau) = E|\sin(\tau X)/(\tau X)|, \qquad (4.2)$$

and X follows the standard normal distribution.

Note that there is an explicit relation between I_n and I_n^G, A_n and A_n^G, which follows from the fact that the functions under the integrals are homogeneous.

Lemma 4.1 *Let $f : R^n \to R$ be a homogeneous function of degree β. Then*

$$\frac{1}{|S^{n-1}|} \int_{S^{n-1}} f(\xi) d\xi = \frac{\Gamma(n/2)}{\Gamma\left(\frac{n+\beta}{2}\right)} (n/2)^{\beta/2} \int_{\mathbb{R}^n} f(x) G(dx).$$

Proof. Writing the integrals in polar coordinates we get

$$\int_{\mathbb{R}^n} f(x)G(dx) = \int_0^\infty \int_{S^{n-1}} f(r\xi)r^{n-1}p_G(r\xi)d\xi dr$$

$$= \int_0^\infty \int_{S^{n-1}} f(\xi)r^{n+\beta-1}(n/2\pi)^{n/2}\exp\{-r^2n/2\}d\xi dr$$

$$= (n/2\pi)^{n/2}\int_0^\infty r^{n+\beta-1}\exp\{-r^2n/2\}dr\int_{S^{n-1}} f(\xi)d\xi$$

$$= (n/2\pi)^{n/2}\frac{\Gamma\left(\frac{n+\beta}{2}\right)}{n}(2/n)^{(n+\beta)/2-1}\frac{2\pi^{n/2}}{\Gamma(n/2)}\frac{1}{|S^{n-1}|}\int_{S^{n-1}} f(\xi)d\xi$$

$$= \frac{\Gamma\left(\frac{n+\beta}{2}\right)}{\Gamma(n/2)}(2/n)^{\beta/2}\frac{1}{|S^{n-1}|}\int_{S^{n-1}} f(\xi)d\xi. \qquad \square$$

The function

$$f(\xi) = \int_0^\infty t^{n-p-1}\prod_{k=1}^n \sin(t\xi_k)/(t\xi_k)dt$$

is homogeneous of degree $\beta = p - n$, and the same is true for the integrals with the absolute values. Now Lemma 4.1 yields

$$I_n = \frac{\Gamma(n/2)}{\Gamma\left(\frac{p}{2}\right)}(n/2)^{(p-n)/2}\, I_n^G, \qquad A_n = \frac{\Gamma(n/2)}{\Gamma\left(\frac{p}{2}\right)}(n/2)^{(p-n)/2}\, A_n^G.$$

4.1 Asymptotic Behavior of the Gaussian Integrals I_n^G

The function g appearing in the expression for the integrals I_n^G can easily be calculated. In fact,

$$(g(\tau)\tau)' = E\cos(\tau X) = e^{-\tau^2/2}.$$

It follows that

$$g(\tau) = \sqrt{\pi/2}(2\Phi(\tau) - 1)\tau^{-1}, \qquad (4.3)$$

where $\Phi(\tau) = P\{X < \tau\}$ denotes the standard normal distribution function. Therefore,

$$g(0) = 1,\ g'(0) = 0,\ g''(0) = -1/3$$

and, as $t \to \infty$,

$$g(\tau) \sim \sqrt{\pi/2}\ \tau^{-1}. \qquad (4.4)$$

In order to study the integrals I_n^G we need to find the point τ_* which maximizes the integrand:

$$t^{n-p-1}g(\frac{t}{\sqrt{n}})^n \to \max.$$

Letting the derivative vanish and using (4.3), we get the equation

$$\frac{\Phi'(\tau_*)\tau_*}{\Phi(\tau_*) - \Phi(0)} = \frac{p+1}{n}. \tag{4.5}$$

Different behavior of τ_* at $n \to \infty$ in three following cases explains the difference of the asymptotics for $AV_p(B_n^\infty)$.

4.1.1 Hypersections, $p = n - d$ with Constant d.

In this case the solution of the equation (4.5) tends to zero and the behavior of I_n^G is determined by the asymptotic of the integrand at zero. Split the domain of integration into the union of $[0, r\sqrt{n}]$ and $(r\sqrt{n}, \infty)$. For the latter interval

$$\int_{r\sqrt{n}}^\infty t^{d-1} g(\frac{t}{\sqrt{n}})^n dt \leq (\max_{\tau > r} g(\tau))^{n-d-1} \int_{r\sqrt{n}}^\infty t^{d-1} g\left(\frac{t}{\sqrt{n}}\right)^{d+1} dt$$

$$\leq (\max_{\tau > r} g(\tau))^{n-d-1} \int_{r\sqrt{n}}^\infty t^{d-1} \left(\frac{c\sqrt{n}}{t}\right)^{d+1} dt$$

$$= (\max_{\tau > r} g(\tau))^{n-d-1} c^{d+1} n^{d/2} r^{-1} \to 0,$$

since the maximum is strictly less than one. On the other hand,

$$\int_0^{r\sqrt{n}} t^{d-1} g(\frac{t}{\sqrt{n}})^n dt = \int_0^{r\sqrt{n}} t^{d-1} \left(1 - \frac{t^2(1+o(r))}{6n}\right)^n dt$$

$$\sim \int_0^\infty \exp\left\{-\frac{t^2(1+o(r))}{6}\right\} t^{d-1} dt.$$

Sending $r \to 0$, we finally obtain for $p = n - d$

$$\lim_n I_n^G = \int_0^\infty \exp\left\{-\frac{t^2}{6}\right\} t^{d-1} dt = \frac{\Gamma(d/2)6^{d/2}}{2}. \tag{4.6}$$

4.1.2 Proportional Sections, $p = \alpha n - 1, 0 < \alpha < 1$.

In this case the equation (4.5) turns into

$$\frac{\Phi'(\tau_*)\tau_*}{\Phi(\tau_*) - \Phi(0)} = \alpha.$$

Hence, the solution τ_* does not depend on n. Make a linear scaling $t = \sqrt{n}\tau_* + u$ and denote by

$$g_\alpha(\tau) = g(\tau)\tau^{1-\alpha}.$$

Then

$$I_n^G = \int_0^\infty t^{(1-\alpha)n} g\left(\frac{t}{\sqrt{n}}\right)^n dt = \int_0^\infty \left[t^{1-\alpha} g\left(\frac{t}{\sqrt{n}}\right)\right]^n dt$$

$$= n^{(1-\alpha)n/2} \int_0^\infty g_\alpha\left(\frac{t}{\sqrt{n}}\right)^n dt = n^{(1-\alpha)n/2} \int_{-\sqrt{n}\tau_*}^\infty g_\alpha\left(\tau_* + \frac{u}{\sqrt{n}}\right)^n du$$

$$= n^{(1-\alpha)n/2} g_\alpha(\tau_*)^n \int_{-\sqrt{n}\tau_*}^\infty \left[\frac{g_\alpha(\tau_* + \frac{u}{\sqrt{n}})}{g_\alpha(\tau_*)}\right]^n du$$

$$\sim n^{(1-\alpha)n/2} g_\alpha(\tau_*)^n \int_{-\sqrt{n}\tau_*}^\infty \left[1 + \frac{g_\alpha''(\tau_*)u^2}{2ng_\alpha(\tau_*)}\right]^n du$$

$$\sim n^{(1-\alpha)n/2} g_\alpha(\tau_*)^n \int_{-\sqrt{n}\tau_*}^\infty \exp\left\{\frac{g_\alpha''(\tau_*)u^2}{2g_\alpha(\tau_*)}\right\} du$$

$$\sim n^{(1-\alpha)n/2} g_\alpha(\tau_*)^n \sqrt{\frac{2\pi g_\alpha(\tau_*)}{|g_\alpha''(\tau_*)|}}$$

$$= n^{(1-\alpha)n/2} g_\alpha(\tau_*)^{n+1/2} \sqrt{\frac{2\pi}{|g_\alpha''(\tau_*)|}}. \tag{4.7}$$

4.1.3 Low-dimensional Sections, $p = const.$ In this case we make a linear scaling $t = \sqrt{n}\tau$,

$$I_n^G = \int_0^\infty t^{n-p-1} g\left(\frac{t}{\sqrt{n}}\right)^n dt = n^{(n-p)/2} \int_0^\infty \tau^{n-p-1} g(\tau)^n d\tau$$

$$= n^{(n-p)/2} \int_0^\infty \tau^{-p-1} [\tau g(\tau)]^n dt$$

$$= n^{(n-p)/2} \int_0^\infty \tau^{-p-1} [\sqrt{\pi/2}(2\Phi(\tau) - 1)]^n d\tau.$$

The solution τ_* of the equation (4.5) now tends to infinity. We have

$$\tau_* \sim \sqrt{2\ln n}, \qquad \exp\{-\tau_*^2/2\} \sim \frac{(p+1)\sqrt{2\pi}}{2\tau_* n} \ll \frac{1}{n}.$$

Therefore, $(2\Phi(\tau_*) - 1)^n \sim 1$ and the main contribution emerges from

$$\int_{\tau_*}^\infty \tau^{-p-1}(2\Phi(\tau) - 1)^n d\tau \sim \tau_*^{-p}/p \sim (2\ln n)^{-p/2}/p.$$

Note that the integral over $[0, \alpha\tau_*]$ does not give any significant contribution for every $\alpha < 1$, since on this interval

$$(2\Phi(\tau)-1)^n \le (2\Phi(\alpha\tau_*)-1)^n \le \exp\left\{-c(\alpha)(\ln n)^{-(1+\alpha^2)/2} n^{1-\alpha^2}\right\} \ll (\ln n)^{-p/2}.$$

The resulting asymptotics turns out to be

$$I_n^G \sim n^{(n-p)/2}(\pi/2)^{n/2}(2\ln n)^{-p/2}/p. \tag{4.8}$$

4.2   Asymptotic Behavior of A_n^G

In order to study the integral A_n^G we need to know the properties of the function h (see (4.2)). Note that the asymptotic behavior of g and h at zero is the same. In fact,

$$|h(\tau) - g(\tau)| \le P\{|X| > \frac{\pi}{\tau}\} << \exp\{-\pi^2/2\tau^2\}.$$

In particular, $h(0) = 1$, $h'(0) = 0$, $h''(0) = -1/3$. But the behavior of h at infinity is different. As $\tau \to \infty$, we have

$$\tau h(\tau) \sim \sqrt{2/\pi} \int_0^\tau \frac{|\sin y|dy}{y} \sim \sqrt{2/\pi} \frac{2}{\pi} \ln \tau = (2/\pi)^{3/2} \ln \tau. \quad (4.9)$$

4.2.1   Hypersections, $p = n - d$ with Constant d. Since in this case only a neighbourhood of zero is important, the result is the same as in (4.6):

$$\lim_n A_n^G = \lim_n I_n^G = \int_0^\infty \exp\{-\frac{t^2}{6}\}t^{d-1}dt = \frac{\Gamma(d/2)6^{d/2}}{2}.$$

4.2.2   Proportional Sections, $p = \alpha n - 1$, $0 < \alpha < 1$. Denote by

$$h_\alpha(\tau) = h(\tau)\tau^{1-\alpha}$$

and let τ_* be the maximal point of this function. Then, as in (4.7),

$$A_n^G \sim n^{(1-\alpha)n/2}h_\alpha(\tau_*)^{n+1/2}\sqrt{\frac{2\pi}{|h_\alpha''(\tau_*)|}}. \quad (4.10)$$

We see that the main term is the same for A_n^G and I_n^G, and they essentially differ by a constant to the power n.

4.2.3   Low-dimensional Sections, $p = const$. In this case we again make the scaling $t = \sqrt{n}\tau$,

$$A_n^G = \int_0^\infty t^{n-p-1}h(\frac{t}{\sqrt{n}})^n dt = n^{(n-p)/2} \int_0^\infty \tau^{n-p-1}h(\tau)^n d\tau$$

$$= n^{(n-p)/2} \int_0^\infty \tau^{-p-1}[\tau h(\tau)]^n d\tau.$$

Taking into account the asymptotics (2.9), we get

$$\int_R^\infty \tau^{-p-1}[\tau h(\tau)]^n d\tau = (1 + o(R))(2/\pi)^{3n/2} \int_R^\infty \tau^{-p-1}[\ln \tau]^n d\tau$$

$$= (1 + o(R))(2/\pi)^{3n/2} \int_{p \ln R}^\infty e^{-u}[u/p]^n \frac{du}{p}$$

$$\sim (2/\pi)^{3n/2}n!p^{-n-1} \sim (2/\pi)^{3n/2}(n/e)^n\sqrt{2\pi n}p^{-n-1}.$$

Therefore, the final answer is

$$A_n^G \sim (2n/\pi)^{3n/2}(pe)^{-n}n^{(1-p)/2}\frac{\sqrt{2\pi}}{p}, \tag{4.11}$$

which is significantly bigger than (4.8).

4.3   Asymptotics for the Average Volume of Sections of the Cube

We consider the sections of the unit balls B_∞^n of the spaces ℓ_∞^n, which are cubes with side 2.

4.3.1   Hypersections, $p = n - d$ with Constant d. In this case, we have

$$I_n = \frac{\Gamma(n/2)}{\Gamma\left(\frac{n-d}{2}\right)}(n/2)^{-d/2}\ I_n^G = (1 + o(1))I_n^G.$$

Therefore, I_n converges to the same constant (4.6) as the Gaussian integral. The same is true for A_n. Now we see from (4.1) and (4.6) that

$$AV_{n-d}(B_\infty^n) \sim 2^{n-d}\pi^{-d}|S^{d-1}|\frac{\Gamma(d/2)6^{d/2}}{2}$$

$$= 2^{n-d}\pi^{-d}\frac{2\pi^{d/2}}{\Gamma(d/2)}\frac{\Gamma(d/2)6^{d/2}}{2}$$

$$= 2^{n-d}(6/\pi)^{d/2} = (6/\pi)^{d/2}\mathrm{vol}_n(B_\infty^n)^{(n-d)/n}.$$

One can see that this result is fairly close to the general estimate (2.9) and to Ball's bound for the maximal section of the cube.

4.3.2   Proportional Sections, $p = \alpha n - 1$, $0 < \alpha < 1$. We have

$$I_n = \frac{\Gamma(n/2)}{\Gamma\left(\frac{\alpha n - 1}{2}\right)}(n/2)^{((\alpha-1)n-1)/2}\ I_n^G.$$

Since

$$\Gamma(n/2) \sim \frac{n^{n/2-1/2}\pi^{1/2}}{(2e)^{n/2-1}e},$$

$$\Gamma\left(\frac{\alpha n - 1}{2}\right) \sim (\pi\alpha n)^{1/2}e^{-3/2}\left(\frac{\alpha n}{2e}\right)^{\alpha n/2-3/2},$$

we have

$$\frac{\Gamma(n/2)}{\Gamma\left(\frac{\alpha n - 1}{2}\right)}(n/2)^{((\alpha-1)n-1)/2} \sim \alpha^{1-\alpha n/2}e^{(\alpha-1)n/2}. \tag{4.12}$$

It follows from (4.7) that

$$I_n \sim n^{(1-\alpha)n/2} g_\alpha(\tau_*)^{n+1/2} \sqrt{\frac{2\pi}{|g_\alpha''(\tau_*)|}} \alpha^{1-\alpha n/2} e^{(\alpha-1)n/2}. \quad (4.13)$$

Combining this expression with (4.1) and using the asymptotics

$$|S^{n-p-1}| \sim 2^{1/2} \left(\frac{2\pi e}{n(1-\alpha)} \right)^{n/2-\alpha n/2} \quad (4.14)$$

we get

$$AV_{\alpha n-1}(B_\infty^n) \sim 2^{n/2+\alpha n/2} \pi^{-1/2-n/2+\alpha n/2} \alpha^{1-\alpha n/2} (1-\alpha)^{\alpha n/2-n/2} \frac{g_\alpha(\tau_*)^{n+1/2}}{\sqrt{|g_\alpha''(\tau_*)|}}.$$

It is worthwhile to note that the main (exponential) term of this expression is

$$\left(\frac{2^{1+\alpha} g_\alpha(\tau_*)^2}{\pi^{1-\alpha} \alpha^\alpha (1-\alpha)^{1-\alpha}} \right)^{n/2}.$$

For example, if $\alpha = 1/2$, solving the equation (4.5) numerically, one can find $\tau_* \approx 1.4$, $\Phi(\tau_*) \approx 0.91924$. Hence, by (4.3)

$$g_\alpha(\tau_*) = \sqrt{\pi/2}(2\Phi(\tau_*) - 1)\tau_*^{-1/2} \approx 0.888,$$

and the main term of the asymptotics of the average volume is

$$\left(\frac{2^{5/2} g_\alpha(\tau_*)^2}{\pi^{1/2}} \right)^{n/2} = (2.516\ldots)^{n/2}.$$

Similarly to (4.12), it follows from (4.10) that

$$A_n \sim n^{(1-\alpha)n/2} h_\alpha(\tau_*)^{n+1/2} \sqrt{\frac{2\pi}{|h_\alpha''(\tau_*)|}} \alpha^{1-\alpha n/2} e^{(\alpha-1)n/2}.$$

4.3.3 Low-dimensional sections, $p = const.$ We have

$$I_n = \frac{\Gamma(n/2)}{\Gamma\left(\frac{p}{2}\right)} (n/2)^{(p-n)/2} I_n^G;$$

Since

$$\Gamma(n/2)(n/2)^{(p-n)/2} \sim \frac{n^{n/2-1/2}\pi^{1/2}}{(2e)^{n/2-1}e} (n/2)^{(p-n)/2} = \frac{n^{p/2-1/2}\pi^{1/2}}{2^{p/2-1}e^{n/2}}, \quad (4.15)$$

we get from (4.8) that

$$I_n \sim \frac{n^{p/2-1/2}\pi^{1/2}}{2^{p/2-1}\Gamma(p/2)e^{n/2}}n^{(n-p)/2}(\pi/2)^{n/2}(2\ln n)^{-p/2}/p$$

$$\sim \frac{n^{(n-1)/2}\pi^{(n+1)/2}}{2^{n/2+p-1}\Gamma(p/2)e^{n/2}(\ln n)^{p/2}\,p}.$$

Now using (4.1) and

$$|S^{n-p-1}| \sim \left(\frac{2\pi}{n}\right)^{n/2-p/2}\sqrt{\frac{n}{\pi}}e^{n/2}, \qquad (4.16)$$

we see that

$$AV_p(B_\infty^n) \sim 2^p\pi^{p-n}|S^{n-p-1}|\frac{n^{(n-1)/2}\pi^{(n+1)/2}}{2^{n/2+p-1}\Gamma(p/2)e^{n/2}(\ln n)^{p/2}\,p}$$

$$\sim \frac{2}{p\Gamma(p/2)}\left(\frac{\pi n}{2\ln n}\right)^{p/2}. \qquad (4.17)$$

It is interesting to compare this formula with the well-known behaviour of the 1-dimensional sections of the cube. The length of such sections is given by

$$\frac{2\left(\sum_{j=1}^n |\xi_j|^2\right)^{1/2}}{\max_{1\le j\le n}|\xi_j|},$$

where ξ is a random vector with spherically symmetric distribution. Taking the standard normal vector as ξ, we see that for big dimensions the numerator is equivalent to $2\sqrt{n}$ (by the law of large numbers) and the denominator is equivalent to $\sqrt{2\ln n}$ (by the well-known behavior of the maximal value of a Gaussian i.i.d. sequence). Therefore, the average section length is precisely $(2n/\ln n)^{1/2}$, as suggested by (4.17) with $p = 1$. Moreover, this example suggests that the volume of sections is highly concentrated near the average, at least for the low-dimensional sections. It also shows that the average differs from the minimal and maximal section lengths which are of the orders 2 and $2\sqrt{n}$, respectively.

Similarly to (4.17), we get from (4.11) that

$$A_n \sim \frac{n^{p/2-1/2}\pi^{1/2}}{\Gamma(p/2)2^{p/2-1}e^{n/2}}(2n/\pi)^{3n/2}(pe)^{-n}n^{(1-p)/2}\frac{\sqrt{2\pi}}{p}$$

$$\sim \frac{n^{3n/2}2^{(3n-p+3)/2}}{\Gamma(p/2)\pi^{3n/2}p^{n+1}e^{3n/2}}.$$

5   Sections of the Balls B_q^n, $0 < q < \infty$

Recall that, by Proposition 3.3,

$$(\|\cdot\|_q^{-p})^\wedge(\xi) = \frac{q}{\Gamma(p/q)} \int_0^\infty t^{n-p-1} \prod_{k=1}^n \gamma_q(\xi_k t)dt,$$

where

$$\gamma_q(t) = \int_{-\infty}^\infty e^{itx} \exp(-|x|^q)\, dx, \qquad t \in R^1.$$

5.1   Hypersections, $p = n - d$

In this case, our calculations are based on the behavior of the function γ_q at zero. In particular, we need

$$\gamma_q(0) = \int_{-\infty}^\infty \exp\{-|x|^q\}dx = \frac{2\Gamma(1/q)}{q} \qquad (5.1)$$

and

$$\gamma_q''(0) = -\int_{-\infty}^\infty x^2 \exp\{-|x|^q\}dx = \frac{-2\Gamma(3/q)}{q}.$$

We normalize the function γ_q at zero by introducing $\bar{\gamma}_q(t) = \gamma_q(t)/\gamma_q(0)$. Then we have

$$(\|\cdot\|_q^{-p})^\wedge(\xi) = \frac{q\gamma_q(0)^n}{\Gamma(p/q)} \int_0^\infty t^{n-p-1} \prod_{k=1}^n \bar{\gamma}_q(\xi_k t)dt.$$

Now we proceed as in Sections 4.1.1 and 4.3.1. The analog of (4.6) reads as

$$\lim_n \frac{1}{|S^{n-1}|} \int_{S^{n-1}} \int_0^\infty t^{n-p-1} \prod_{k=1}^n \bar{\gamma}_q(\xi_k t)dtd\xi = \int_0^\infty \exp\{-\frac{t^2|\bar{\gamma}_q''(0)|}{2}\}t^{d-1}dt$$

$$= \frac{\Gamma(d/2)}{2(|\bar{\gamma}_q''(0)|/2)^{d/2}} = \frac{\Gamma(d/2)2^{d/2-1}\Gamma(1/q)^{d/2}}{\Gamma(3/q)^{d/2}}.$$

We end up with

$$\frac{1}{|S^{n-1}|} \int_{S^{n-1}} (\|\cdot\|_q^{-p})^\wedge(\xi)\, d\xi \sim \frac{(2\Gamma(1/q)/q)^n}{\Gamma(p/q)} \; \frac{q\, \Gamma(d/2)2^{d/2-1}\Gamma(1/q)^{d/2}}{\Gamma(3/q)^{d/2}}$$

and since

$$\Gamma(p/q) = \Gamma\left(\frac{n-d}{q}\right) \sim \sqrt{2\pi}(n/q)^{\frac{n-d}{q}-1/2}e^{-n/q},$$

we get

$$\frac{1}{|S^{n-1}|} \int_{S^{n-1}} (\|\cdot\|_q^{-p})^\wedge(\xi)\, d\xi \sim \frac{(2\Gamma(1/q)/q)^n e^{n/q}}{(n/q)^{\frac{n-d}{q}-1/2}} \; \frac{q\, \Gamma(d/2)2^{d/2-1}\Gamma(1/q)^{d/2}}{\sqrt{2\pi}\Gamma(3/q)^{d/2}}.$$

The main term $n^{-n/q}$ shows a significant difference with the ℓ_∞-case when $q \to \infty$.

Now applying Lemma 3.1 we get

$$AV_{n-d}(B_q^n) \sim n^{-1}(2\pi)^{-d}|S^{d-1}| \frac{(2\Gamma(1/q)/q)^n e^{\frac{n}{q}}}{(n/q)^{\frac{n-d}{q}-1/2}} \frac{q}{\sqrt{2\pi}\Gamma(3/q)^{\frac{d}{2}}} \frac{\Gamma(d/2)2^{\frac{d}{2-1}}\Gamma(1/q)^{\frac{d}{2}}}{\sqrt{2\pi}\Gamma(3/q)^{\frac{d}{2}}}$$

$$= (2\pi)^{-(d+1)/2} \left(\frac{\Gamma(1/q)}{\Gamma(3/q)}\right)^{d/2} \frac{(2\Gamma(1/q)/q)^n e^{n/q}}{(n/q)^{\frac{n-d}{q}+1/2}}.$$

Note that if $q = 2$ we deal with the Euclidean balls of dimension p. Our asymptotics gives in this case

$$2^{-d/2}\pi^{-d/2-1/2}n^{d/2-1/2}\left(\frac{2\pi e}{n}\right)^{n/2},$$

which is equivalent to the volume of the Euclidean ball, i.e. to the expression $\mathrm{vol}_{n-d}(B_2^{n-d}) = (n-d)^{-1}|S^{n-d-1}|$.

5.2 Proportional Sections, $p = \alpha n - 1$, $0 < \alpha < 1$

Using Lemma 3.1, Proposition 3.3, Lemma 4.1, formulae (4.12), (4.14), (5.1) and repeating the calculation from subsection 4.1.2 we obtain

$$AV_{\alpha n-1}(B_n^q) \sim \frac{2^{n/2+\alpha n/2}\Gamma(1/q)^n g_{\alpha,q}(\tau_*)^{n+1/2}}{\sqrt{\pi|g_{\alpha,q}''(\tau_*)|n}\pi^{n/2-\alpha/2}(1-\alpha)^{n/2-\alpha n/2}\alpha^{\alpha n/2}q^{n-1}\Gamma(p/q)},$$

where $g_{\alpha,q}(\tau) = \tau^{1-\alpha}E\bar{\gamma}(\tau X)$ and τ_* is the argmax of this function. Therefore, using the asymtotics

$$\Gamma(p/q) \sim \sqrt{2\pi}(\alpha n/q)^{\alpha n/q-1/2-1/q}e^{-\alpha n/q},$$

we get

$$AV_{\alpha n-1}(B_n^q) \sim$$

$$\frac{2^{n/2+\alpha n/2-1/2}\Gamma(1/q)^n g_{\alpha,q}(\tau_*)^{n+1/2}e^{\alpha n/q}}{\sqrt{|g_{\alpha,q}''(\tau_*)|n}\pi^{(1-\alpha)n/2+1}(1-\alpha)^{(1-\alpha)n/2}\alpha^{(1/2+1/q)(\alpha n-1)}(n/q)^{(\alpha n-1)/q+1/2}}.$$

The main term of this asymptotics is $n^{-\alpha n/q}$.

5.3 Low-dimensional Sections, $p = const$

Beginning again with general Lemma 3.1 and Proposition 3.3, then using specific asymptotic expressions (4.15) and (4.16), we arrive at

$$AV_p(B_q^n) \sim \frac{q\gamma_q(0)^n}{n^{n/2-p}p2^{n/2-1}\pi^{n/2-p/2}\Gamma(p/q)\Gamma(p/2)} I_n^G \qquad (5.2)$$

with the Gaussian integral

$$I_n^G = \int_0^\infty t^{n-p-1}[E\bar{\gamma}_q(tX/\sqrt{n})]^n dt$$

$$= n^{n/2-p/2}\gamma_q(0)^{-n}\int_0^\infty \tau^{-p-1}[\tau\, E\gamma_q(\tau X)]^n d\tau. \qquad (5.3)$$

Note that we have

$$\tau E\gamma_q(\tau X) = \tau\int_{-\infty}^\infty Ee^{i\tau Xx}e^{-|x|^q}dx = \tau\int_{-\infty}^\infty e^{-\tau^2x^2/2-|x|^q}dx$$

$$= \int_{-\infty}^\infty e^{-u^2/2-|u/\tau|^q}dx.$$

Replacing for large τ the expression $e^{-|u/\tau|^q}$ by $1-|u/\tau|^q$ we get

$$\tau E\gamma_q(\tau X) = \sqrt{2\pi}(1 - E_q(\tau^{-q} + o(\tau^{-q})))$$

$$= \sqrt{2\pi}\exp(-E_q(\tau^{-q} + o(\tau^{-q}))), \quad \tau \to \infty,$$

where

$$E_q = E|X|^q = 2^{q/2}\pi^{-1/2}\Gamma\left(\frac{q+1}{2}\right).$$

Therefore,

$$\int_0^\infty \tau^{-p-1}[\tau E\gamma(\tau X)]^n d\tau \sim (2\pi)^{n/2}\int_0^\infty \tau^{-p-1}\exp\{-E_q\tau^{-q}\}d\tau$$

$$= (2\pi)^{n/2}q^{-1}\Gamma(p/q)(E_q n)^{-p/q}.$$

Chaining this estimate with (5.2) and (5.3), we get the final answer,

$$AV_p(B_q^n) \sim \frac{\pi^{p/2}n^{p/2-p/q}}{\Gamma(p/2+1)E_q^{p/q}}. \qquad (5.4)$$

In the simplest case where $p = 1$, this behavior corresponds to the intuitive picture. Namely, the length of the one dimensional central section in direction ξ is

$$\frac{\left(\sum_{k=1}^n \xi_k^2\right)^{1/2}}{\left(\sum_{k=1}^n |\xi_k|^q\right)^{1/q}}.$$

Considering ξ as a standard Gaussian vector in R^n and using twice the law of large numbers, we see that the typical value of the above fraction is $2n^{1/2}/(nE_q)^{1/q} = 2n^{1/2-1/q}E_q^{-1/q}$, just as stated in (5.4).

6   Exponential Concentration of Hyperplane Section Volumes

Let $\xi \in S^{n-1}$. Combining (3.1) with Proposition 3.2 or Proposition 3.3, respectively, we obtain the following formula for the volume of the central section of the ℓ_q^n-ball by the hyperplane orthogonal to ξ,

$$\mathrm{vol}_{n-1}(B_q^n \cap \xi^\perp) = a_n V(\xi), \tag{6.1}$$

where

$$V(\xi) = \int_0^\infty \prod_{k=1}^n \bar{\gamma}(\xi_k t) dt$$

and

$$\bar{\gamma}(\tau) = \sin \tau / \tau, \quad a_n = \frac{2^n}{\pi}, \qquad q = \infty,$$

or

$$\bar{\gamma}(\tau) = \frac{q}{2\Gamma(1/q)} \int_{-\infty}^\infty e^{i\tau x - |x|^q} dx, \quad a_n = \frac{2^n \Gamma(1/q)^n}{\pi(n-1)\Gamma(\frac{n-1}{q})q^{n-1}}, \qquad q < \infty.$$

In this section we show that the "typical" volume of a hyperplane section is, in fact, very close to the average volume calculated in Sections 4.3.1 and 5.1, respectively. To quantify this statement we prove the exponential concentration property of $V(\cdot)$ considering it as a random variable defined on the unit sphere S^{n-1} equipped with the unit Haar measure $mes(\cdot)$. Our study of the concentration properties of V was inspired by the following classical result (functional form of Lévy's isoperimetric inequality). Denote by $Lip(C)$ the class of functions on the sphere S^{n-1} satisfying the Lipschitz condition

$$|f(\xi) - f(\xi')| \le C d(\xi, \xi')$$

with respect to the geodesic distance $d(\cdot, \cdot)$.

Theorem 6.1 (cf. [Le], Section 2) *Let $f \in Lip(C)$ and let m_f be the median of f. Then for all $r \ge 0$*

$$mes\{\xi : \ |f(\xi) - m_f| \ge r\} \le 2\exp\{-r^2(n-1)/2\,C^2\}. \tag{CI}$$

With this main ingredient we are able to prove the following concentration inequality for V.

Theorem 6.2 *There exist n_0 and c depending on the function $\bar{\gamma}$ such that for all $n > n_0$ and all $r \in (0, 1)$,*

$$mes\{|V - m_V| > r\} \le 4\exp\{-c\,r^2(n-1)\}.$$

Proof. Define some parameters, depending on the function $\bar{\gamma}$, namely $\tau_0, \sigma, \alpha, I$, as follows. We choose $\tau_0, \sigma > 0$ so that for all τ satisfying $|\tau| \leq \tau_0$

$$|\bar{\gamma}(\tau)| \leq \exp\{-\sigma^2\tau^2\}.$$

Next, fix

$$\alpha = \sup_{|\tau| \geq \tau_0} \bar{\gamma}(\tau) < 1$$

and let

$$I = \int_0^\infty \tau^2 |\bar{\gamma}(\tau)|^4 d\tau < \infty.$$

The reason for considering I comes from the following Hölder-type estimate: for all real $x_1 \ldots, x_4$

$$\int_0^\infty t^2 \prod_{k=1}^4 |\bar{\gamma}(x_k t)| dt \leq \prod_{k=1}^4 \left(\int_0^\infty t^2 |\bar{\gamma}(x_k t)|^4 dt \right)^{1/4} = \prod_{k=1}^4 \left(|x_k|^{-3} I \right)^{1/4}$$

$$= I \prod_{k=1}^4 |x_k|^{-3/4} \leq I \left(\min_{1 \leq k \leq 4} |x_k| \right)^{-3}. \qquad (6.2)$$

Furthermore, for every $\xi = (\xi_k) \in S^{n-1}$ introduce the sum of four maximal squares,

$$\Lambda(\xi) = \max_{k_1 < k_2 < k_3 < k_4} |\xi_{k_1}|^2 + \cdots + |\xi_{k_4}|^2.$$

We try to show that the functional V is Lipschitz on the sphere in order to apply the concentration inequality (CI). Unfortunately, our estimate for the derivative V' works only for ξ with $\Lambda(\xi)$ not approaching 1.

The partial derivative of V by ξ_j is equal to

$$V_j'(\xi) = \int_0^\infty t\, \bar{\gamma}'(\xi_j t) \prod_{k \leq n, k \neq j} \bar{\gamma}(\xi_k t) dt.$$

Using an elementary estimate

$$|\bar{\gamma}'(\tau)| \leq D_q |\tau|, \qquad (6.3)$$

where $D_\infty = 1/3$ and $D_q = \Gamma(3/q)/\Gamma(1/q)$ for $0 < q < \infty$, we get

$$|V_j'(\xi)| \leq D_q |\xi_j| \int_0^\infty t^2 \prod_{k \leq n, k \neq j} |\bar{\gamma}(\xi_k t)| dt. \qquad (6.4)$$

Here is the key estimate for the integral in the right-hand side of (6.4).

Lemma 6.3 *There exist $n_0 \doteq n_0(\bar{\gamma})$, $C = C(\bar{\gamma})$ so that for all $n > n_0$, all $j \le n$, and all $\xi \in S^{n-1}$ satisfying $\Lambda(\xi) \le 1/3$ one has $|V(\xi)| \le C$ and*

$$\int_0^\infty t^2 \prod_{k \le n, k \ne j} |\bar{\gamma}(\xi_k t)| dt \le C. \tag{6.5}$$

Proof. Before proceeding with rather technical proof of the general case, let us indicate a short proof for $q \in (0,2)$. In this case $\bar{\gamma}(\cdot)$ is positive and the function $\log \bar{\gamma}(\sqrt{\cdot})$ is logarithmically convex, as stated in [K1, Lemma 3]. It follows immediately from the log-convexity that for every $t > 0$ and all positive $\eta_1 \le \xi_1 \le \xi_2 \le \eta_2$ with $\eta_1^2 + \eta_2^2 = \xi_1^2 + \xi_2^2$ one has

$$\bar{\gamma}(t\xi_1)\bar{\gamma}(t\xi_2) \le \bar{\gamma}(t\eta_1)\bar{\gamma}(t\eta_2).$$

Considering now the products from (6.5) and modifying step by step the pairs of coordinates one can easily show that

$$\prod_{k \le n, k \ne j} \bar{\gamma}(\xi_k t) \le \prod_{k \le n} \bar{\gamma}(\eta_k t)$$

where $\eta_1 \ge \cdots \ge \eta_4 \ge 1/6$, $\eta_5 = \cdots = 0$. It follows by (6.1) that

$$\int_0^\infty t^2 \prod_{k \le n, k \ne j} |\bar{\gamma}(\xi_k t)| dt \le 6^3 I.$$

Now we give a proof of the general case, which does not rely on log-convexity. Without lost of generality, assume that $n > 5$, $j = n$ and $|\xi_1| \le \cdots \le |\xi_{n-1}|$. Let $l = [\frac{3\ln|\xi_{n-4}|}{\ln \alpha}] + 1$, $\kappa = n - 4 - l$, $\tau_* = \frac{\tau_0}{|\xi_n|}$. All these parameters depend on ξ. To ensure that κ is well defined note that

$$|\xi_{n-4}| \ge \frac{1}{n-4} \sum_{k=1}^{n-4} |\xi_k|^2 \ge \frac{1 - \Lambda(\xi)}{n-4} \ge \frac{2}{3n}.$$

Hence,

$$l \le \frac{3\ln(3n/2)}{|\ln \alpha|} + 2 \ll n$$

and, therefore, κ is positive, say for $n > n_0(\bar{\gamma})$. The dependence of n_0 on $\bar{\gamma}$ comes via the parameter α.

Finally, let $\beta = \beta(\bar{\gamma}) < 1$ be so small that

$$\sup_{0 \le x \le \beta} \frac{3x \ln x}{\ln \alpha} \le \frac{1}{6}. \tag{6.6}$$

Now consider two cases.

1) If $|\xi_{n-4}| \geq \beta$, we may use the estimate (6.2) and write

$$\int_0^\infty t^2 \prod_{k \leq n-1} |\bar{\gamma}(\xi_k t)| dt \leq \int_0^\infty t^2 \prod_{k=n-4}^{n-1} |\bar{\gamma}(\xi_k t)| dt \leq I|\xi_{n-4}|^{-3} \leq I\beta^{-3},$$

(6.7)

thus proving (6.5).

2) If $|\xi_{n-4}| \leq \beta$, write separate estimates for two integration domains. For the domain $0 \leq t \leq \tau_*$ first note that by the estimate (6.6) and the assumption on $\Lambda(\xi)$

$$\sum_{k=1}^\kappa \xi_k^2 = 1 - \sum_{k=\kappa+1}^{n-4} \xi_k^2 - \sum_{k=n-3}^n \xi_k^2 \geq 1 - l|\xi_{n-4}| - \Lambda(\xi) \geq 1/2,$$

and then write (applying the definitions of τ_0, σ, τ_*)

$$\int_0^{\tau_*} t^2 \prod_{k \leq n-1} |\bar{\gamma}(\xi_k t)| dt \leq \int_0^{\tau_*} t^2 \prod_{k \leq \kappa} |\bar{\gamma}(\xi_k t)| dt \leq \int_0^\infty t^2 \prod_{k \leq \kappa} \exp\{-\sigma^2 \xi_k^2 t^2\} dt$$

$$\leq \int_0^\infty t^2 \exp\{-\sigma^2 t^2/2\} dt = \sqrt{\pi/2}\sigma^{-3}.$$

For the domain $\tau_* \leq t < \infty$, apply the definition of α, τ_* and the second inequality in (6.7) to see that

$$\int_{\tau_*}^\infty t^2 \prod_{k \leq n-1} |\bar{\gamma}(\xi_k t)| dt \leq \max_{t \geq \tau_*} \prod_{k=\kappa}^{n-5} |\bar{\gamma}(\xi_k t)| \int_0^\infty t^2 \prod_{k=n-4}^{n-1} |\bar{\gamma}(\xi_k t)| dt$$

$$\leq \alpha^l I|\xi_{n-4}|^{-3}.$$

Moreover, by the definition of l

$$\alpha^l |\xi_{n-4}|^{-3} \leq 1.$$

By summing up two estimates, we get

$$\int_0^\infty t^2 \prod_{k \leq n-1} |\bar{\gamma}(\xi_k t)| dt \leq \sqrt{\pi/2}\sigma^{-3} + I,$$

and we have proved (6.5) in this case either. We omit the proof of the inequality $|V(\xi)| \leq C$ since it follows the same lines but is even easier. $\quad\square$

Note that, in the case of the cube, the inequality $|V(\xi)| \leq C$ was proved in [H] and [B1].

Corollary 6.4 *Under the assumptions of Lemma 6.3, we have the Lipschitz property*

$$|V_j'(\xi)| \leq D_q C .$$

(6.8)

Remark 6.5 We could obtain the same result under a weaker assumption $\Lambda(\xi) \leq 1 - \delta$, but then the constant C depends on δ and tends to infinity when δ tends to zero.

We now finish the *proof of Theorem 6.2*. Note that every sum of squares of coordinates belongs to $Lip(2)$. Hence, the function $\Lambda(\cdot)$ being the maximum of such sums also belongs to $Lip(2)$. Let $\theta(\cdot)$ be a smooth decreasing function on the real line such that $\theta(r) = 1$ for $r \leq 1/5$ and $\theta(r) = 0$ for $r \geq 1/4$. Then define a smoothened functional

$$W(\xi) = \theta(\Lambda(\xi))V(\xi).$$

The point is that the first term kills V in the domain where the Lipschitz properties of V are unknown and, moreover, $V = W$ on a big set $\{\xi : \Lambda(\xi) \leq 1/5\}$ that contains the set $\{\xi : \max |\xi_k|^2 \leq 1/20\}$. It follows from the definition that W belongs locally to a Lipschitz class $Lip(C')$, hence it belongs to this class globally on the sphere. Applying (CI) to W and to the maximum of coordinate functionals $M(\xi) = \max_k |\xi_k| \in Lip(1)$, we observe that for some c, all $n > n_0$ and all r

$$mes\{|V - m_W| > r\} \leq mes\{V \neq W\} + mes\{|W - m_W| > r\}$$
$$\leq mes\{\xi : M(\xi)^2 > 1/20\} + 2\exp\{-cr^2(n-1)\}$$
$$\leq 2\exp\{-(n-1)(\sqrt{1/20} - m_M)^2/2\} + 2\exp\{-cr^2(n-1)\}. \qquad (6.9)$$

It follows from this estimate that

$$\lim_{n \to \infty} |m_V - m_W| = 0.$$

Hence, slightly changing n_0 and c, we can replace in (6.9) m_W by m_V. Since for $r \leq 1$ the second term dominates the first one, we get, as claimed,

$$mes\{|V - m_V| > r\} \leq 4\exp\{-cr^2(n-1)\}.\ \square$$

Remark 6.6 One can see from explicit formulae in Sections 4 and 5 that the volumes of one-dimensional sections also exhibit the concentration behavior. Therefore, one could conjecture that the same is true for all intermediate dimensions of sections. To address this question, one has to consider the isoperimetry on more complicated Grassman manifolds than the unit sphere, which could be a subject of another research.

Remark 6.7 Another general question arises naturally from our result. Which systems of bodies $K_n \subset R^n$ exhibit the same strong concentration effect for hyperplane section volumes?

7   Average Volume of Non-central Sections of the Cube

We investigate here the behavior of the average volume of non-central sections of the cube B_∞^n and encounter the Gaussian dependence of this volume on the displacement of the slicing hyperplane.

7.1   Approximation of Spheric Averages by Gaussian Averages

As in Section 6, denote $\bar\gamma(v) = \frac{\sin v}{v}$. Obviously, $|\bar\gamma(v)| \le 1$ and, by (6.3), $|\bar\gamma'(\tau)| \le \frac{|\tau|}{3}$.

Consider the function $\Pi(t,\xi) = \prod_{k=1}^n \bar\gamma(t\xi_k)$, $t > 0$, $\xi \in R^n$. Then

$$|\Pi_t'(t,\xi)| \le \sum_{k=1}^n |\bar\gamma'(t\xi)|\,|\xi_k| \prod_{j \ne k} |\bar\gamma(t\xi_j)| \le \frac{t}{3}\sum_{k=1}^n \xi_k^2 = \frac{t\,|\xi|^2}{3}. \qquad (7.1)$$

In what follows we make use of the "Gaussian" functions $g(\cdot)$ and $h(\cdot)$ introduced in (4.2). Consider the following spheric and Gaussian averages:

$$I_n(t) = \frac{1}{|S^{n-1}|}\int_{S^{n-1}} \Pi(t,\xi)d\xi \quad \text{and} \quad I_n^G(t) = E_G \Pi(t,\xi) = g(t/\sqrt{n})^n.$$

Note that by (7.1) we have $|I_n'(t)| \le |t|/3$. We show now that $I_n(t)$ and $I_n^G(t)$ are close.

Theorem 7.1 *For every bounded function f on $[0,\infty)$, we have*

$$\lim_{n\to\infty}\left(\int_0^\infty f(t)I_n(t)dt - \int_0^\infty f(t)I_n^G(t)dt\right) = 0.$$

Proof. First, we show that I_n^G is an integral transform of I_n. Using the equality $\Pi(t,r\xi) = \Pi(rt,\xi)$ and representing the Gaussian measure in polar coordinates, we get

$$I_n^G(t) = \int_0^\infty I_n(rt)p_n(r)dr, \qquad (7.2)$$

where

$$p_n(r) = \left(\frac{n}{2\pi}\right)^{n/2}|S^{n-1}|\,r^{n-1}e^{-r^2 n/2} \qquad (7.3)$$

is the probability density of the random variable $\sqrt{\sum_{k=1}^n X_k^2/n}$ with X_k i.i.d. standard normal. Take any $\delta \in (0,1)$ and recall that $1-\delta < \sqrt{1-\delta} <$

$\sqrt{1+\delta} < 1+\delta$. By our probabilistic interpretation and Chebyshev's inequality,

$$\int_{|r-1|>\delta} p_n(r)dr \le P\left\{\sum_{k=1}^n X_k^2/n \le 1-\delta\right\} + P\left\{\sum_{k=1}^n X_k^2/n \ge 1+\delta\right\}$$

$$\le \frac{Var X^2}{\delta^2 n} = \frac{2}{\delta^2 n}.$$

Therefore, by (7.2),

$$|I_n(t) - I_n^G(t)| \le \int_0^\infty |I_n(t) - I_n(rt)|p_n(r)dr$$

$$\le 2\int_{|r-1|>\delta} p_n(r)dr + \sup_{(1-\delta)t \le u \le (1+\delta)t} |I_n(t) - I_n(u)|$$

$$\le \frac{4}{\delta^2 n} + 2\delta \sup_{u \le (1+\delta)t} |I_n'(u)| \le \frac{4}{\delta^2 n} + \frac{2\delta(1+\delta)\,t}{3}.$$

Letting $\delta = n^{-1/3}$ we have

$$|I_n(t) - I_n^G(t)| \le (4 + \frac{4\,t}{3})n^{-1/3}. \tag{7.4}$$

(More precise probabilistic tools would yield approximation of the order $n^{-1/2}$).

We see from (7.4) that $I_n(t)$ and $I_n^G(t)$ are asymptotically uniformly close on compacts. We also need some estimates to handle big values of t. Let

$$A_n(t) = \frac{1}{|S^{n-1}|}\int_{S^{n-1}} |\Pi(t,\xi)|d\xi \quad \text{and} \quad A_n^G(t) = E_G|\Pi(t,\xi)| = h(t/\sqrt{n})^n.$$

Then, as in (7.2),

$$A_n^G(t) = \int_0^\infty A_n(rt)p_n(r)dr.$$

We focus our attention on an integration domain near $r = 1$. Let $a_n = 1 - 1/\sqrt{2n}$, $b_n = 1 + 1/\sqrt{2n}$. Then the local central limit theorem suggests (and it is possible to check directly from the definition) that there exists a constant $c > 0$ such that for all n and all $u \in [a_n, b_n]$ we have $p_n(u) \ge c\sqrt{n}$. Then

$$A_n^G(t) \ge c\sqrt{n}\int_{a_n}^{b_n} A_n(rt)dr = \frac{c\sqrt{n}}{t}\int_{a_n t}^{b_n t} A_n(u)du.$$

For every $T > 0$,

$$\int_T^\infty A_n^G(t)dt \ge c\sqrt{n}\int_T^\infty\int_{a_n t}^{b_n t} A_n(u)du\frac{dt}{t} \ge c\sqrt{n}\int_{b_n T}^\infty A_n(u)\int_{u/b_n}^{u/a_n}\frac{dt}{t}du$$

$$= c\sqrt{n}(\ln b_n - \ln a_n)\int_{b_n T}^\infty A_n(u)du \ge c\int_{2T}^\infty A_n(u)du. \tag{7.5}$$

We get

$$\int_T^\infty |I_n(t)| dt \le \int_T^\infty A_n(t) dt$$

$$\le c^{-1} \int_{T/2}^\infty A_n^G(t) dt = c^{-1} \int_{T/2}^\infty h(t/\sqrt{n})^n dt.$$

Since we know the behavior of the function h, we can evaluate the latter integral. Let $\delta > 0$ and choose $\sigma = \sigma(h(\cdot), \delta)$ that satisfies

$$|h(\tau)| \le 1 - \sigma\tau^2, \qquad 0 < \tau < \delta.$$

Introduce also a constant $\beta = \beta(h(\cdot), \delta)$ by

$$\beta = \sup_{\tau \ge \delta} h(\tau) < 1.$$

Then

$$\int_{T/2}^\infty h(t/\sqrt{n})^n dt \le \int_{T/2}^{\delta\sqrt{n}} (1 - \sigma t^2/n)^n dt + \beta^{n-2} \int_{\delta\sqrt{n}}^\infty h(t/\sqrt{n})^2 dt$$

$$\le \int_{T/2}^\infty \exp\{-\sigma t^2\} dt + \beta^{n-2} \sqrt{n} \int_0^\infty h(\tau)^2 d\tau. \qquad (7.6)$$

Since $\beta < 1$, we get from (7.5) and (7.6) that

$$\lim_{T\to\infty} \limsup_{n\to\infty} \int_T^\infty |I_n(t)| dt = 0. \qquad (7.7)$$

The same property of I_n^G is now obvious.

As a consequence of (7.4) and (7.7), we get the statement of the Theorem 7.1. □

7.2   An Application to Non-central Sections

For every $\xi \in S^{n-1}$, denote by

$$A_\xi(r) = \mathrm{vol}_{n-1}(B_\infty^n \cap \{r\xi + \xi^\perp\}), \ r \in [0, \infty)$$

the parallel section function of B_∞^n in the direction of ξ.

Let χ be the indicator function of the interval $[-1, 1]$. The Fourier transform of the function $A_\xi(\cdot)$ is equal to

$$\hat{A}_\xi(t) = \int_{-\infty}^\infty A_\xi(z) e^{-izt} dt = \int_{\mathbb{R}^n} \chi(\|x\|_\infty) e^{-i(x,\xi)t} dx$$

$$= \prod_{k=1}^n \int_{-\infty}^\infty \chi(x_k) e^{-ix_k\xi_k t} dt = 2^n \Pi(t, \xi).$$

Inverting the Fourier transform we get

$$A_\xi(r) = \frac{2^n}{2\pi} \int_{-\infty}^{\infty} e^{irt} \Pi(t,\xi)dt = \frac{2^n}{\pi} \int_{0}^{\infty} \cos(rt)\Pi(t,\xi)dt. \qquad (7.8)$$

Respectively, the average volume of sections at distance r from the origin is

$$\frac{1}{|S^{n-1}|} \int_{S^{n-1}} A_\xi(r)d\xi = \frac{2^n}{\pi} \int_{0}^{\infty} \cos(rt)I_n(t)dt.$$

Using Theorem 7.1 and arguing as in Section 4.1.1, we get

$$\lim_{n\to\infty} \int_{0}^{\infty} \cos(rt)\, I_n(t)dt = \lim_{n\to\infty} \int_{0}^{\infty} \cos(rt)I_n^G(t)dt$$

$$= \int_{0}^{\infty} \cos(rt)\, \exp\{-t^2/6\}dt = \sqrt{3\pi/2}\exp\{-3r^2/2\}.$$

This shows that the average volume of non-central hyperplane sections of B_∞^n at distance r from the origin is asymptotically Gaussian with respect to r. Namely,

$$\lim_{n\to\infty} \frac{1}{\mathrm{vol}_n(B_\infty^n)|S^{n-1}|} \int_{S^{n-1}} \mathrm{vol}_{n-1}(B_\infty^n \cap (r\xi+\xi^\perp))d\xi = \sqrt{3/2\pi}\exp\{-3r^2/2\}.$$

Remark 7.2 Having at hand the representation (7.8) for the average volume it is natural to ask for the concentration properties of the volume of non-central section as a function of direction ξ (with fixed displacement r). Without any changes, all the reasonings of Section 6 hold in this case, starting from (7.8) instead of (6.1). We get the same exponential inequality as in Theorem 6.2. Perhaps, one might wish to obtain an inequality which becomes stronger when the displacement parameter r increases.

Remark 7.3 We conjecture that the results of this section also hold for non-central sections of B_q^n with finite q. Unfortunately, in this case we don't see such a convenient starting point, as (7.8).

References

[ABP]   Anttila M., Ball K., Perissinaki I. The central limit problem for convex bodies. Preprint

[A]   Artin E. (1964) The Gamma Function. Athena Series: Selected Topics in Mathematics. Holt, Rinehart and Winston, New York-Toronto-London

[Ba1]   Ball K. (1986) Cube slicing in $\mathbb{R}^n$. Proc. Amer. Math. Soc. 97:465–473

[Ba2]   Ball K. (1988) Logarithmically concave functions and sections of convex sets in $\mathbb{R}^n$. Studia Math. 87:69–84

[Ba3]   Ball K. (1989) Volumes of sections of cubes and related problems. Geometric Aspects of Functional Analysis (1987–88), Lecture Notes in Math., 1376, Springer, Berlin-New York, 251–260

[BV]   Brehm U., Voigt J. Asymptotics of cross sections for convex bodies. Preprint

[DF]   Diaconis P., Freedman D. (1987) A dozen de Finetti-style results in search of a theory. Ann. Inst. H. Poincare 23:397–423

[H]   Hensley D. (1979) Slicing the cube in $\mathbb{R}^n$ and probability bounds for the measure of a central cube slice in $\mathbb{R}^n$ by probability methods. Proc. Amer. Math. Soc. 73:95–100

[K1]   Koldobsky A. (1998) An application of the Fourier transform to sections of star bodies. Israel J. Math. 106:157–164

[K2]   Koldobsky A. (1998) Intersection bodies in $\mathbb{R}^4$. Adv. Math. 136:1–14

[K3]   Koldobsky A. (1999) A generalization of the Busemann-Petty problem on sections of convex bodies. Israel J. Math. 110:75–91

[K4]   Koldobsky A. A functional analytic approach to intersection bodies. GAFA, to appear

[La]   Laplace P.S. (1812) Th'eorie analytique des probabilit'es, Paris

[Le]   Ledoux M. (1996) Isoperimetry and Gaussian analysis. Lecture Notes Math. 1648:165–296

[Lu]   Lutwak E. (1975) Dual cross-sectional measures. Rend. Acad. Naz. Lincei 58:1–5

[MeP]   Meyer M., Pajor A. (1988) Sections of the unit ball of ℓ_q^n. J. Funct. Anal. 80:109–123

[MiP]   Milman V.D., Pajor A. (1989) Isotropic position and inertia ellipsoids and zonoids of the unit ball of a normed n-dimensional space. In: Lindenstrauss J., Milman V.D. (Eds.) Geometric Aspects of Functional Analysis, Lecture Notes in Mathematics, 1376, Springer, Heidelberg, 64–104

[P]   Polya G. (1913) Berechnung eines bestimmten integrals. Math. Ann. 74:204–212

[R]   Romik D. Randomized central limit theorems. Preprint

[S]   Sudakov V.N. (1978) Typical distributions of linear functionals in finite-dimensional spaces of higher dimension. Soviet Math. Dokl. 19:1578–1582

[Va]   Vaaler J.D. (1979) A geometric inequality with applications to linear forms. Pacific J. Math. 83:543–553

[Vo]   Voigt J. (2000) A concentration of mass property for isotropic convex bodies in high dimensions. Israel J. Math. 115:235–251

[W]   von Weizsäcker H. (1997) Sudakov's typical marginals, random linear functionals and a conditional central limit theorem. Prob. Theory and Related Fields 107:313–324

Between Sobolev and Poincaré

R. Latała and K. Oleszkiewicz

Institute of Mathematics, Warsaw University, Banacha 2, 02-097 Warszawa, Poland

Abstract. Let $a \in [0,1]$ and $r \in [1,2]$ satisfy relation $r = 2/(2-a)$. Let $\mu(dx) = c_r^n \exp(-(|x_1|^r + |x_2|^r + \ldots + |x_n|^r))dx_1 dx_2 \ldots dx_n$ be a probability measure on the Euclidean space $(R^n, \|\cdot\|)$. We prove that there exists a universal constant C such that for any smooth real function f on R^n and any $p \in [1,2)$

$$E_\mu f^2 - (E_\mu |f|^p)^{2/p} \leq C(2-p)^a E_\mu \|\nabla f\|^2.$$

We prove also that if for some probabilistic measure μ on R^n the above inequality is satisfied for any $p \in [1,2)$ and any smooth f then for any $h : R^n \longrightarrow R$ such that $|h(x) - h(y)| \leq \|x - y\|$ there is $E_\mu |h| < \infty$ and

$$\mu(h - E_\mu h > \sqrt{C} \cdot t) \leq e^{-Kt^r}$$

for $t > 1$, where $K > 0$ is some universal constant.

Let us begin with a few definitions.

Definition 1. Let (Ω, μ) be a probability space and let f be a measurable, square integrable non-negative function on Ω. For $p \in [1,2)$ we define the p-variance of f by

$$Var(p)_\mu(f) = \int_\Omega f(x)^2 \mu(dx) - \left(\int_\Omega f(x)^p \mu(dx)\right)^{2/p} = E_\mu f^2 - (E_\mu f^p)^{2/p}.$$

Note that $Var(1)_\mu(f) = D_\mu^2(f) = Var_\mu(f)$ coincides with classical notion of variance, while

$$\lim_{p \to 2^-} \frac{Var(p)_\mu(f)}{2-p} = \frac{1}{2}(E_\mu f^2 \ln(f^2) - E_\mu f^2 \cdot \ln(E_\mu f^2)) = \frac{1}{2} Ent_\mu(f^2),$$

where Ent_μ denotes a classical entropy functional (see [L] for a nice introduction to the subject).

Definition 2. Let $\mathcal{E}$ be a non-negative functional on some class $\mathcal{C}$ of non-negative functions from $L^2(\Omega, \mu)$. We will say that $f \in \mathcal{C}$ satisfies

- the Poincaré inequality with constant C
  if $Var_\mu(f) \leq C \cdot \mathcal{E}(f)$,

Research partially supported by KBN Grant 2 P03A 043 15.

- the logarithmic Sobolev inequality with constant C
  if $Ent_\mu(f^2) \leq C \cdot \mathcal{E}(f)$,
- the inequality $I_\mu(a)$ (for $0 \leq a \leq 1$) with constant C
  if $Var(p)_\mu(f) \leq C \cdot (2-p)^a \cdot \mathcal{E}(f)$ for all $p \in [1,2)$.

Lemma 1. *For a fixed $f \in C$ and $p \in [1,2)$ let*

$$\varphi(p) = \frac{Var(p)_\mu(f)}{1/p - 1/2}.$$

Then φ is a non-decreasing function.

Proof. Hölder's inequality yields that $\alpha(t) = t \ln(E_\mu f^{1/t})$ is a convex function for $t \in (1/2, 1]$. Hence also $\beta(t) = e^{2\alpha(t)} = (E_\mu f^{1/t})^{2t}$ is convex and therefore $\frac{\beta(t) - \beta(1/2)}{t - 1/2}$ is non-decreasing on $(1/2, 1]$. Observation that

$$\varphi(p) = \frac{\beta(1/2) - \beta(1/p)}{1/p - 1/2}$$

completes the proof.   □

Corollary 1. *For $f \in C$ the following implications hold true:*

- *f satisfies the Poincaré inequality with constant C*
  if and only if f satisfies $I_\mu(0)$ with constant C,
- *if f satisfies the logarithmic Sobolev inequality with constant C*
  then f satisfies $I_\mu(1)$ with constant C,
- *if f satisfies $I_\mu(1)$ with constant C*
  then f satisfies the logarithmic Sobolev inequality with constant $2C$,
- *if f satisfies $I_\mu(a)$ with constant C and $0 \leq \alpha \leq a \leq 1$*
  then f satisfies $I_\mu(\alpha)$ with constant C.

Proof.

- To prove the first part of Corollary 1 it suffices to note that $p \longmapsto Var(p)_\mu(f)$ is a non-increasing function.
- The second part of Corollary 1 follows easily from the fact that

$$\lim_{p \to 2^-} \frac{Var(p)_\mu(f)}{2 - p} = \frac{1}{2} \cdot Ent_\mu(f^2).$$

- To prove the third part of Corollary 1 use Lemma 1 and note that for $p \in [1,2)$ we have

$$\frac{Var(p)_\mu(f)}{2 - p} = \frac{\varphi(p)}{2p} \leq \frac{\lim_{p \to 2^-} \varphi(p)}{2} = Ent_\mu(f^2).$$

- The last part of statement is trivial.   □

Corollary 1 shows that inequalities $I_\mu(a)$ interpolate between Poincaré and logarithmic Sobolev inequalities. Note that $I_\mu(a)$ for $a < 0$ would be equivalent to the Poincaré inequality and the only functions satisfying $I_\mu(a)$ for $a > 1$ would be the constant functions (because in this case $I_\mu(a)$ would imply the logarithmic Sobolev inequality with constant 0). Therefore restriction to $a \in [0, 1]$ is natural.

Definition 3. Given probability space (Ω, μ), a class $C \subseteq L^2_+(\Omega, \mu)$ and non-negative functional $\mathcal{E}$ on C we will say that a pair $(\mu, \mathcal{E})$ satisfies $I(a)$ (respectively the Poincaré or the logarithmic Sobolev) inequality if every $f \in C$ satisfies $I_\mu(a)$ (resp. the Poincaré or the logarithmic Sobolev) inequality with constant C (for these particular μ and $\mathcal{E}$). For the sake of brevity we will assume that μ identifies probability space and $\mathcal{E}$ carries information about C.

An obvious modification of Corollary 1 for pairs $(\mu, \mathcal{E})$ follows. In some cases we can establish the precise relation between best possible constants in $I(1)$ and logarithmic Sobolev inequalities.

Let $m : (-a, a) \longrightarrow R$ be an even, strictly positive continuous density of some probability measure μ on $(-a, a)$, where $0 < a \leq \infty$ and assume that $\int_{-a}^{a} x^2 m(x)dx < \infty$. For $f \in C_0^\infty(-a, a)$ put

$$(Lf)(x) = xf'(x) - u(x)f''(x),$$

where $u(x) = \dfrac{\int_x^a tm(t)\, dt}{m(x)} \geq 0$. General theory (see [KLO] for detailed references and some related results) yields that L can be extended to a positive definite self-adjoint operator (denoted by the same symbol), defined on a dense subspace $Dom(L)$ of $L^2((-a, a), \mu)$, whose spectrum $\sigma(L)$ is contained in $\{0\} \cup [1, \infty)$. Moreover $P_t = e^{-tL}$ $(t \geq 0)$ is a Markov semigroup with invariant measure μ. Put $\mathcal{E}(f) = \|L^{1/2}f\|_2^2$ (we accept $\mathcal{E}(f) = +\infty$ for f which do not belong to $Dom(L^{1/2})$) and take $C = L^2_+((-a, a), \mu)$.

Lemma 2. *Under the above assumptions the following equivalence holds true:*
$(\mu, \mathcal{E})$ satisfies the inequality $I(1)$ with constant C
if and only if
$(\mu, \mathcal{E})$ satisfies the logarithmic Sobolev inequality with constant $2C$.

Proof. If $(\mu, \mathcal{E})$ satisfies the inequality $I(1)$ with constant C then by Corollary 1 it satisfies the logarithmic Sobolev inequality with constant $2C$. Now let us assume that $(\mu, \mathcal{E})$ satisfies the logarithmic Sobolev inequality with constant $2C$. Then for any $f \in L^2((-a, a), \mu)$ we have

$$Ent_\mu(f^2) = Ent_\mu(|f|^2) \leq 2C\mathcal{E}(|f|) \leq 2C\mathcal{E}(f)$$

(the last inequality is a well known property of Dirichlet forms of Markov semigroups - see for example Theorem 1. 3. 2 of [D]). Therefore classical hypercontractivity result [G] yields

$$\|P_{t(p)}f\|_2 \leq \|f\|_p,$$

where $t(p) = \frac{C}{2}\ln(\frac{1}{p-1})$ for $p \in [1,2)$; if $p = 1$ then we put $t(p) = \infty$ and $P_\infty(f) = E_\mu f$. Hence

$$Efe^{-2t(p)L}f \le (Ef^p)^{2/p}$$

or equivalently

$$Ef^2 - (Ef^p)^{2/p} \le Ef(Id - e^{-2t(p)L})f$$

for any $f \in C$. Now it suffices to prove that for any $\lambda \in \sigma(L)$ we have

$$1 - e^{-2t(p)\lambda} \le (2-p)C\lambda,$$

i.e.

$$1 - (2-p)C\lambda \le (p-1)^{C\lambda}.$$

For $\lambda = 0$ and $p \in (1,2)$ the inequality is trivial. It is known that if $(\mu, \mathcal{E})$ satisfies the logarithmic Sobolev inequality with constant $2C$ then (under the assumptions of Lemma 2) $C \ge 1$ - to see this consider the logarithmic Sobolev inequality for functions of the form $f(x) = |1 + \varepsilon x|$ with ε tending to zero (this is a special case of more general observation which says that, for functionals $\mathcal{E}$ satisfying certain natural conditions, if $(\mu, \mathcal{E})$ satisfies the logarithmic Sobolev inequality with constant $2C$ then it also satisfies the Poincaré inequality with constant C). We can restrict our considerations to the case $\lambda \ge 1$ since $\sigma(L) \setminus \{0\} \subseteq [1, \infty)$. Therefore $(p-1)^{C\lambda}$ is a convex function of p and to prove that

$$h(p) = (p-1)^{C\lambda} + (2-p)C\lambda - 1 \ge 0$$

for $p \in [1,2)$ it suffices to check that $h(2) = h'(2) = 0$ which is obvious. The case $p = 1$ (omitted when $\lambda = 0$ because $(p-1)^{C\lambda}$ was not well defined) follows easily since the function $p \longmapsto (Ef^p)^{2/p}$ is continuous for $p \in [1,2]$. $\square$

Corollary 2. *If μ is a $\mathcal{N}(0,1)$ Gaussian measure on real line, $\mathcal{E}(f) = E_\mu(f')^2$ and C is a class of non-negative smooth functions then $(\mu, \mathcal{E})$ satisfies $I(1)$ with constant 1.*

Proof. If μ is a $\mathcal{N}(0,1)$ Gaussian measure and operator L is defined as before then

$$E_\mu f L f = E_\mu(f')^2.$$

The assertion follows from Lemma 2 and well known fact ([G]) that Gaussian measures satisfy the logarithmic Sobolev inequality with constant 2. $\square$

Remark 1. Method used in Lemma 2 seems applicable also in more general situation (see [O] for possible directions of generalization). Let us mention just one interesting application. If $\Omega = \{-1, 1\}$, $\mu(\{-1\}) = \mu(\{1\}) = 1/2$ and $\mathcal{E}(f) = (\frac{f(1)-f(-1)}{2})^2$ then $(\mu, \mathcal{E})$ satisfies $I(1)$ with constant 1.

Remark 2. Let μ be a non-symmetric two-point distribution on $\{-1,1\}$, $\mu(\{1\}) = 1 - \mu(\{-1\}) = \alpha$ with $\alpha \in (0,1/2) \cup (1/2,1)$. Then for any $p \in [1,2)$ and any $f : \{-1,1\} \to R_+$ the inequality

$$E_\mu f^2 - (E_\mu f^p)^{2/p} \le C_\alpha(p)(f(1) - f(-1))^2$$

holds with

$$C_\alpha(p) = \frac{\alpha^{1-2/p} - (1-\alpha)^{1-2/p}}{\alpha^{-2/p} - (1-\alpha)^{-2/p}}$$

and the constant cannot be improved.

Proof (sketch). To check the optimality of $C_\alpha(p)$ put $f(-1) = \alpha^{2/p}$ and $f(1) = (1-\alpha)^{2/p}$. To prove the inequality observe that for $p \in (1,2)$, $\varphi(y) = ((1+\sqrt{y})^p + (1-\sqrt{y})^p)^{2/p}$ is a strictly convex function of $y \in (0,1)$, since

$$\varphi'(y) = [(1+\sqrt{y})^p + (1-\sqrt{y})^p]^{\frac{2}{p}-1} \frac{(1+\sqrt{y})^{p-1} - (1-\sqrt{y})^{p-1}}{\sqrt{y}}$$

$$= (2 \sum_{k=0}^{\infty} \binom{p}{2k} y^k)^{\frac{2}{p}-1} 2 \sum_{k=0}^{\infty} \binom{p-1}{2k+1} y^k$$

is clearly increasing (note that $\binom{p}{2k}$ and $\binom{p-1}{2k+1}$ are positive for $k = 0, 1, \ldots$). Hence for each $y_0 \in (0,1)$ and $p \in (1,2)$ there exist unique real numbers A and B such that

$$\varphi(y^2) = ((1+y)^p + (1-y)^p)^{2/p} \ge A + By^2 \text{ for all } y \in (-1,1)$$

with equality holding for $|y| = y_0$ only. By the homogeneity we may assume that $f(-1) = (1-\alpha)^{-1/p}(1+y)$ and $f(1) = \alpha^{-1/p}(1-y)$. Putting $y_0 = \frac{(1-\alpha)^{1/p} - \alpha^{1/p}}{(1-\alpha)^{1/p} + \alpha^{1/p}}$, using the above inequality after some elementary, but a little involved computations one proves the assertion. $\qquad\square$

Definition 4. Let us denote by $\varPhi$ the class of all continuous functions $\varphi : [0,\infty) \longrightarrow R$ having strictly positive second derivative and such that $1/\varphi''$ is a concave function. Let us additionally include in $\varPhi$ all functions φ of the form $\varphi(x) = ax + b$, where a and b are some real constants.

Although it is not obvious, functions belonging to $\varPhi$ form a convex cone. There are some interesting questions connected with the class $\varPhi$ and its generalizations but we postpone them till the end of the note.

Lemma 3. *For any $\varphi \in \varPhi$ and $t \in [0,1]$ the function $F_t : [0,\infty) \times [0,\infty) \longrightarrow R$ defined by*

$$F_t(x,y) = t\varphi(x) + (1-t)\varphi(y) - \varphi(tx + (1-t)y)$$

is non-negative and convex.

Proof. Non-negativity of F_t is an easy consequence of convexity of φ. Obviously F_t is continuous on $[0, \infty) \times [0, \infty)$ and twice differentiable on $(0, \infty) \times (0, \infty)$. Therefore it suffices to prove that $Hess\, F_t$ (second derivative matrix) is positive definite on $(0, \infty) \times (0, \infty)$. We skip the trivial case of φ being an affine function. Note that from the positivity of φ'' and the concavity of $1/\varphi''$ it follows that

$$\frac{1}{\varphi''(tx + (1-t)y)} \geq \frac{t}{\varphi''(x)} + \frac{1-t}{\varphi''(y)} \geq \frac{t}{\varphi''(x)}.$$

Therefore

$$\frac{\partial^2 F_t}{\partial x^2}(x, y) = t\varphi''(x) - t^2\varphi''(tx + (1-t)y) \geq 0.$$

In a similar way we prove that $\frac{\partial^2 F_t}{\partial y^2}(x, y) \geq 0$. Now it is enough to prove that $\det(Hess\, F_t) \geq 0$ i.e. that

$$\frac{\partial^2 F_t}{\partial x^2}(x, y) \cdot \frac{\partial^2 F_t}{\partial y^2}(x, y) \geq \left(\frac{\partial^2 F_t}{\partial x \partial y}(x, y)\right)^2$$

which is equivalent to

$$(t\varphi''(x) - t^2\varphi''(tx + (1-t)y))((1-t)\varphi''(y) - (1-t)^2\varphi''(tx + (1-t)y))$$

$$\geq (-t(1-t)\varphi''(tx + (1-t)y))^2$$

or

$$\varphi''(x)\varphi''(y) \geq t\varphi''(y)\varphi''(tx + (1-t)y) + (1-t)\varphi''(x)\varphi''(tx + (1-t)y).$$

After dividing by $\varphi''(x)\varphi''(y)\varphi''(tx + (1-t)y)$ the last inequality follows from concavity of $1/\varphi''$ and the proof is complete. $\square$

Lemma 4. *For a non-negative real random variable Z defined on probability space (Ω, μ) and having finite first moment, and for $\varphi \in \Phi$ let*

$$\Psi_\varphi(Z) = E_\mu \varphi(Z) - \varphi(E_\mu Z).$$

Then for any non-negative real random variables X and Y defined on (Ω, μ) and having finite first moment, and for any $t \in [0, 1]$ the following inequality holds:

$$\Psi_\varphi(tX + (1-t)Y) \geq t\Psi_\varphi(X) + (1-t)\Psi_\varphi(Y);$$

in other words Ψ_φ is a convex functional on the convex cone of integrable non-negative real random variables defined on (Ω, μ).

Proof. Let us note that (under notation of Lemma 3)

$$\Psi_\varphi(tX + (1-t)Y) - t\Psi_\varphi(X) - (1-t)\Psi_\varphi(Y) =$$

$$(E_\mu \varphi(tX + (1-t)Y) - tE_\mu \varphi(X) - (1-t)E_\mu \varphi(Y)) -$$
$$(\varphi(tE_\mu X + (1-t)E_\mu Y) - t\varphi(E_\mu X) - (1-t)\varphi(E_\mu Y))$$
$$= E_\mu F_t(X,Y) - F_t(E_\mu X, E_\mu Y) = E_\mu F_t(X,Y) - F_t(E_\mu(X,Y)).$$

We are to prove that it is a non-negative expression and this follows easily from Jensen inequality. For the sake of clarity we present a detailed argument.

Let $x_0 = E_\mu X$ and $y_0 = E_\mu Y$. Lemma 3 yields that F_t is convex, so that there exist constants $a, b, c \in R$ such that

$$F_t(x,y) \geq ax + by + c$$

for any $x, y \in [0, \infty)$ and

$$F_t(x_0, y_0) = ax_0 + by_0 + c.$$

Therefore

$$E_\mu F_t(X,Y) \geq E_\mu(aX + bY + c) = ax_0 + by_0 + c = F_t(x_0, y_0) = F_t(E_\mu X, E_\mu Y)$$

and the proof is finished. $\qquad\qquad\square$

Lemma 5. *Let (Ω_1, μ_1) and (Ω_2, μ_2) be probability spaces and let $(\Omega, \mu) = (\Omega_1 \times \Omega_2, \mu_1 \otimes \mu_2)$ be their product probability space. For any non-negative random variable Z defined on (Ω, μ) and having finite first moment and for any $\varphi \in \Phi$ the following inequality holds true:*

$$E_\mu \varphi(Z) - \varphi(E_\mu Z) \leq E_\mu \big([E_{\mu_1} \varphi(Z) - \varphi(E_{\mu_1} Z)] + [E_{\mu_2} \varphi(Z) - \varphi(E_{\mu_2} Z)]\big).$$

Proof. For $\omega_2 \in \Omega_2$ let $Z_{(\omega_2)}$ be a non-negative random variable defined on (Ω_1, μ_1) by the formula

$$Z_{[\omega_2]}(\omega_1) = Z(\omega_1, \omega_2).$$

By Lemma 4 used for the probability space (Ω_1, μ_1) and Jensen inequality used for the family of random variables $(Z_{[\omega_2]})_{\omega_2 \in \Omega_2}$ (this time we skip the detailed argument which the reader can easily repeat after the proof of Lemma 4) we get

$$E_{\mu_2}\big(E_{\mu_1}\varphi(Z) - \varphi(E_{\mu_1} Z)\big) \geq E_{\mu_1}\varphi(E_{\mu_2} Z) - \varphi\big(E_{\mu_1}(E_{\mu_2} Z)\big)$$

which is equivalent to the assertion of Lemma 5. $\qquad\qquad\square$

By an easy induction argument we obtain

Corollary 3. *Let $(\Omega_1, \mu_1), (\Omega_2, \mu_2), \ldots, (\Omega_n, \mu_n)$ be probability spaces and let $(\Omega, \mu) = (\Omega_1 \times \Omega_2 \times \ldots \times \Omega_n, \mu_1 \otimes \mu_2 \otimes \ldots \otimes \mu_n)$ be their product probability space. Let Z be any integrable non-negative real random variable defined on (Ω, μ). Then for any $\varphi \in \Phi$ the following inequality holds:*

$$E_\mu \varphi(Z) - \varphi(E_\mu Z) \leq \sum_{k=1}^{n} E_\mu\big(E_{\mu_k}\varphi(Z) - \varphi(E_{\mu_k} Z)\big).$$

Let us observe that the function φ defined by $\varphi(x) = x^{2/p}$ belongs to the class Φ if $p \in [1, 2]$. Therefore by applying Corollary 3 to the random variable $Z = f^p$, where $f \in L^2_+(\Omega, \mu)$, we obtain

Corollary 4. *Under the notation of Corollary 3 for any $f \in L^2_+(\Omega, \mu)$ we have*

$$E_\mu f^2 - (Ef^p)^{2/p} \leq \sum_{k=1}^{n} E_\mu \big(E_{\mu_k} f^2 - (E_{\mu_k} f^p)^{2/p} \big).$$

This sub-additivity property of functional $Var(p)_\mu$ immediately yields the following

Corollary 5. *Assume that pairs $(\mu_1, \mathcal{E}_1), (\mu_2, \mathcal{E}_2), \ldots (\mu_n, \mathcal{E}_n)$ satisfy the inequality $I(a)$ with some constant C. Let $\mu = \mu_1 \otimes \mu_2 \otimes \ldots \otimes \mu_n$ and $\mathcal{E}(f) = E_\mu(\mathcal{E}_1(f_1) + \mathcal{E}_2(f_2) + \ldots + \mathcal{E}_n(f_n))$, where*

$$f_i(x) = f(x_1, \ldots, x_{i-1}, x, x_{i+1}, \ldots, x_n)$$

for given $x_1, \ldots, x_{i-1}, x_{i+1}, \ldots, x_n$. Class $\mathcal{C}$ can be chosen in any way which assures that $f \in \mathcal{C}$ implies $f_i \in \mathcal{C}_i$, for example $\mathcal{C} = \mathcal{C}_1 \otimes \mathcal{C}_2 \otimes \ldots \otimes \mathcal{C}_n$. Then the pair $(\mu, \mathcal{E})$ also satisfies the inequality $I(a)$ with constant C.

The case we will concentrate on is $\mathcal{E}(f) = E_\mu \|\nabla f\|^2$.

Proposition 1. *Let $\mu_1, \mu_2, \ldots \mu_n$ be probability measures on R. Let $C > 0$ and $a \in [0, 1]$. Assume that for any smooth function $f : R \longrightarrow [0, \infty)$ the inequality*

$$E_{\mu_i} f^2 - (E_{\mu_i} f^p)^{2/p} \leq C(2-p)^a E_{\mu_i} (f')^2$$

holds true for $p \in [1, 2)$ and $i = 1, 2, \ldots n$. Then for $\mu = \mu_1 \otimes \mu_2 \otimes \ldots \otimes \mu_n$ the inequality

$$E_\mu f^2 - (E_\mu f^p)^{2/p} \leq C(2-p)^a E_\mu \|\nabla f\|^2,$$

where $\| \cdot \|$ denotes standard Euclidean norm, is satisfied for $p \in [1, 2)$ and any smooth function $f : R^n \longrightarrow [0, \infty)$.

Proof. Use Corollary 5 and note that

$$E_\mu \|\nabla f\|^2 = E_\mu \Big[\big(\frac{\partial f}{\partial x_1}\big)^2 + \ldots + \big(\frac{\partial f}{\partial x_1}\big)^2 \Big] = E_\mu [(f'_1)^2 + \ldots + (f'_n)^2]$$

$$= E_\mu [E_{\mu_1}(f'_1)^2 + \ldots + E_{\mu_n}(f'_n)^2]. \qquad \square$$

Now let us demonstrate that the inequality $I(a)$ for the $\mathcal{E}(f) = E_\mu \|\nabla f\|^2$ functional implies concentration of Lipschitz functions.

Theorem 1. *Let μ be a probability measure on R^n. Assume that there exist constants $C > 0$ and $a \in [0,1]$ such that the inequality*

$$E_\mu f^2 - (E_\mu f^p)^{2/p} \leq C(2-p)^a E_\mu \|\nabla f\|^2$$

is satisfied for any smooth function $f : R^n \longrightarrow [0,\infty)$ and $p \in [1,2)$. Let $h : R^n \longrightarrow R$ be a Lipschitz function with Lipschitz constant 1, i.e. $|h(x) - h(y)| \leq \|x - y\|$ for any $x, y \in R^n$, where $\|\cdot\|$ denotes a standard Euclidean norm. Then $E_\mu |h| < \infty$ and

- *for any $t \in [0,1]$*

$$\mu(h - E_\mu h \geq t\sqrt{C}) \leq e^{-Kt^2}$$

- *for any $t \geq 1$*

$$\mu(h - E_\mu h \geq t\sqrt{C}) \leq e^{-Kt^{\frac{2}{2-a}}}$$

where K is some universal constant.

Proof. Our proof will work for $K = 1/3$ but we do not know optimal constants (it is also interesting what the optimal K is for given value of parameter a). Note that it is essential part of the assumptions that we study the limit behaviour when $p \to 2$. For any fixed $p \in (1,2)$ the inequality

$$E_\mu f^2 - (E_\mu f^p)^{2/p} \leq C(2-p)^a E_\mu \|\nabla f\|^2$$

is weaker than the Poincaré inequality with constant $C(2-p)^a$ and therefore it cannot imply anything stronger than the exponential concentration.

We will follow the approach of [AS]. Assume first that h is bounded and smooth. Then $\|\nabla h\| \leq 1$. Define $H(\lambda) = E_\mu e^{\lambda h}$ for $\lambda \geq 0$. Assumptions of Theorem 1 for $f = e^{\lambda h/2}$ give

$$H(\lambda) - H\left(\frac{p}{2}\lambda\right)^{2/p} \leq \frac{C\lambda^2}{4}(2-p)^a E_\mu \|\nabla h\|^2 e^{\lambda h} \leq \frac{C\lambda^2}{4}(2-p)^a H(\lambda).$$

Hence

$$H(\lambda) \leq \frac{H(\frac{p}{2}\lambda)^{2/p}}{1 - \frac{C}{4}(2-p)^a \lambda^2}$$

for any $p \in [1,2)$ and $0 \leq \lambda \leq \frac{2}{\sqrt{C}}(2-p)^{-a/2}$. Applying the same inequality for $\frac{p}{2}\lambda$ and iterating, after m steps we get

$$H(\lambda) \leq \frac{H((\frac{p}{2})^m \lambda)^{(2/p)^m}}{\prod_{k=0}^{m-1}(1 - \frac{C\lambda^2}{4}(2-p)^a \cdot (\frac{p}{2})^{2k})^{(2/p)^k}}.$$

Note that

$$1 - \frac{C\lambda^2}{4}(2-p)^a \cdot (\frac{p}{2})^{2k} \geq \left(1 - \frac{C\lambda^2}{4}(2-p)^a\right)^{(p/2)^{2k}}$$

since $(\frac{p}{2})^{2k} < 1$. Hence

$$H(\lambda) \leq H\left((\tfrac{p}{2})^m \lambda\right)^{(2/p)^m} \left(1 - \frac{C\lambda^2}{4}(2-p)^a\right)^{-\sum_{k=0}^{m-1}(p/2)^k}.$$

As $\lim_{m \to \infty}(\frac{p}{2})^m = 0$ we get

$$\lim_{m \to \infty} H\left((\tfrac{p}{2})^m \lambda\right)^{(2/p)^m} = e^{\lambda E_\mu h}.$$

Therefore

$$E_\mu e^{\lambda(h - E_\mu h)} \leq \left(1 - \frac{C\lambda^2}{4}(2-p)^a\right)^{-\frac{2}{2-p}}$$

and

$$\mu(h - E_\mu h \geq t\sqrt{C}) \leq e^{-\lambda t\sqrt{C}} \cdot \left(1 - \frac{C\lambda^2}{4}(2-p)^a\right)^{-\frac{2}{2-p}}.$$

- Putting $p = 1$ and $\lambda = \frac{t}{\sqrt{C}}$ we get for any $t \in [0, 2)$

$$\mu(h - E_\mu h \geq t\sqrt{C}) \leq e^{-t^2} \cdot \left(1 - \frac{t^2}{4}\right)^{-2}.$$

In particular for $t \in [0, 1]$ we have $1 - \frac{t^2}{4} > e^{-t^2/3}$ and

$$\mu(h - E_\mu h \geq t\sqrt{C}) \leq e^{-t^2/3}.$$

- If $t \geq 1$, let us put $p = 2 - t^{-\frac{2}{2-a}}$ and $\lambda = t^{\frac{a}{2-a}}/\sqrt{C}$. Then we arrive at

$$\mu(h - E_\mu h \geq t\sqrt{C}) \leq e^{-t^{\frac{2}{2-a}}} \cdot \left(1 - \frac{1}{4}\right)^{-2t^{\frac{2}{2-a}}} = \left(\frac{16}{9e}\right)^{t^{\frac{2}{2-a}}}$$

which completes the proof (if h is bounded and smooth) since $\frac{16}{9e} \leq e^{-1/3}$.

Therefore by a standard approximation argument we prove the assertion for any bounded h which satisfies assumptions of Theorem 1. Finally for general h define its bounded truncations $(h_N)_{N=1}^\infty$ putting $h_N(x) = h(x)$ if $|x| \leq N$ and $h_N(x) = N \cdot sgn(x)$ if $|x| \geq N$. One can easily check that if h satisfies the assumptions of Theorem 1 then $|h_N|$ is also a Lipschitz function with a Lipschitz constant 1 and therefore using Theorem 1 for a bounded function $|h_N|$ we arrive at

$$\mu(|h_N| - E_\mu|h_N| \geq 4\sqrt{C}) \leq \left(\frac{16}{9e}\right)^{4^{\frac{2}{2-a}}} \leq \left(\frac{16}{9e}\right)^4 \leq \frac{1}{5}.$$

Similarly

$$\mu(|h_N| - E_\mu|h_N| \leq -4\sqrt{C}) = \mu(-|h_N| - E_\mu(-|h_N|) \geq 4\sqrt{C}) \leq \frac{1}{5}.$$

Hence

$$\mu(|\,|h_N| - E_\mu|h_N|\,| \geq 4\sqrt{C}) \leq \frac{2}{5}$$

and

$$\mu(|\,|h| - E_\mu|h_N|\,| \geq 4\sqrt{C}) \leq \frac{2}{5} + \mu(|h| > N).$$

Therefore

$$\mu(|E_\mu|h_N| - E_\mu|h_M|\,| \geq 8\sqrt{C}) \leq$$

$$\mu(|\,|h| - E_\mu|h_N|\,| \geq 4\sqrt{C}) + \mu(|\,|h| - E_\mu|h_M|\,| \geq 4\sqrt{C}) \leq$$

$$\frac{4}{5} + \mu(|h| > N) + \mu(|h| > M) \longrightarrow \frac{4}{5} < 1$$

as $\min(N, M) \longrightarrow \infty$, which means that the sequence $(E_\mu|h_N|)_{N=1}^\infty$ is bounded. As $|h_N|$ grows monotonically to $|h|$, by Lebesgue Lemma we get $E_\mu|h| < \infty$ and $E_\mu h_N \longrightarrow E_\mu h$ as $N \longrightarrow \infty$. Now an easy approximation argument completes the proof. □

In order to prove that the order of concentration implied by Theorem 1 cannot be improved we will need the following

Theorem 2. *Let $a \in [0,1]$ and $r \in [1,2]$ satisfy $r = 2/(2-a)$. Put $c_r = \frac{1}{2\Gamma(1+1/r)} = \frac{r}{2\Gamma(1/r)}$. Then $\mu_r(dx) = c_r^n \exp(-(|x_1|^r + |x_2|^r + \ldots + |x_n|^r))dx_1 dx_2 \ldots dx_n$ is a probability measure on R^n and there exists a universal constant $C > 0$ (not depending on a or n) such that*

$$E_{\mu_r} f^2 - (E_{\mu_r} f^p)^{2/p} \leq C(2-p)^a E_{\mu_r} \|\nabla f\|^2$$

for any smooth non-negative function f on R^n and any $p \in [1,2)$.

Proof. Proposition 1 shows that it is enough to prove Theorem 2 in the case $n = 1$. Therefore the assertion easily follows from the two following propositions. □

Proposition 2. *Let $a \in [0,1]$ and $r \in [1,2]$ satisfy $r = 2/(2-a)$. Put $c_r = \frac{1}{2\Gamma(1+1/r)}$, so that $\mu_r(dx) = c_r \exp(-|x_1|^r)dx$ is a probability measure on R. Let $\lambda(dx) = \frac{1}{2}e^{-|x|}$ be a symmetric exponential probability measure on R. Under these assumptions the following implications hold true:*

- *If $C > 0$ is a constant such that for any smooth function $f : R \longrightarrow [0,\infty)$ and any $p \in [1,2)$ there is*

$$E_{\mu_r} f^2 - (E_{\mu_r} f^p)^{2/p} \leq C(2-p)^a E_{\mu_r} (f')^2$$

then for any smooth function $g : R \longrightarrow [0,\infty)$ and any $p \in [1,2)$ there is

$$\int_R g(x)^2 \lambda(dx) - \left(\int_R g(x)^p \lambda(dx) \right)^{2/p}$$

$$\leq 600C(2-p)^a \int_R \max(1, |x|^a)g'(x)^2 \lambda(dx).$$

- *Conversely, if $C > 0$ is such a constant that for any smooth function $g : R \longrightarrow [0, \infty)$ and any $p \in [1, 2)$ there is*

$$\int_R g(x)^2 \lambda(dx) - \left(\int_R g(x)^p \lambda(dx) \right)^{2/p} \leq C(2-p)^a \int_R \max(1, |x|^a) g'(x)^2 \lambda(dx)$$

then for any smooth function $f : R \longrightarrow [0, \infty)$ and any $p \in [1, 2)$ there is

$$E_{\mu_r} f^2 - (E_{\mu_r} f^p)^{2/p} \leq 50 C(2-p)^a E_{\mu_r}(f')^2.$$

Proposition 3. *There exists a universal constant C such that for any $a \in [0,1]$, any $p \in [1,2)$ and any smooth function $g : R \longrightarrow [0, \infty)$ the following inequality holds*

$$\int_R g(x)^2 \lambda(dx) - \left(\int_R g(x)^p \lambda(dx) \right)^{2/p} \leq C(2-p)^a \int_R \max(1, |x|^a) g'(x)^2 \lambda(dx).$$

We will start with proof of Proposition 2. The proof of Proposition 3 will be postponed to the end of the paper.

Proof of Proposition 2. Let us define the function $z_r : R \longrightarrow R$ by

$$\frac{1}{2} \int_{z_r(x)}^{\infty} e^{-|t|} dt = c_r \int_x^{\infty} e^{-|t|^r} dt,$$

where $c_r = \frac{r}{2\Gamma(1/r)} = \frac{1}{2\Gamma(1+1/r)}$. It is easy to see that z_r is a homeomorphism of R onto itself and

$$z_r'(x) = 2c_r e^{|z_r(x)| - |x|^r}.$$

Therefore z_r is a C^1–diffeomorphism of R onto itself. Binding f and g by relation $f(x) = g(z_r(x))$ and using standard change of variables formula we reduce the proof of Proposition 2 to the following lemma. $\qquad \square$

Lemma 6. *Under notation introduced above*

$$\frac{1}{50} \max(1, |x|^a) \leq \left(z_r'(z_r^{-1}(x)) \right)^2 \leq 600 \max(1, |x|^a)$$

for any $x \in R$.

Proof. First let us note that $1/3 \leq c_r \leq e/2$. Indeed,

$$\Gamma(1/r) = \int_0^{\infty} x^{\frac{1}{r}-1} e^{-x} dx \leq \int_0^1 x^{\frac{1}{r}-1} dx + \int_1^{\infty} e^{-x} dx = r + 1/e.$$

Hence $c_r \geq \frac{r}{2r+2/e} \geq 1/3$. On the other hand

$$\Gamma(1/r) = \int_0^{\infty} x^{\frac{1}{r}-1} e^{-x} dx \geq \frac{1}{e} \int_0^1 x^{\frac{1}{r}-1} dx = r/e.$$

Therefore $c_r \leq e/2$. Let us also notice that by obvious symmetry we can consider only the case $x > 0$. Now let us estimate from below $z_r^{-1}(1)$. We have

$$\frac{e}{2} z_r^{-1}(1) \geq c_r z_r^{-1}(1) \geq c_r \int_0^{z_r^{-1}(1)} e^{-t^r} dt = \frac{1}{2} \int_0^1 e^{-t} dt = \frac{1}{2}(1 - 1/e)$$

and therefore $z_r^{-1}(1) \geq \frac{e-1}{e^2} \geq 1/5$. Note that by definition of $z_r(x)$ for $x > 0$ we have

$$\frac{1}{2} e^{-z_r(x)} = c_r \int_x^\infty e^{-t^r} dt \leq c_r \int_x^\infty \frac{rt^{r-1}}{rx^{r-1}} e^{-t^r} dt = \frac{c_r e^{-x^r}}{rx^{r-1}}$$

and therefore

$$z_r'(x) = 2c_r e^{z_r(x)-x^r} \geq rx^{r-1}.$$

Hence also $z_r(x) \geq x^r$ and $z_r^{-1}(x) \leq x^{1/r}$ for all positive x. If $x \geq 1/5$ then

$$\int_x^\infty e^{-t^r} dt \geq \int_x^{6x} e^{-t^r} dt \geq \frac{1}{r(6x)^{r-1}} \int_x^{6x} rt^{r-1} e^{-t^r} dt =$$

$$6^{1-r} \frac{e^{-x^r} - e^{-6^r x^r}}{rx^{r-1}} \geq \frac{1}{12} \frac{e^{-x^r}}{rx^{r-1}},$$

since $6^r x^r \geq x^r + 1$ for $x \geq 1/5$ and $r \in [1, 2]$. Therefore for $x \geq z_r^{-1}(1) \geq 1/5$ we have

$$z_r'(x) \leq 12rx^{r-1} \leq 24x^{r-1}$$

and

$$z_r(x) \leq z_r\left(z_r^{-1}(1)\right) + 12 \int_{z_r^{-1}(1)}^x rt^{r-1} dt$$

$$= 1 + 12\left(x^r - [z_r^{-1}(1)]^r\right) \leq 1 + 12x^r \leq 37x^r.$$

Hence $z_r^{-1}(x) \geq (x/37)^{1/r}$ for $x \geq z_r^{-1}(1)$. If $x \geq 1$ then $z_r^{-1}(x) \geq 1/5$ and therefore

$$z_r'\left(z_r^{-1}(x)\right) \leq 24\left[z_r^{-1}(x)\right]^{r-1} \leq 24x^{\frac{r-1}{r}} = 24x^{a/2}.$$

Also if $x \geq 1$ then $z_r^{-1}(x) \geq z_r^{-1}(1)$ and

$$z_r'\left(z_r^{-1}(x)\right) \geq r\left[z_r^{-1}(x)\right]^{r-1} \geq (x/37)^{\frac{r-1}{r}} \geq 37^{\frac{1}{r}-1} x^{a/2} \geq \frac{1}{7} x^{a/2}.$$

This proves Lemma 6 for $|x| \geq 1$. For any $x \geq 0$ we have

$$z_r'\left(z_r^{-1}(x)\right) = 2c_r e^{x - z_r^{-1}(x)^r} \geq 2c_r \geq 2/3.$$

We used the previously proved fact that $z_r^{-1}(x) \leq x^{1/r}$. Now it remains only to establish upper estimate on $z_r'(z_r^{-1}(x))$ for $x \in [0, 1]$. Note that if $x \leq z_r^{-1}(1)$ then

$$c_r \int_x^\infty e^{-t^r} dt = \frac{1}{2} \int_{z_r(x)}^\infty e^{-t} dt \geq \frac{1}{2} \int_1^\infty e^{-t} dt = \frac{1}{2e}$$

and therefore

$$z_r'(x) = \frac{2c_r e^{-x^r}}{2c_r \int_x^\infty e^{-t^r} dt} \le \frac{c_r}{c_r \int_x^\infty e^{-t^r} dt} \le 2ec_r \le e^2 \le 8.$$

Hence $z_r'(z_r^{-1}(x)) \le 8$ for any $|x| \le 1$ and the proof is finished.   □

Lemma 7. *For $s \in (1, 2]$ and $x, y \ge 0$ put*

$$\rho_s(x, y) = \left(\frac{x^s + y^s}{2} - \left(\frac{x + y}{2} \right)^s \right)^{1/2}.$$

Then ρ_s is a metric on $[0, \infty)$.

Proof. Since $k_t(a, b) = e^{-(a+b)t}$ is obviously positive definite integral kernel and $K(a, b) = s(s - 1)(a + b)^{s-2} = \frac{s(s-1)}{\Gamma(2-s)} \int_0^\infty t^{1-s} k_t(a, b)\, dt$ we get, by Schwartz inequality (applied to a scalar product defined by the kernel $K(a, b)$), that for any $y \ge x \ge 0$ and $z \ge t \ge 0$ the following inequality is true:

$$\int_{x/2}^{y/2} \int_{t/2}^{z/2} K(a, b)\, da\, db$$

$$\le \left(\int_{x/2}^{y/2} \int_{x/2}^{y/2} K(a, b)\, da\, db \right)^{1/2} \left(\int_{t/2}^{z/2} \int_{t/2}^{z/2} K(a, b)\, da\, db \right)^{1/2}.$$

Now, as

$$K(a, b) = \frac{\partial^2}{\partial a\, \partial b}(a + b)^s,$$

we get by integration by parts

$$\left(\frac{y + z}{2} \right)^s + \left(\frac{x + t}{2} \right)^s - \left(\frac{x + z}{2} \right)^s - \left(\frac{y + t}{2} \right)^s \le$$

$$\left(x^s + y^s - 2\left(\frac{x + y}{2} \right)^s \right)^{1/2} \left(z^s + t^s - 2\left(\frac{z + t}{2} \right)^s \right)^{1/2}.$$

Putting $t = y$ we arrive at

$$\left(\frac{x + y}{2} \right)^s + \left(\frac{y + z}{2} \right)^s - \left(\frac{x + z}{2} \right)^s - y^s \le 2\rho_s(x, y)\rho_s(y, z)$$

which is equivalent to

$$\rho_s(x, z)^2 - \rho_s(x, y)^2 - \rho_s(y, z)^2 \le 2\rho_s(x, y)\rho_s(y, z).$$

Hence $\rho_s(x, z) \le \rho_s(x, y) + \rho_s(y, z)$. For $x \le y \le z$ we have also easily $\rho_s(x, z) \ge \rho_s(x, y)$ and $\rho_s(x, z) \ge \rho_s(y, z)$, so that $\rho_s(x, y) \le \rho_s(x, z) + \rho_s(z, y)$ and $\rho_s(y, z) \le \rho_s(y, x) + \rho_s(x, z)$. This completes the proof of triangle inequality for $s < 2$. Other metric properties of ρ_s as well as the case $s = 2$ are trivial.   □

Remark 3. In a similar way one can prove that $\rho_s(x, y) = |\frac{x^s + y^s}{2} - (\frac{x+y}{2})^s|^{1/2}$ is a metric on $(0, \infty)$ for $s \in (-\infty, 0) \cup (0, 1)$. It was pointed out to the authors by B. Maurey that Lemma 7 follows also from isometrical immersion of $([0, \infty), \rho_s)$ into $L^2([0, \infty), \kappa_s^{-1}t^{-s-1}dt)$, where $x \in [0, \infty)$ is sent to the function $e^{-xt} - 1$ and $\kappa_s = 2^{s+1} \int_0^\infty (e^{-u} - 1 + u)u^{-s-1}du$.

Lemma 8. *Let* $s \in [1, 2]$, $t \in [0, 1]$ *and* c, d, x *be nonnegative numbers. The following inequality holds*

$$(1 - t)c^s + td^s - ((1 - t)c + td)^s \le$$

$$K[(1 - t)c^s + td^s + x^s - ((1 - t)c + tx)^s - (td + (1 - t)x)^s]. \qquad (1)$$

under anyone of the following additional assumptions

- x *lies outside the open interval* (c, d) *and* $K = 1$
- $t = \frac{1}{2}$ *and* $K = 2$
- $t \le \frac{1}{2}$, $c \ge d$ *and* $K = 12$

Proof. Let us remind that

$$F_t(x, y) = tx^s + (1 - t)y^s - (tx + (1 - t)y)^s$$

is a convex function on $[0, \infty) \times [0, \infty)$. Note that the inequality of Lemma 8 is equivalent to

$$F_t(d, c) \le K[F_t(d, x) + F_t(x, c)].$$

- As

$$\frac{\partial}{\partial x}[F_t(d, x) + F_t(x, c)]$$

$$= s[(1 - t)(x^{s-1} - (td + (1 - t)x)^{s-1}) + t(x^{s-1} - (tx + (1 - t)c)^{s-1})],$$

we see that the right-hand side of the inequality as a function of x is increasing on $(\max(c, d), \infty)$ and decreasing on $[0, \min(c, d))$. For $x = \max(c, d)$ and $x = \min(c, d)$ the inequality is trivially satisfied with $K = 1$. This completes the case of x which does not lie between c and d.

- The second part of Lemma 8 follows easily by Lemma 7, as

$$F_{1/2}(d, c) = \rho_s(d, c)^2 \le (\rho_s(d, x) + \rho_s(x, c))^2 \le$$

$$2[\rho_s(d, x)^2 + \rho_s(x, c)^2] = 2[F_{1/2}(d, x) + F_{1/2}(x, c)].$$

- To prove the last part of the statement we will use convexity of F_t. Since $F_t(d, x) + F_t(x, c) \ge F_t(\frac{d+x}{2}, \frac{x+c}{2})$, it suffices to prove that $F_t(d, c) \le 12F_t(\frac{d+x}{2}, \frac{x+c}{2})$. Thanks to the first part of Lemma 8 we can restrict our considerations to the case $x \in [d, c]$. Note that

$$\frac{\partial}{\partial x}F_t\left(\frac{d+x}{2}, \frac{x+c}{2}\right)$$

$$= \frac{s}{2}\left[t(\frac{d+x}{2})^{s-1}+(1-t)(\frac{x+c}{2})^{s-1}-(t(\frac{d+x}{2})+(1-t)(\frac{x+c}{2}))^{s-1}\right] \leq 0,$$

since the function $\varphi(u) = u^{s-1}$ is concave. Therefore it is enough to prove that

$$F_t(d,c) \leq 12F_t\left(\frac{d+c}{2},c\right).$$

Using the homogeneity of the above formula we can reduce our task to proving that

$$F_t(1-u,1) \leq 12F_t(1-u/2,1)$$

for any $u \in [0,1]$ and $t \in [0,1/2]$.
Using the Taylor expansion we get

$$F_t(1-u,1) = t(1-u)^s + 1 - t - (1-tu)^s =$$

$$s(s-1)u^2t(1-t) \cdot \left[\frac{1}{2} + \sum_{k=1}^{\infty} \frac{u^k}{(k+1)(k+2)} \sum_{m=0}^{k} t^m \cdot \prod_{l=1}^{k}(1-\frac{s-1}{l})\right].$$

Therefore

$$F_t(1-u/2,1) \geq \frac{1}{2}s(s-1)(u/2)^2t(1-t)$$

and

$$F_t(1-u,1) \leq s(s-1)u^2t(1-t) \cdot \left[\frac{1}{2} + 2\sum_{k=1}^{\infty} \frac{1}{(k+1)(k+2)}\right]$$

$$= \frac{3}{2}s(s-1)u^2t(1-t)$$

because $\sum_{m=0}^{\infty} t^m \leq 2$. Hence

$$F_t(1-u,1) \leq 12F_t(1-u/2,1)$$

which completes the proof. $\square$

Lemma 9. *Let* $a \in [0,1]$, $0 \leq x_1 < x_2$ *and* g *be a smooth function on* $[x_1,x_2]$ *such that* $g(x_1) = y_1, g(x_2) = y_2$. *Then*

$$\int_{x_1}^{x_2} \max(1,x^a)g'(x)^2d\lambda(x) \geq \frac{(y_2-y_1)^2}{4(e^{x_2}-e^{x_1})}\max(1,x_2^a). \qquad (2)$$

Proof. By the Schwartz inequality

$$|y_2 - y_1| \leq \int_{x_1}^{x_2} |g'(x)|dx$$

$$\leq \left(\int_{x_1}^{x_2} \max(1,x^a)g'(x)^2d\lambda(x)\right)^{1/2}\left(2\int_{x_1}^{x_2} \min(1,x^{-a})e^x dx\right)^{1/2}.$$

Therefore to show (2) it is enough to prove that

$$f_1(x_2) = \int_{x_1}^{x_2} \min(1, x^{-a}) e^x \, dx \leq 2 \min(1, x_2^{-a})(e^{x_2} - e^{x_1}) = f_2(x_2).$$

For $x_2 \leq 2$ this is obvious because for $0 < x < x_2 \leq 2$ we have $\min(1, x^{-a}) \leq 1 \leq 2 \min(1, x_2^{-a})$, and for $x \geq 2$ we have

$$f_2'(x) = 2x^{-a}\left(e^x - ax^{-1}(e^x - e^{x_1})\right) \geq x^{-a} e^x = f_1'(x). \qquad \square$$

Lemma 10. *Let* $0 \leq y_1 < y_2$, $0 \leq x_1 < x_2$ *and* g *is defined on* $(-\infty, x_2)$ *by the formula*

$$g(x) = \begin{cases} y_1 & \text{for } x \leq x_1 \\ y_1 + (e^x - e^{x_1}) \frac{y_2 - y_1}{e^{x_2} - e^{x_1}} & \text{for } x \in (x_1, x_2] \end{cases}$$

Then

$$\int_{-\infty}^{x_2} g'(x)^2 \, d\lambda(x) = \frac{(y_2 - y_1)^2}{2(e^{x_2} - e^{x_1})}. \tag{3}$$

and for all $p \geq 1$

$$\int_{-\infty}^{x_2} g(x)^p \, d\lambda(x) \leq \lambda(-\infty, x_2)\left[(1 - \frac{x_2}{2} e^{-x_2}) y_1^p + \frac{x_2}{2} e^{-x_2} y_2^p\right]. \tag{4}$$

Proof. Equation (3) follows by direct calculations. It is easy to see that $g(x)$ is maximal (for fixed values of x_2, y_1 and y_2) when $x_1 = 0$, so to prove (4) we may and will assume that this is the case. To easy the notation we will denote x_2 by x. First we will consider $p = 1$. After some standard calculations (4) is equivalent in this case to

$$\frac{e^x(x - 1 + e^{-x})}{(2e^x - 1)(e^x - 1)} \leq \frac{1}{2} x e^{-x} \text{ for all } x > 0,$$

that is

$$2 + 3x \leq x e^{-x} + 2e^x \text{ for all } x > 0,$$

which immediately follows from well known estimates $e^{-x} \geq 1 - x$ and $e^x \geq 1 + x + x^2/2$.

Now, for arbitrary $p \geq 1$ notice that $g(x) = (1 - \theta(x)) y_1 + \theta(x) y_2$ with $0 \leq \theta(x) \leq 1$. Therefore we have by the convexity of x^p

$$\int_{-\infty}^{x_2} g(x)^p \, d\lambda(x) \leq \int_{-\infty}^{x_2} \left((1 - \theta(x)) y_1^p + \theta(x) y_2^p\right) d\lambda(x) \leq$$

$$\lambda(-\infty, x_2)\left[(1 - \frac{x_2}{2} e^{-x_2}) y_1^p + \frac{x_2}{2} e^{-x_2} y_2^p\right],$$

where the last inequality follows by the previously established case $p = 1$. $\square$

Lemma 11. *Suppose that* $s \in (1, 2]$, $t \in (0, 1)$, $u = \frac{s}{4(s-1)}e^{-s/2(s-1)}$ *and positive numbers* $a, b, c, d, \tilde{a}, \tilde{c}, x$ *satisfy the following conditions*

$$c < x < d, c^s \leq a, d^s \leq b, \tilde{c}^s \leq \tilde{a}, \tilde{c} \leq (1-u)c + ux.$$

Then

$$(1-t)a + tb - ((1-t)c + td)^s \leq$$

$$8\big[(1-t)\tilde{a} + tb - ((1-t)\tilde{c} + td)^s + (1-t)a + tx^s - ((1-t)c + tx)^s\big].$$

$$(5)$$

Proof. Without loss of generality we may assume that $a = c^s, b = d^s, \tilde{a} = \tilde{c}^s$. Since the function $y \to (1-t)y^s - ((1-t)y + td)^s$ is nonincreasing on $[0, d]$, it is enough to show that

$$(1-t)c^s + td^s - ((1-t)c + td)^s \leq$$

$$3\big[(1-t)((1-u)c + ud)^s + td^s - ((1-t)(1-u)c + (t + (1-t)u)d)^s\big].$$

By the homogeneity we may and will assume that $d = 1$. We are then to show that

$$f((1-c)) \leq 8f((1-u)(1-c)),$$

$$(6)$$

where

$$f(x) = (1-t)(1-x)^s + t - (1-(1-t)x)^s = \sum_{i=2}^{\infty}(-1)^i \binom{s}{i}(1-t)(1-(1-t)^{i-1})x^k.$$

We use the following simple observation: if a_i, b_i are two summable sequences of positive numbers such that for any $i > j$, $a_i/a_j \geq b_i/b_j$ then for any nondecreasing nonnegative sequence c_i

$$\frac{\sum a_i c_i}{\sum a_i} \geq \frac{\sum b_i c_i}{\sum b_i}.$$

We apply the above to the sequences $a_i = (-1)^i \binom{s}{i}(1-t)(1-(1-t)^{i-1})x^i$, $b_i = (i-1)(-1)^i \binom{s}{i}$ and $c_i = (1-u)^i$, $i = 2, 3, \ldots$ and notice that

$$h(y) := \sum_{i=2}^{\infty} b_i y^i = 1 - (1-y)^{s-1}(1 + (s-1)y) \text{ for } y \in [0, 1]$$

Therefore we get

$$f((1-u)x) \geq \frac{h(1-u)}{h(1)} = \big(1 - u^{s-1}(1 + (s-1)(1-u))\big)f(x)$$

Inequality (6) follows if we notice that

$$u^{s-1}(1 + (s-1)(1-u)) \leq su^{s-1} = \frac{s^s}{4^{s-1}}e^{-s/2}\left(\frac{1}{s-1}\right)^{s-1} \leq 1e^{-1/2}e^{1/e} \leq \frac{7}{8}\square$$

Proposition 4. *Suppose that for all* $p \in [1, 2)$ *and all nonnegative smooth functions* g *we have*

$$\int_R g^2 d\lambda - \left(\int_R g^p d\lambda \right)^{2/p} \leq$$

$$K_1 (2 - p)^i \int_R (g'(x))^2 \max(1, |x|^i) d\lambda(x) \text{ for } i = 0, 1, \tag{7}$$

where K_1 *is a universal constant. Then for all* p *and* g *as above we have*

$$\int_R g^2 d\lambda - \left(\int_R g^p d\lambda \right)^{2/p} \leq$$

$$K_2 (2 - p)^a \int_R (g'(x))^2 \max(1, |x|^a) d\lambda(x) \text{ for } a \in (0, 1), \tag{8}$$

where $K_2 \leq 32 K_1$ *is some universal constant.*

Proof. An easy approximation argument shows that (7) holds for any continuous function g, continuously differentiable everywhere except possibly finitely many points.

First we assume that g is constant on R^- or R^+, without loss of generality say it is R^-, and we show that (8) holds with $K_2 = 16K_1$. Let us fix $p \in [1, 2)$ and define

$$x_p = (2 - p)^{-1}, y = g(x_p), t = \lambda(x_p, \infty), s = \frac{2}{p},$$

$$a = \frac{1}{1 - t} \int_{-\infty}^{x_p} g^2 d\lambda, b = \frac{1}{t} \int_{x_p}^{\infty} g^2 d\lambda$$

$$c = \frac{1}{1 - t} \int_{-\infty}^{x_p} g^p d\lambda \text{ and } d = \frac{1}{t} \int_{x_p}^{\infty} g^p d\lambda.$$

Notice that by Hölder's inequality we have

$$a \geq c^s \text{ and } b \geq d^s. \tag{9}$$

We will consider two cases

Case 1. y^p *lies outside* (c, d) *or* $c > d$.

We first apply inequality (7) for $i = 1$ and a function $g I_{(-\infty, x_p)} + y I_{[x_p, \infty)}$ to get

$$(1 - t)a + ty^2 - ((1 - t)c + ty^p)^s \leq K_1 (2 - p) \int_0^{x_p} (g'(x))^2 \max(1, |x|) d\lambda(x) \leq$$

$$K_1 (2 - p)^a \int_0^{x_p} (g'(x))^2 \max(1, |x|^a) d\lambda(x).$$

In a similar way using the case of $i = 0$ for the function $yI_{(-\infty, x_p)} + gI_{[x_p, \infty)}$ we get

$$tb + (1-t)y^2 - (td + (1-t)y^p)^s \le K_1 \int_{x_p}^{\infty} (g'(x))^2 d\lambda(x) \le$$

$$K_1(2-p)^a \int_{x_p}^{\infty} (g'(x))^2 \max(1, |x|^a) d\lambda(x).$$

Notice also that

$$\int_R g^2 d\lambda - \left(\int_R g^p d\lambda \right)^{2/p} = (1-t)a + tb - ((1-t)c + td)^s \le$$

$$12\left[(1-t)a + ty^2 - ((1-t)c + ty^p)^s + tb + (1-t)y^2 - (td + (1-t)y^p)^s \right] \le$$

$$12K_1(2-p)^a \int_R (g'(x))^2 \max(1, |x|^a) d\lambda(x).$$

The middle inequality follows by Lemma 8 with $x = y^p$ together with estimates (9).

Case 2. $c < y^p < d$, we can then find $0 < x_0 < x_p$ such that $g(x_0) = c^{1/p}$. Define new function f by the formula

$$f(x) = \begin{cases} g(x) & \text{for } x > x_p \\ c^{1/p} + \frac{y-c^{1/p}}{e^{x_p} - e^{x_0}}(e^x - e^{x_0}) & \text{for } x \in [x_0, x_p] \\ c^{1/p} & \text{for } x < x_0. \end{cases}$$

Let

$$\tilde{a} = \frac{1}{1-t} \int_{-\infty}^{x_p} f^2 d\lambda \text{ and } \tilde{c} = \frac{1}{1-t} \int_{-\infty}^{x_p} f^p d\lambda.$$

By Lemma 9 and 10 we have

$$\int_R f'(x)^2 d\lambda(x) \le 2(2-p)^a \int_R \max(1, |x|^a) g'(x)^2 d\lambda(x).$$

Therefore by (7) with $i = 0$, used for the function f, we have

$$(1-t)\tilde{a} + tb - ((1-t)\tilde{c} + td)^s \le 2K_1(2-p)^a \int \max(1, |x|^a) g'(x)^2 d\lambda(x).$$

We conclude as in the previous case using Lemmas 10 and 11 instead of Lemma 8.

Finally suppose that g is arbitrary. A similar argument as in case 1 (but now with $x_p = 0$ and $t = 1/2$) together with the already proved case of g constant on R_- or R_+ proves the assertion in this case. $\qquad\square$

Proof of Proposition 3. We need only to prove that assumptions of Proposition 4 are satisfied. But in view of Proposition 2 they are equivalent to the Poincaré inequality for symmetric exponential probability measure ($i = 0$) and the logarithmic Sobolev inequality for the centered $\mathcal{N}(0, \sqrt{2}/2)$ Gaussian measure ($i = 1$) which are well known to hold with some universal constants. This completes the proof. $\quad\square$

In the end of the paper we would like to come back to the class Φ introduced in Definition 4. It is easy to check that if Lemma 5 holds for some function $\varphi \in C^2((0, \infty)) \cap C([0, \infty))$ for any $(\Omega_1, \mu_1), (\Omega_2, \mu_2)$ and any Z then $\varphi \in \Phi$. Indeed, it is even true if we restrict our consideration to (Ω_1, μ_1) and (Ω_2, μ_2) being two-point probability spaces whose atoms have $1/2$ measures. This gives a natural characterization of the class Φ.

One can try to generalize the definition of Φ. Let U be an open, convex subset of R^d. We will say that a continuous function $f : U \longrightarrow R$ belongs to the class $C_n(U)$ if for any probability spaces $(\Omega_1, \mu_1), \ldots, (\Omega_n, \mu_n)$ and any integrable random variable Z with values in U, defined on $(\Omega, \mu) = (\Omega_1 \times \ldots \times \Omega_n, \mu_1 \otimes \ldots \otimes \mu_n)$ the following inequality is satisfied:

$$\sum_{K \subseteq \{1,2,\ldots,n\}} (-1)^{|K|} E_{K^c} f(E_K Z) \geq 0,$$

where E_K denotes expectation with respect to μ_k for all $k \in K$. One can easily see that $C_1(U)$ is just a set of all convex functions on U, while $C_2((0, \infty))$ is closely related to the class Φ. In fact $f \in C_2((0, \infty))$ if and only if it is an affine function or it has a continuous strictly positive second derivative such that $1/f''$ is a concave function. One can prove that always $C_{n+1}(U) \subseteq C_n(U)$ and therefore it is natural to define $C_\infty(U)$ as an intersection of all $C_n(U)$. Then it appears that $f \in C_\infty(U)$ if and only if f is given by the formula $f(x) = Q(x, x) + x^*(x) + y$, where Q is a non-negative definite symmetric quadratic form, x^* is a linear functional on R^d and y is a constant. The above inclusions do not need to be strict. For example it is easy to see that $C_2(R) = C_\infty(R)$. It would be interesting to know some nice characterization of $C_2(U)$ for general U and $C_n((0, \infty))$ for $n > 2$. It is not clear what applications of C_n for $n > 2$ could be found but it is easy to see that this class has some tensorization property. By now, we do not know even the answer to the following question: For which $p \in [1, 2]$ does $f(x) = x^p$ belong to $C_n((0, \infty))$? We can only give some estimates.

These problems will be discussed in a separate paper.

Remark 4. Recently some new results were announced to the authors by F. Barthe (private communication) - he proved (using Theorem 2 above) that if a log-concave probability measure μ on the Euclidean space $(R^n, \|\cdot\|)$ satisfies inequality $\mu(\{x \in R^n ; \|x\| > t\}) \leq ce^{-(t/c)^r}$ for some constants $c > 0, r \in [1, 2]$ and any $t > 0$ then it satisfies also inequality

$$E_\mu f^2 - (E_\mu f^p)^{2/p} \leq C(c, n, r)(2 - p)^a E_\mu \|\nabla f\|^2$$

for any non-negative smooth function f on R^n and $p \in [1, 2)$, where $C(c, n, r)$ is some positive constant depending on c, n and r only and $a = 2 - 2/r$.

Acknowledgements

The article was inspired by the questions of Prof. Stanisław Kwapień and Prof. Gideon Schechtman. This work was done while the first named author was visiting Southeast Applied Analysis Center at School of Mathematics, Georgia Institute of Technology and was partially supported by NSF Grant DMS 96-32032. The research of the second named author was performed at the Weizmann Institute of Science in Rehovot, Israel and Equipe d'Analyse, Université Paris VI.

References

[AS]   Aida S., Stroock D. (1994) Moment estimates derived from Poincaré and logarithmic Sobolev inequalities. Math. Res. Lett. 1:75–86

[D]   Davies E.B. (1989) Heat Kernels and Spectral Theory. Cambridge Tracts in Mathematics, 92, Cambridge University Press

[G]   Gross L. (1975) Logarithmic Sobolev inequalities. Amer. J. Math. 97:1061–1083

[KLO]   Kwapień S., Latała R., Oleszkiewicz K. (1996) Comparison of moments of sums of independent random variables and differential inequalities. J. Funct. Anal. 136:258–268

[L]   Ledoux M. Concentration of measure and logarithmic Sobolev inequalities. Séminaire de Probabilités XXXIII, Lecture Notes in Math., Springer, to appear

[O]   Oleszkiewicz K. (1999) Comparison of moments via Poincaré-type inequality. In: Contemporary Mathematics, 234, Advances in Stochastic Inequalities, AMS, Providence, 135–148 ▪

Random Aspects of High-dimensional Convex Bodies

A.E. Litvak and N. Tomczak-Jaegermann

Department of Mathematical Sciences, University of Alberta, Edmonton, Alberta, Canada T6G 2G1

1   Introduction

In this paper we study geometry of compact, not necessarily centrally symmetric, convex bodies in $\mathbb{R}^n$. Over the years, local theory of Banach spaces developed many sophisticated methods to study centrally symmetric convex bodies; and already some time ago it became clear that many results, if valid for arbitrary convex bodies, may be of interest in other areas of mathematics.

In recent years many results on non-centrally symmetric convex bodies were proved and a number of papers have been written (see e.g., [BLPS], [GGM], [LMP], [MP], [R1], [R2] among others). The present paper concentrates on random aspects of compact convex bodies and investigates some invariants fundamental in the local theory of Banach spaces, restricted to random sections and projections of such bodies. It turns out that, loosely speaking, such random operations kill the effect of non-symmetry in the sense that resulting estimates are very close to their centrally symmetric counterparts (this is despite the fact that random sections might be still far from being symmetric (see Section 5 below)). At the same time these estimates might be in a very essential way better than for general bodies.

We are mostly interested in two directions. One is connected with so-called MM^*- estimate, and related inequalities. For a centrally symmetric convex body $K \subset \mathbb{R}^n$, an estimate $M(K)M(K^0) \leq c \log n$ (see the definitions in Section 2 below) is an important technical tool intimately related to the K-convexity constant. It follows by combining works by Lewis and by Figiel and Tomczak-Jaegermann, with deep results of Pisier on Rademacher projections (see e.g., [Pi]). Although the symmetry can be easily removed from the first two parts, Pisier's argument use it in a very essential way. In Section 4 we show, in particular, that every convex body K has a position K_1 (i.e., $K_1 = uK - a$ for some operator u and $a \in \mathbb{R}^n$) such that a random projection, PK_1, of dimension $[n/2]$ satisfies $M(PK_1)M(K_1^0) \leq C \log n$, where C is an absolute constant. Moreover, there exists a unitary operator u such that $M(K_1 + uK_1)M(K_1^0) \leq C \log n$. Our proof is based essentially on symmetric

---

Current address of first author: Department of Mathematics, Technion, Haifa, Israel 32000. The second author held the Canada Council Killam Research Fellowship in 1997/99.

considerations, a non-symmetric part is reduced to classical facts and simple lemmas. Similar estimates were recently proved by a different method by Rudelson ([R2]).

In Section 3 we develop some tools. We introduce the concept of random Gelfand numbers, which formalises the phenomenon of lower estimates by the Euclidean norm on random subspaces. Theorem 3.2 shows that, roughly speaking, for any convex body $K \subset \mathbb{R}^n$, good lower estimates on (non-random) subspaces of proportional dimensions imply similar estimates valid on random subspaces of comparable dimensions. The results of this section are new even for centrally symmetric bodies.

Our second source of invariants is related to distances between convex bodies. For symmetric convex bodies this theory has been much studied (see [T] and references therein); in contrast, for non-symmetric convex bodies very little is known in this direction (see [R2] for a few recent results). In Section 5 we consider, for example, a natural way to measure a "non-symmetry" of a convex body, as the distance of a body to the set of all centrally symmetric bodies, and we investigate the behaviour of this distance for random sections (and projections) of convex bodies. It turns out that for random projections of rank k of a simplex $S \subset \mathbb{R}^n$, this distance can be asymptotically estimated from below by $\sqrt{k/\ln n}$ (Theorem 5.1). On the other hand, the case of simplex is the worst (up to a logarithmic factor). This follows from Theorem 5.3 related to an "isomorphic" version of Dvoretzky's theorem ([MS2], [MS3]).

Finally, in Section 6 we discuss the so-called proportional Dvoretzky-Rogers factorization in the non-symmetric setting. The proof essentially follows the known symmetric argument ([ST]), but it has a few delicate points and it needs some extra work (based on [BS]). An interesting standard application says that every convex body has k-dimensional projection and section ($k = [n/2]$) whose distances to ℓ_1^k and to ℓ_∞^k, respectively, are smaller than $C\sqrt{n}$ (where C is an absolute constant). Another application, Theorem 6.7, extends the proofs of "isomorphic" version of Dvoretzky's theorem by Milman and Schechtman ([MS2], [MS3]) to the non-symmetric case (cf. [GGM]).

2   Definitions and Notation

Let n be a positive integer. Denote the canonical Euclidean norm on $\mathbb{R}^n$ by $|\cdot|$ and the Euclidean unit ball by D. Given a set $A \subset \mathbb{R}^n$, denote the volume of A by $|A|$. Given a subspace $E \subset \mathbb{R}^n$, denote the orthogonal projection on E by P_E.

By a convex body $K \subset \mathbb{R}^n$ we shall always mean a compact convex set with a non-empty interior, and without loss of generality, we shall assume that $0 \in \text{Int } K$. The gauge of K denoted by $\|\cdot\|_K$ is a positive convex homogeneous functional; and $X = (\mathbb{R}^n, \|\cdot\|_K)$ is an n-dimensional linear space, corresponding to the functional. This space will be also denoted by $(\mathbb{R}^n, K)$. For the Banach space notation which corresponds to the case of centrally symmetric convex bodies we refer the reader to [MS1], [Pi] and [T]. In particular, an operator always means a linear operator.

For an arbitrary $z \in \text{Int } K$, by K^z we denote the polar body with respect to z, that is, $K^z = \{y + z \mid (y, x - z) \leq 1 \text{ for all } x \in K\}$. So, K^0 is the standard polar of K. Recall that for every convex body K there is a unique point $z \in \text{Int } K$, called a Santaló point, such that $|K| \, |K^z| \leq |D|^2$. We shall say that $z \in \text{Int } K$ is an isomorphic Santaló's point for K (with constant C), if $|K| \, |K^z| \leq C^n |D|^2$.

Let us recall the notation connected with distances.

Given convex bodies K, L in $\mathbb{R}^n$, we define the geometric distance by

$$\tilde{d}(K, L) = \inf\{\alpha\beta \mid (1/\beta)L \subset K \subset \alpha L\}.$$

The Banach–Mazur distance is defined by

$$d(K, L) = \inf\{\tilde{d}(u(K - z), L - x) \mid x, z \in \mathbb{R}^n; u : \mathbb{R}^n \to \mathbb{R}^n \text{ an isomorphism}\}$$

and it corresponds to the notion of the Banach–Mazur distance in the centrally symmetric case, i.e. if K and L are centrally symmetric convex bodies then to attain the infimum in the definition it is enough to take $x = z = 0$. Given convex body K in $\mathbb{R}^n$ and centrally symmetric convex body L in $\mathbb{R}^n$ we define the distance corresponding to a center z by

$$d_z(K, L) = \inf\{\tilde{d}(u(K - z), L) \mid u : \mathbb{R}^n \to \mathbb{R}^n \text{ an isomorphism}\}.$$

Clearly, $d(K, L) \leq 2\inf\{d_z(K, L) \mid z \in \mathbb{R}^n\}$.

For simplicity below we will use $d(K, L)$ for $\inf\{d_z(K, L) \mid z \in \mathbb{R}^n\}$ in the case $L = -L$. For a convex body $K \subset \mathbb{R}^n$ we use shorthand notation $d_K = d(K, D)$ and $d_{K,z} = d_z(K, D)$ for $z \in \mathbb{R}^n$.

Many well known definitions for centrally symmetric bodies are naturally extended to the non-symmetric case by the same formulas. Let $K \subset \mathbb{R}^n$ be a convex body. Recall that

$$M = M_K = M(K) = \int_{S^{n-1}} \|x\|_K \, d\nu(x),$$

where S_{n-1} denotes the Euclidean unit sphere and ν the normalized Haar measure on S_{n-1}. Let $M^* = M_K^* = M^*(K) = M(K^0)$. It is easy to check that

$$M(K) \leq M(K \cap -K) \leq 2M(K). \tag{1}$$

Let

$$\ell(K) = \mathbf{E} \left\| \sum_{i=1}^{n} g_i e_i \right\|_K,$$

where g_i are independent standard Gaussian random variables, and $\{e_i\}$ is the canonical basis of $\mathbb{R}^n$. It is well known that $\ell(K) = c_n M(K)\sqrt{n}$, where $c_n < 1$ and $c_n \to 1$ as $n \longrightarrow \infty$.

For a centrally symmetric convex body $B \subset \mathbb{R}^n$ the K-convexity constant of $(\mathbb{R}^n, B)$ is denoted by $\kappa(B)$. It is well known that there exists an isomorphism $u : \mathbb{R}^n \longrightarrow \mathbb{R}^n$ such that the body $\widetilde{B} = u(B)$ satisfies $M(\widetilde{B})M(\widetilde{B}^0) \leq C\kappa(B) \leq C' \log d_B$, where C and C' are absolute constants. (u is determined by the so-called ℓ-ellipsoid for B, and the latter estimate is the well-known estimate by Pisier, cf. e.g., [MS1], [Pi], [T].)

For $1 \leq k \leq n$, by $\mu_{n,k}$ we denote the normalized Haar measure on the Grassmann manifold $G_{n,k}$ of all k-dimensional subspaces of $\mathbb{R}^n$.

Let us recall the so-called "lower M^*-estimate".

Theorem 2.1 *Let $K \subset \mathbb{R}^n$ be a convex body. For every $1 \leq k \leq n$, the set of all subspaces $E \subset \mathbb{R}^n$ with* codim $E = k - 1$ *such that*

$$\sqrt{k/n}\ |x| \leq 2\,M^*(K)\,\|x\|_K \qquad \text{for all } x \in E,$$

has measure larger than $1 - \exp\left(-\alpha_0 k\right)$, *where* $\alpha_0 > 0$ *is an absolute constant.*

The first estimate of this type was proved by Milman ([M1]) with a certain function $f(\lambda)$ replacing $\sqrt{\lambda}$ on the left hand side, where $\lambda = k/n \in (0, 1)$, and the factor 2 replaced by an absolute constant. Then f was improved to a polynomial by Milman ([M2], [M3]), and to the present form (which is asymptotically optimal) by Pajor and Tomczak-Jaegermann ([PT2]). Subsequently, Gordon ([Go]) improved the factor on the right hand side (see also [M4]). The non-symmetric case has been known for a long time and is obtained by all the methods above.

Let us finally mention that following the practise typical for the local theory of Banach spaces, we write most of our results in terms of convex bodies rather than operators acting between two such bodies. Passing to the operator language is completely standard.

3   Random Gelfand Numbers

The concept of random subspaces is fundamental in large parts of the theory. In this paper, given a property of m-dimensional subspaces of $\mathbb{R}^n$, we say that this property is satisfied by a random m-dimensional subspace if the measure $\mu_{n,m}$ of the subset of $G_{n,m}$ of all subspaces satisfying this property is larger than $1 - \exp\left(-\alpha_0(n - m + 1)\right)$.

For sake of generality, in the next definitions we consider arbitrary operators. Let $K, L \subset \mathbb{R}^n$ be convex bodies and let $u : (\mathbb{R}^n, K) \longrightarrow (\mathbb{R}^n, L)$ be an operator. Let $1 \leq k \leq n$. First we recall a classical definition of Gelfand numbers, $c_k(u)$. Let

$$c_k(u) = \inf\left\{ \max_{x \in E} \|ux\|_L\,/\,\|x\|_K \ \mid\ E \subset \mathbb{R}^n,\ \text{codim } E < k \right\}.$$

Definition 3.1 We define the k-th random Gelfand number by $cr_k(u) =$ $\inf a$, where $a > 0$ is a real number such that the inequality

$$a > \max_{x \in E} \|ux\|_L / \|x\|_K$$

is satisfied for a random subspace E of codim $E = k - 1$.

We shall write $c_k(K)$ and $cr_k(K)$ to denote $c_k(id)$ and $cr_k(id)$, respectively, where $id : (\mathbb{R}^n, K) \longrightarrow (\mathbb{R}^n, D)$ is the formal identity operator.

Theorem 2.1 can then be reformulated as follows.

Theorem 2.1' *Let $K \subset \mathbb{R}^n$ be a convex body. For every $1 \leq k \leq n$ we have*

$$\sqrt{k}\, cr_k(K) \leq 2\ell(K^0).$$

Let us recall a definition well-known in the symmetric case. For a convex body $K \subset \mathbb{R}^n$, and $1 \leq k \leq n$, the volume number $v_k(K)$ is defined by

$$v_k(K) = \sup\left\{(|PK|/|PD|)^{1/k} \mid P \text{ is an orthogonal projection of rank } k\right\}.$$

Theorem 3.2 *Let $K \subset \mathbb{R}^n$ be a convex body. For every $1 \leq k \leq n$ we have*

$$\sqrt{k}\, cr_k(K) \leq C \sum_{j=m}^{n} c_j(K)/\sqrt{j},$$

where $m = [ck]$, and $c > 0$ and $C > 1$ are absolute constants. Moreover, if 0 is an isomorphic Santaló point for K (with constant C'), then we also have

$$\sum_{j=m}^{n} c_j(K)/\sqrt{j} \leq C''(\lambda) \sum_{j=[m/2]}^{n} v_j(K)/\sqrt{j},$$

where $\lambda = k/n$ and $C''(\lambda)$ depends on C' and λ.

Proof. The moreover part in the symmetric case is (1.13) of [PT3]. Since this inequality follows from Milman's quotient-subspace theorem (see the proof in [MaT]), it holds in the non-symmetric case, as long as the choice of the center ensures the validity of Milman's theorem. This in turn was shown in [MP] and [R2] to depend on the inequality (7) below to be satisfied for K. Finally, if 0 is an isomorphic Santaló point for K, then (7) holds for K.

The first part requires an additional notation. For a convex body $K \subset \mathbb{R}^n$ and $\rho > 0$, set $K_\rho = K \cap \rho D$, and by $\|x\|_\rho$ denote the gauge

$$\|x\|_{K_\rho} = \max\{\|x\|_K, |x|/\rho\}.$$

The following lemma is analogous to (4) of [PT1].

Lemma 3.3 *Let $K \subset \mathbb{R}^n$ be a convex body and let $\rho = \inf \beta$, over all β such that*

$$\mu_{n,n-k+1}\left(\left\{E \mid \operatorname{codim} E = k - 1, \; \beta > \max_{x \in E} \frac{|x|}{\|x\|_\beta}\right\}\right) \geq 1 - \exp(-\alpha_0 k).$$

Then

$$\rho = cr_k(K_\rho) = cr_k(K).$$

Proof. Clearly, $cr_k(K_\rho) \leq cr_k(K)$. Write $\|\cdot\|$ for $\|\cdot\|_K$, and μ for $\mu_{n,n-k+1}$, and let $\alpha_k = 1 - \exp(-\alpha_0 k)$.

Given β denote

$$A_\beta := \left\{E \mid \operatorname{codim} E = k - 1, \; \beta > \max_{x \in E} \frac{|x|}{\|x\|_\beta}\right\}$$

and

$$A_\beta' := \left\{E \mid \operatorname{codim} E = k - 1, \; \beta > \max_{x \in E} \frac{|x|}{\|x\|}\right\}.$$

Since $A_\beta' \subset A_\beta$, we have $\mu(A_\beta') \leq \mu(A_\beta)$; and, by definition, for every $\beta > \rho$ if $E \in A_\beta$ then $\|x\|_\beta = \|x\|$ for every $x \in E$. Therefore $\mu(A_\beta') = \mu(A_\beta)$. Thus,

$$cr_k(K) = \inf\left\{\beta \mid \mu(A_\beta') \geq \alpha_k\right\} \leq \inf\left\{\beta > \rho \mid \mu(A_\beta') \geq \alpha_k\right\} = \rho.$$

On the other hand for every $\beta < \rho$ one has

$$\mu\left(\left\{E \mid \operatorname{codim} E = k - 1, \; \beta > \max_{x \in E} \frac{|x|}{\|x\|_\rho}\right\}\right) \leq \mu(A_\beta) < \alpha_k,$$

which means $cr_k(K_\rho) \geq \rho$. $\qquad\square$

Let us also recall the result based on the Dudley theorem (see e.g., [Pi], Chapter 5; one can easily check that the argument works in the non-symmetric case as well). It says that if $K \subset \mathbb{R}^n$ is a convex body then

$$\ell(K^0) \leq C_0 \sum_{j=1}^{n} c_j(K)/\sqrt{j}, \tag{2}$$

where $C_0 > 1$ is an absolute constant.

Returning to the proof of Theorem 3.2, let ρ be as in the lemma. Then by Theorem 2.1' and (2) one has

$$\rho = cr_k(K) = cr_k(K_\rho) \leq \frac{2C_0}{\sqrt{k}} \sum_{j=1}^{n} c_j(K_\rho)/\sqrt{j}.$$

Since for any j

$$c_j(K_\rho) \leq \|id : (\mathbb{R}^n, K_\rho) \longrightarrow (\mathbb{R}^n, D)\| = \rho,$$

we have for every m

$$\rho \leq \frac{2C_0}{\sqrt{k}} \left(\sum_{j=1}^{m} \rho/\sqrt{j} + \sum_{j=m+1}^{n} c_k(K_\rho)/\sqrt{j} \right)$$

$$\leq \frac{4C_0}{\sqrt{k}} \sqrt{m}\rho + \frac{2C_0}{\sqrt{k}} \sum_{j=m+1}^{n} c_k(K)/\sqrt{j}.$$

Choosing $m = [k/(12C_0)^2]$ we complete the proof of the theorem, for appropriate constants c and C. $\qquad\square$

Remark 3.4 It is well known that in centrally symmetric case $C''(\lambda)$ in Theorem 3.2 can be taken as $c/(\lambda \ln \lambda)$, where c is an absolute constant. In the general case that function can be taken as $c/(\lambda \ln^a \lambda)$, where c and a are absolute constants ([R2]).

Theorem 3.5 *Let $K \subset \mathbb{R}^n$ be a convex body such that 0 is an isomorphic Santaló point for K (with constant C') and let $1 \leq k \leq n$. For a random subspace E with codim $E = k - 1$ one has*

$$\ell(P_E K^0) \leq C'(\lambda) \sum_{j=[ck]}^{n} v_j(K)/\sqrt{j},$$

where $c > 0$ is an absolute constant, $\lambda = k/n$ and $C'(\lambda) > 1$ depends on C and λ.

Proof. Let ρ be as in Lemma 3.3 and $\beta \leq 2\rho$ such that $\mu_{n,n-k+1}(A_\beta) \geq \alpha_k$, where A_β and α_k were defined in the proof of Lemma 3.3. Then

$$\rho = cr_k(K_\rho) \leq cr_k(K_\beta) \leq cr_k(K) = \rho.$$

Repeating the argument of Theorem 3.2 and setting $K_\beta^0 = (K_\beta)^0$, we get

$$\ell(K_\beta^0) \leq C_0 \sum_{j=1}^{n} c_j(K_\beta)/\sqrt{j} \leq \frac{4C_0}{\sqrt{k}} \sum_{j=m}^{n} c_j(K)/\sqrt{j}$$

$$\leq C(\lambda) \sum_{j=m/2}^{n} v_k(K)/\sqrt{j},$$

where $m = [k/(24C_0)^2]$ (since $\beta \leq 2\rho$).

Moreover, for every $E \in A_\beta$ we have

$$\left(P_E K_\beta^0\right)_E^0 = K_\beta \cap E = K \cap E,$$

where $\left(P_E K_\beta^0\right)_E^0$ denotes polar to $P_E K_\beta^0$ with respect to the subspace E. This means that

$$P_E K_\beta^0 = P_E K^0.$$

Thus, by the ideal property of the ℓ-functional we get

$$\ell\left(P_E K^0\right) = \ell\left(P_E K_\beta^0\right) \le \ell\left(K_\beta^0\right),$$

which completes the proof. $\square$

4 $M M^*$ Estimates for Convex Bodies

For a non-symmetric convex body the usual meaning of a position (that is, an image of the body by a linear operator) has to be modified to reflect the importance of the choice of the center.

Given a convex body $K \subset \mathbb{R}^n$, we say that a body K_1 is a position of K if for some absolute constant C we have:

1. $K_1 = uK - a$ for some isomorphism u and some $a \in \mathbb{R}^n$,
2. $d_0(K_1, D) \le C d_K$,
3. $|K_1| \cdot |K_1^0| \le C^n |D|^2$.

From Lemmas 4.4 and 4.5 below it follows that the set of $a \in \mathbb{R}^n$ satisfying conditions 2 and 3 is non-empty.

The main result in this section is the following theorem.

Theorem 4.1 *Let $K \subset \mathbb{R}^n$ be a convex body. There exists a position K_1 of K such that for every $0 < \lambda < 1$, a random subspace $E \subset \mathbb{R}^n$ with $\dim E = [\lambda n]$ satisfies*

$$M\left(P_E(K_1 \cap -K_1)\right) M(K_1^0) \le C(\lambda)\kappa(K - K) \le C'(\lambda)\log\left(1 + d_K\right),$$

where $C(\lambda)$ and $C'(\lambda)$ depend on λ only.

This theorem can be reformulated in a global form.

Theorem 4.1' *Let $K \subset \mathbb{R}^n$ be a convex body. There exists a position K_1 of K such that, letting $B = K_1 \cap -K_1$, there exists a unitary operator U satisfying*

$$M(B + UB) M(K_1^0) \le C\kappa(K - K),$$

where $C > 1$ is an absolute constant.

Proof. Let $E \subset \mathbb{R}^n$ be a random $[n/2]$-dimensional subspace such that the both projections $P = P_E$ and $P' = P_{E^\perp}$ satisfy the conclusion of Theorem 4.1. Set $U = P - P'$. Since B is a symmetric body, we clearly have $PB + P'B \subset B + UB$; and so $M(B + UB) \leq M(PB + P'B)$. Since PB and $P'B$ are contained in mutually orthogonal subspaces of $\mathbb{R}^n$, the latter quantity is less than or equal to $2\left(M(PB) + M(P'B)\right)$. The conclusion follows by the choice of E. $\qquad\square$

The above proof is modeled on an argument from [LMP].

Before going on, let us discuss several consequences of Theorem 4.1. The first one means that if we pass to random subspace E of $\mathbb{R}^n$ then the body $\widetilde{K} = P_E K$ satisfies the same estimate for MM^* as in the symmetric case. (This is despite the fact that $\widetilde{K}$ may happen to be quite far from being symmetric.)

Corollary 4.2 *Let $K \subset \mathbb{R}^n$ be a convex body. There exists a position K_1 of K such that for every $0 < \lambda < 1$, a random subspace $E \subset \mathbb{R}^n$ with $\dim E = [\lambda n]$ satisfies*

$$M(P_E K_1)\, M(K_1^0 \cap E) \leq C(\lambda) \kappa(K - K) \leq C'(\lambda) \log\left(1 + d_K\right),$$

where $C(\lambda)$ and $C'(\lambda)$ depend on λ only.

The second corollary allows to pass from a projection of a body to a section, but in doing so we lose randomness.

Corollary 4.3 *Let $K \subset \mathbb{R}^n$ be a convex body. There exists a position K_1 of K such that for every $0 < \lambda < 1$, there exists a subspace $E \subset \mathbb{R}^n$ with $\dim E = [\lambda n]$ such that*

$$M(K_1 \cap E)\, M(K_1^0) \leq C(\lambda) \kappa(K - K) \kappa(K_1 \cap -K_1) \leq C'(\lambda) \log^2\left(1 + d_K\right),$$

where $C(\lambda)$ and $C'(\lambda)$ depend on λ only.

This easily follows from Theorem 4.1, by using the lifting property of the ℓ-functional (see Lemma 9.4 of [Pi] and the remark afterwards).

Corollary 4.3 was proved by a different method by Rudelson ([R2], see also [R1], for the main technical step), who obtained a better dependence of $C(\lambda)$ on λ.

We now start the proof of Theorem 4.1, which requires some preparation.

Lemma 4.4 *Let $K \subset \mathbb{R}^n$ be a convex body (with $0 \in \mathrm{Int}\ K$). Let $z \in (1/2)K$. Then the polar K^z with respect to z satisfies $|K^z| \leq 2^n |K^0|$.*

Proof. By definition

$$K^z - z = \{y \mid (y, x - z) \leq 1 \text{ for all } x \in K\}.$$

However if $(y, x) \leq 1 + (y, z)$ then $\|y\|_{K^o} \leq 1 + (1/2)\|y\|_{K^o}$, because $z \in (1/2)K$. That means $K^z - z \subset 2K^0$ which implies the result. $\square$

Lemma 4.5 *Let $K \subset \mathbb{R}^n$ be a convex body (with $0 \in \text{Int } K$). Let $a \in K$ and let $0 < \theta < 1$. Then*

$$d_{K,\theta a} \leq \left(\frac{2}{\theta} - 1\right) d_{K,a}.$$

Proof. Denote $d_{K,a}$ by d. Let $\mathcal{E}$ be an ellipsoid centered at 0 such that $a + \mathcal{E} \subset K \subset a + d\mathcal{E}$. Clearly $\theta a + \theta\mathcal{E} \subset \theta K \subset K$. On the other hand, since $0 \in K \subset a + d\mathcal{E}$ then $-a \in d\mathcal{E}$, hence $a \in d\mathcal{E}$. Thus $K \subset \theta a + (2 - \theta)d\mathcal{E}$. Therefore, $d_{K,\theta a} \leq ((2 - \theta)/\theta)d$. $\square$

The key ingredient of our approach is contained in the following proposition for symmetric convex bodies.

Proposition 4.6 *Let $B_1 \subset B_2 \subset \mathbb{R}^n$ be symmetric convex bodies. For every $0 < \lambda < 1$, a random subspace $E \subset \mathbb{R}^n$ with $\dim E = [\lambda n]$ satisfies*

$$\ell(P_E B_1) \leq C(\lambda) \left(\frac{|B_2|}{|B_1|}\right)^{c/k} \ell(B_2),$$

where $k = n - [\lambda n]$, $C(\lambda)$ depend on λ only, and c is an absolute constant.

The proof requires the definition of covering and entropy numbers. Recall that for arbitrary subsets K_1, K_2 of $\mathbb{R}^n$ the covering number $N(K_1, K_2)$ is defined as the smallest number N such that there exist N points $y_1, ..., y_N$ in $\mathbb{R}^n$ satisfying

$$K_1 \subset \bigcup_{i=1}^{N}(y_i + K_2).$$

The entropy numbers are defined by

$$e_k(K_1, K_2) = \inf\left\{\varepsilon > 0 \mid N(K_1, \varepsilon K_2) \leq 2^{k-1}\right\}.$$

It is known (and easy to check) that for an arbitrary convex body $K \subset \mathbb{R}^n$ one has $v_k(K) \leq 2e_k(K, D)$ (see e.g., [Pi], ch. 9).

Proof of Proposition 4.6. By Theorem 3.5 for a random subspace $E \subset \mathbb{R}^n$ with codim $E = k = n - [\lambda n]$ one has

$$\ell(P_E B_1) \leq C(\lambda) \sum_{m=[ck]}^{n} \frac{v_m(B_1^0)}{\sqrt{m}} \tag{3}$$

where c is numerical constants and $v_m(B_1^0)$ are the volume numbers of B_1^0. Let $\varepsilon = e_n(B_2^0, D)$ (entropy number). Since $B_2^0 \subset B_1^0$ and by definition of entropy numbers we have

$$N := N(B_1^0, \varepsilon D) \leq N(B_1^0, B_2^0) N(B_2^0, \varepsilon D) \leq 3^n \frac{|B_1^0|}{|B_2^0|} 2^n \qquad (4)$$

(the latter inequality follows from Lemma 4.16 of [Pi] and the choice of ε). Therefore there are N points $x_1, ..., x_N$ such that

$$B_1^0 \subset \bigcup_{i=1}^{N}(x_i + \varepsilon D).$$

Thus for every projection P of rank m we get

$$|PB_1^0| \leq N\varepsilon^m |PD|,$$

which means by (4)

$$v_m(B_1^0) \leq \left(6^n \frac{|B_1^0|}{|B_2^0|}\right)^{1/m} e_n(B_2^0, D). \qquad (5)$$

Using Sudakov's inequality (see e.g., Theorem 5.5 of [Pi]) and (3), (5) we obtain

$$\frac{\ell(P_E B_1)}{\ell(B_2)} \leq C(\lambda) \frac{1}{\sqrt{n} e_n(B_2^0, D)} \sum_{m=[ck]}^{n} \frac{(6^n |B_1^0|/|B_2^0|)^{1/m} e_n(B_2^0, D)}{\sqrt{m}}$$

$$\leq 4C(\lambda) (6^n |B_1^0|/|B_2^0|)^{1/ck}.$$

The result follows from the Santaló and the inverse Santaló inequalities. $\square$

Corollary 4.7 *Let $K \subset \mathbb{R}^n$ be a convex body and let $a \in \mathbb{R}^n$ be an isomorphic Santaló point for K (with constant C'). Let $K_1 = K - a$. For every $0 < \lambda < 1$, a random subspace $E \subset \mathbb{R}^n$ with $\dim E = [\lambda n]$, satisfies*

$$M(P_E(K_1 \cap -K_1)) \leq C(\lambda) M(K_1 - K_1), \qquad (6)$$

where $C(\lambda)$ depends on C' and λ.

Proof. By Proposition 4.6 it is sufficient to show that for some absolute constant C we have

$$|K_1 - K_1| \leq C^n |K_1 \cap -K_1|. \qquad (7)$$

This follows from the Rogers–Sheppard inequality, the definition of an isomorphic Santaló point and the symmetric inverse Santaló inequality. $\square$

Remark 4.8 A global form of (6) says that there exists a unitary operator U such that $M(B + UB) \leq CM(K_1 - K_1)$, where $B = K_1 \cap -K_1$ and C is an absolute constant.

Remark 4.9 Inequality (7) was obtained in [MP], and in [R2].

Proof of Theorem 4.1. We start with general remarks on positions of a given convex body $K \subset \mathbb{R}^n$. Fix an arbitrary image $K_2 = u(K)$ of K under an operator u. Let z be the Santaló point of K_2 and set $K_3 = K_2 - z$, so that 0 is the Santaló point for K_3. Pick $b \in \text{Int } K_3$ such that $d_K = d_{K_2} = d_{K_3,b}$ and set $a = b/2$. By Lemma 4.4, the polar K_3^a with respect to a satisfies $(|K_3|\,|K_3^a|)^{1/n} \leq 2|D|^{2/n}$. By Lemma 4.5, $d_{K_3,a} \leq 3d_K$. Now let $K_1 = K_3 - a$. Then $d_0(K_1, D) = d_{K_3,a} \leq 3d_K$ and K_1 is clearly a position of K. Moreover, $K_1 - K_1 = u(K - K)$.

As recalled in the introduction, for the centrally symmetric convex body $K - K$ there exists a linear operator u such that the body $B = u(K - K)$ satisfies $M(B)M(B^0) \leq C\kappa(B) = C\kappa(K - K)$. Let $K_2 = u(K)$. Now, let K_1 be a position of K constructed above, starting with K_2.

Since $B = K_1 - K_1$, the first inequality follows immediately by Corollary 4.7 and (1). Since the Banach–Mazur distance satisfies $d(K - K, D) \leq d_0(K_1, D) \leq 3d_K$, the second estimate follows.   □

5   Distances between Random Projections of Convex Bodies

A natural way to measure a "non-symmetry" of convex bodies in $\mathbb{R}^n$ can be defined as the distance from a convex body $K \subset \mathbb{R}^n$ to the set of all symmetric bodies, which we shall denote by $\partial(K)$. That is, $\partial(K) = \inf d(K, B)$, where the infimum is taken over all symmetric convex bodies $B \subset \mathbb{R}^n$ (see [Gr] for different ways to measure the "non-symmetry"). By compactness, there exist $a \in K$ and a symmetric convex body $B \subset \mathbb{R}^n$ such that $\partial(K) = \tilde{d}(K - a, B)$. Observe that we also have

$$\partial(K) = \tilde{d}(K - a, (K - a) \cap -(K - a))$$
$$= \tilde{d}(K - a, \text{conv}\{(K - a) \cup -(K - a)\}). \tag{8}$$

That is, $(K - a) \cap -(K - a)$ and conv $\{(K - a) \cup -(K - a)\}$ are two symmetric bodies closest to K. Also note, that for any $a \in K$

$$\frac{1}{2}(K - K) \subset \text{conv}\{(K - a) \cup -(K - a)\} \subset K - K.$$

By John's theorem, $\partial(K) \leq d(K, D) \leq n$ for all $K \subset \mathbb{R}^n$. If S is a regular simplex in $\mathbb{R}^n$, it is well known and easy to see that $d(S, D) = \partial(S) = n$. So the simplex is as far as possible from being symmetric and as far as possible from the Euclidean ball. Let us also mention that the simplex is the only body with the distance n to the Euclidean ball ([Pa]). On the other hand, it was observed in [GGM] that S has a $k = [(n+1)/2]$-dimensional projection which is symmetric (in fact, it is isometric to the unit ball of ℓ_1^k). In contrast, as our results show, orthogonal projections of S on "random" k-dimensional subspaces are still very far from being symmetric.

Theorem 5.1 *Let S be a regular simplex in $\mathbb{R}^n$ having 0 as the center of mass, and let $p > 1$. There exist constants $c > 0$ and $C > 0$ such that for $k \geq C \cdot p \cdot \log n$, the set of subspaces $E \in G_{n,k}$ such that projections $P = P_E$ satisfy*

$$c\sqrt{\frac{k}{p \cdot \log n}} \leq \partial(PS), \tag{9}$$

has measure larger than $1 - n^{-p}$.

Denote

$$A = \sqrt{\frac{n}{k}} \int\limits_{S^{n-1}} \left(\sum_{i=1}^{k} x_i^2\right)^{1/2} d\nu(x) = \frac{\sqrt{n}\ \Gamma\left(\frac{k+1}{2}\right)\ \Gamma\left(\frac{n}{2}\right)}{\sqrt{k}\ \Gamma\left(\frac{k}{2}\right)\ \Gamma\left(\frac{n+1}{2}\right)},$$

where ν is the normalized rotationally invariant measure on the Euclidean sphere S^{n-1} and $\Gamma(\cdot)$ is the Gamma function. Then $A = A(n,k) < 1$ and $A \longrightarrow 1$ as $n, k \longrightarrow \infty$. We need a lemma, which follows from usual concentration inequalities on the sphere (we use it in the formulation from [JL]).

Lemma 5.2 *There exist two constants $c_1 > 0$ and $c > 0$ such that for any N and vectors $y_1, ..., y_N \in S^{n-1}$, any $\varepsilon > 0$, and any integer $0 < k < n$, the set of subspaces $E \in G_{n,k}$ such that projections $P = P_E$ satisfy*

$$\forall j : \left| |Py_j| - A\sqrt{k/n} \right| \leq A\varepsilon\sqrt{k/n}, \tag{10}$$

has measure larger than $1 - c_1 N \exp\left(-c\varepsilon^2 k\right)$.

Proof of Theorem 5.1. Fix an arbitrary subspace $E \in G_{n,k}$. Using (8) it is easy to check that

$$\partial(PS) = \inf_{a \in E} \tilde{d}(PS - a, (PS - a) \cap -(PS - a))$$

$$= \inf_{a \in \mathbb{R}^n} \tilde{d}(P(S - a), P(S - a) \cap -P(S - a)).$$

We shall estimate from below the expression under the latter infimum, for every $a \in \mathbb{R}^n$.

Assume first that $a = 0$. Denote the vertices of the simplex by v_i, with $|v_i| = 1$, for $1 \leq i \leq n + 1$. Then $\sum v_i = 0$ and $\langle v_i, v_j \rangle = -1/n$ for $i \neq j$.

Applying Lemma 5.2 for the set $\{v_i - v_j\}_{i,j=0}^{n+1}$, where $v_0 = 0$, we obtain that with probability larger than $1 - c_1(n+2)^2 \exp\left(-c\varepsilon^2 k\right)$

$$A(1-\varepsilon)\sqrt{k/n}\,|v_i - v_j| \le |Pv_i - Pv_j| \le A(1+\varepsilon)\sqrt{k/n}|v_i - v_j|$$

for every $0 \le i, j \le n+1$ and some absolute constants $c_1,\ c > 0$.

Since for every x and y one has $2\langle x, y\rangle = |x|^2 + |y|^2 - |x - y|^2$, then $|v_i - v_j|^2 = 2 + 2/n$ for $i \ne j$, and so

$$|\langle Pv_i, Pv_j\rangle| \le A^2 \frac{k}{n}\left(4\varepsilon + (1+\varepsilon)^2/n\right) \le A^2\frac{k}{n}8\varepsilon,$$

for $\varepsilon \in (1/n, 1)$.

Since $PS = \operatorname{conv}\{Pv_1, \ldots, Pv_{n+1}\}$ and $\sum Pv_i = 0$, to calculate the norm $\|-Pv_{n+1}\|_{PS}$ we clearly need to use only vectors Pv_i for $1 \le i \le n$, that is,

$$\|-Pv_{n+1}\|_{PS} = \inf\left\{\sum_{i=1}^n \lambda_i \ \Big|\ -Pv_{n+1} = \sum_{i=1}^n \lambda_i Pv_i,\ \lambda_i \ge 0\right\}.$$

However, if $-Pv_{n+1} = \sum_{i=1}^n \lambda_i Pv_i$ for $\lambda_i \ge 0$ then

$$A^2(1-\varepsilon)^2\frac{k}{n} \le |\langle Pv_{n+1}, Pv_{n+1}\rangle| \le \sum_{i=1}^n \lambda_i |\langle Pv_{n+1}, Pv_i\rangle| \le A^2\frac{k}{n}8\varepsilon\sum_{i=1}^n \lambda_i.$$

Thus we obtain $\|-Pv_{n+1}\|_{PS} \ge (1-\varepsilon)^2/(8\varepsilon)$ with probability larger than $1 - \exp\left(\ln\left(c_1(n+2)^2\right) - \varepsilon^2 ck\right)$. The choice

$$\varepsilon = \sqrt{\frac{2p \cdot \ln\left(c_1(n+2)^2\right)}{ck}}$$

implies an estimate analogous to (9) for n large enough (with a different absolute constant c). For small n the estimate is trivial.

Now let $a \ne 0$. Of course, it is enough to consider the case $a \in S$ only. Arrange vertices of the simplex in such a way that $a \in \operatorname{conv}\{0, v_1, \ldots, v_n\}$. Then an obvious modification of the previous argument gives the same estimate as for $a = 0$. This completes the proof. $\qquad\square$

The estimate in Theorem 5.1 is optimal, up to a logarithmic factor. This is a consequence of the following general upper estimate.

Theorem 5.3 *Let $K \subset \mathbb{R}^n$ be a convex body such that D is the ellipsoid of minimal volume for K. There exist constants $c > 0$ and $C > 0$ such that for $k \ge c(M_K^*)^2 n$, the set of subspaces $E \in G_{n,k}$ such that projections $P = P_E$ satisfy*

$$d(PK, PD) \le \tilde{A}\sqrt{k \log n}, \tag{11}$$

where $\tilde{A} \le C\sqrt{n/(n-k)}$, has measure larger than $1 - \exp(-ck) - \exp(-c(n-k))$.

By Dvoretzky's theorem, for $k \leq c(M_K^*)^2 n$ an analogous distance is bounded by an absolute constant.

It should be noted that Theorem 5.3 is closely related to an isomorphic version of Dvoretzky's theorem recently proved by Milman and Schechtman ([MS2], [MS3]) for normed spaces, and later by Gordon, Guédon and Meyer ([GGM]) for arbitrary convex bodies (for $k \leq n/\log n$). The upper estimates proved in these papers are better than (11) by a logarithmic factor, which is slightly different in different ranges of k. However, except for the case of $k \leq n/\log n$ ([MS3], [GGM]), the proofs do not give randomness.

Before we proceed with the proof of Theorem 5.3, let us recall a convenient notation, which will be used in the rest of this section. Given a convex body $K \subset \mathbb{R}^n$, let a_K and b_K denote the smallest real numbers for which the inclusions hold,

$$(1/b_K)D \subset K \subset a_K D.$$

In particular, $a_K = b_{K^\circ}$ and $b_K = a_{K^\circ}$.

Remark 5.4 In this notation Theorem 2.1 means that for every convex body $K \subset \mathbb{R}^n$, a projection P on a random k-dimensional subspace satisfies $b_{PK} \leq 2\sqrt{n/(n-k)}\, M_K$.

Theorem 5.3 follows from Theorem 2.1 (see the remark above) and the following two lemmas. The first lemma is well known. In the dual form, this is essentially an upper estimate in Dvoretzky's theorem (valid without restrictions that $\varepsilon > 0$ must be sufficiently small) see e.g., [MS1] (Theorem 4.2 and its proof).

Lemma 5.5 Let $K \subset \mathbb{R}^n$ be a convex body. There exist absolute constants $c > 0$ and $C > 0$ such that for any $1 \leq k \leq n$, the set of subspaces $E \in G_{n,k}$ such that projections $P = P_E$ satisfy $PK \subset rPD$, has measure larger than $1 - \exp(-ck)$; here $r = ca_K \sqrt{k/n}$ for $k \geq (M_K^*/a_K)^2 n$ and $r = cM_K^*$ for $k \leq (M_K^*/a_K)^2 n$.

The next lemma comes from [BLPS].

Lemma 5.6 Let $K \subset D$ be a convex body such that D is ellipsoid of minimal volume for K. Then

$$M_K \leq C\sqrt{n \log n},$$

where C is an absolute constant.

It is worth to mention that Barthe ([Ba]) has recently shown that the maximal value of M_K, over all convex bodies K for which D is the ellipsoid of minimal volume, is attained for the regular simplex S. Recall that M_S is approximately equal to $\sqrt{n \log n}$.

Remark 5.7 If K in $\mathbb{R}^n$ is a body such that 0 is the Santaló point of K (or, equivalently, 0 is the baricenter of K^0) and D is the minimal volume ellipsoid for K with such choice of origin then, using recent results of Guédon ([Gu], Theorems 2.2 and 2.3), the logarithmic factor in (11) can be removed. Indeed, dual form of Theorem 3 in [LMP] says that a projection P on a random k-dimensional subspace satisfies

$$b_{PK} \leq c(n/(n-k))^{3/2}\, \widetilde{M}_{K^0} \quad \text{where} \quad \widetilde{M}_L = \frac{1}{|L|} \int_L |x|\, dx$$

for a convex body L (cf. the previous remark). Recently Guédon has shown that if 0 is baricenter of L and D is the ellipsoid of maximal volume for L with respect to baricenter than $\widetilde{M}_L < \sqrt{n}$ ([Gu], Theorem 2.2). Applying this estimate he obtained (in dual formulation) that $b_{PK} \leq \widetilde{A}_0\sqrt{n}$ for $\widetilde{A}_0 = c(n/(n-k))^{3/2}$ ([Gu], Theorem 2.3). Combining the last estimate with Lemma 5.5 we get $d(PK, PD) \leq \widetilde{A}_0\sqrt{k}$ instead of (11).

The following theorem is a standard application of estimates from Section 4. In the symmetric case it was proved in [BM]. It shows that random sections and random projections of convex bodies are much closer together then it would follow from general estimates, and the distances between them admit the same estimates as for centrally symmetric bodies.

Theorem 5.8 *Let K_1 and K_2 be convex bodies in $\mathbb{R}^n$ such that 0 is an isomorphic Santaló point for each of them (with constant c_1), and that $M(K_i - K_i)M^*(K_i - K_i) \leq c_2\kappa(K_i - K_i)$, for $i = 1, 2$. Let $0 < \lambda < 1$. Then for two random subspaces E_1 and E_2 with $\dim E_1 = \dim E_2 = [\lambda n]$ one has*

$$d(P_{E_1}K_1, K_2^0 \cap E_2) \leq C(\lambda)\sqrt{n}\, \kappa(K_1')\kappa(K_2')$$
$$\leq C(\lambda)\sqrt{n}\, \ln(1 + d_{K_1'})\ln(1 + d_{K_2'}),$$

where $K_i' = K_i - K_i$, for $i = 1, 2$, and $C(\lambda)$ depends on c_1, c_2 and λ.

Proof. We shall use the estimate that follows from Chevet's inequality ([BG], in the symmetric case, see also [T]; and a general case follows by a standard modification). If K and L are convex bodies in $\mathbb{R}^n$, then

$$d(K, L^0) \leq c(a_K M_L^* + a_L M_K^*)(b_K M_L + b_L M_K), \tag{12}$$

where c is an absolute constant.

By the proof of Theorem 4.1, for two random subspaces F_1 and F_2 of dimension $[\sqrt{\lambda}n]$ we have

$$M(P_{F_i}K_i)M(K_i^0) \leq C(\lambda)\kappa(K_i - K_i),$$

for $i = 1, 2$, where $C(\lambda)$ depends on c_1, c_2 and λ.

By Theorem 2.1, for random subspaces $E_i \subset F_i$ of dimension $[\lambda n]$, $i = 1, 2$, we have

$$b_{P_i K_i} \leq 3 \left(1 - \sqrt{\lambda}\right)^{-1/2} M(P_i K_i),$$

where $P_i = P_{E_i}$.

Since $a_{P_i K_i} \leq c' \sqrt{k} M^*(P_i K_i)$ (where c' is an absolute constant), then (12) implies

$$d(P_1 K_1, (P_2 K_2)^0)$$
$$\leq c(1 - \sqrt{\lambda})^{-1/2} \sqrt{k} M^*(P_1 K_1) M^*(P_2 K_2) M(P_{F_1} K_1) M(P_{F_2} K_2)$$
$$\leq C'(\lambda) \sqrt{n} \, \kappa(K_1 - K_1) \kappa(K_2 - K_2),$$

where $C'(\lambda)$ depends on c_1, c_2 and λ. This concludes the proof. □

6   The Proportional Dvoretzky–Rogers Factorization

Results of this section are completely analogous of the theorem by Bourgain and Szarek ([BS], see also [ST]) for centrally symmetric convex bodies.

Theorem 6.1 *Let $K \subset \mathbb{R}^n$ be a convex body, such that D is the ellipsoid of minimal volume containing K. Let $\varepsilon \in (0, 1)$ and set $k = [(1 - \varepsilon)n]$. There exist vectors $x_1, x_2, ..., x_k$ in K, and an orthogonal projection P in $\mathbb{R}^n$ with rank $P \geq k$ such that for all scalars $t_1, ..., t_k$*

$$c\varepsilon^3 \left(\sum_{j=1}^{k} |t_j|^2\right)^{1/2} \leq \left\|\sum_{j=1}^{k} t_j P x_j\right\|_{PK} \leq \frac{6}{\varepsilon} \sum_{j=1}^{k} |t_j|,$$

where $c > 0$ is a universal constant.

Remark 6.2 For a centrally symmetric body K, the conclusion of the theorem can be equivalently expressed in terms of a factorization of the identity operator $i_{1,2}^k : \ell_1^k \to \ell_2^k$ through K. This is no longer true for non-symmetric bodies: the inequalities above imply the factorization of $i_{1,2}^k$ through PK only.

In particular we have two immediate corollaries.

Corollary 6.3 *Let $K \subset \mathbb{R}^n$ be a convex body such that D is the ellipsoid of minimal volume containing K. For every $\varepsilon \in (0, 1)$ there exist an orthogonal projection P in $\mathbb{R}^n$ with $k = \operatorname{rank} P \geq (1 - \varepsilon)n$ and an ellipsoid $\mathcal{D}$ on $P(\mathbb{R}^n)$ such that $(\varepsilon/6)\mathcal{B}_1^k \subset PK \subset (C/\varepsilon^3)\mathcal{D}$, where $\mathcal{B}_1^k$ is the l_1^k-ball corresponding to $\mathcal{D}$. By duality, there exists a subspace $E \subset \mathbb{R}^n$ with $\dim E \geq k$ and an ellipsoid $\mathcal{D}$ on E such that $c\varepsilon^3 \mathcal{D} \subset K \cap E \subset (6/\varepsilon)\mathcal{B}_\infty^k$, where $\mathcal{B}_\infty^k$ is the l_∞^k-ball corresponding to $\mathcal{D}$.*

Corollary 6.4 *Let* $K \subset \mathbb{R}^n$ *be a convex body such that* D *is the ellipsoid of minimal volume containing* K. *For* $\varepsilon \in (0, 1)$, *let* $k = [(1 - \varepsilon)n]$. *Then there exist an orthogonal projection* P *with* $\operatorname{rank} P = k$ *and a subspace* $E \subset \mathbb{R}^n$ *with* $\dim E = k$ *such that the Banach-Mazur distance satisfies*

$$\max \left(d(PK, B_1^k), d(K \cap E, B_\infty^k) \right) \leq (C/\varepsilon^4)\sqrt{n},$$

where C *is an absolute constant and* B_q^k *denotes the unit ball of* ℓ_q^k.

Remark 6.5 The interest in the second corollary is that the estimates of the distance of certain sections and projections of a (non-symmetric) convex body to the cube and the octahedron are of the same order in n as for symmetric bodies; and they are better than general estimates.

Remark 6.6 The second part of Corollary 6.3 immediately shows that the proofs from [MS2] and [MS3] work in the non-centrally symmetric case as well. Thus we have the following theorem.

Theorem 6.7 *For every* $\log n \leq k \leq n/2$, *for every convex body* K *in* $\mathbb{R}^n$ *there is a* k-*dimensional subspace* E *of* $\mathbb{R}^n$ *such that* $d_{E \cap K} \leq C\sqrt{k/\log(1 + n/k)}$ *with an absolute constant* C.

This result in the non-centrally symmetric case was shown to be valid for $k \leq Cn/(\log n)^2$ in [GGM] by a different method.

The proof of Theorem 6.1 essentially follows the argument from Szarek and Talagrand's paper [ST], which however needs to be modified into a non-symmetric setting.

Lemma 6.8 *Let* $K \subset \mathbb{R}^n$ *be a compact convex body such that* D *is the ellipsoid of minimal volume containing* K. *Let* $\varepsilon \in (0, 1)$. *There exist* $k \geq (1 - \varepsilon)n$ *and contact points* $x_1, x_2, ..., x_k$ *of* K *and* D *such that*

$$\operatorname{dist} \left(x_j, \operatorname{span} \{ x_i \mid i \neq j, 1 \leq i \leq k \} \right) \geq \sqrt{\varepsilon},$$

for $j = 1, ..., k$.

Sketch of the proof. For sake of future reference we outline the known proof. By John's theorem there exist contact points $x_1, ..., x_m$ of K and D and positive scalars $c_1, ..., c_m$ which give a resolution of the identity, $x = \sum_i c_i(x_i, x)x_i$ for all $x \in \mathbb{R}^n$ and $\sum_i c_i = n$ (additionally, $\sum_i c_i x_i = 0$, but we shall not use this). Then for any orthogonal projection Q with $\operatorname{rank} Q = \ell$ one has $\max_{1 \leq i \leq m} |Qx_i| \geq (\ell/n)^{1/2}$. Indeed, $\ell = \operatorname{tr} Q = \sum_i c_i(x_i, Qx_i) = \sum_i c_i |Qx_i|^2 \leq \max_i |Qx_i|^2 \sum_i c_i = n \max_i |Qx_i|^2$.

Let $k = [(1 - \varepsilon)n] + 1$. For every subset $\sigma \subset \{1, \ldots, m\}$ with $|\sigma| = k$, consider $B_\sigma = \mathrm{conv}\,\{\pm x_i : i \in \sigma\}$ and pick a subset σ_0 for which the k-dimensional volume of B_{σ_0} is maximal. Then for all $j \in \sigma_0$ one has

$$\mathrm{dist}\,(x_j, \mathrm{span}\,\{x_i \mid i \neq j, i \in \sigma_0\})$$
$$= |Qx_j| = \max\{|Qx_i| \mid i \notin \sigma_0 \text{ or } i = j\} \geq \sqrt{\varepsilon},$$

where Q is the orthogonal projection with $\ker Q = \mathrm{span}\,\{x_i \mid i \neq j, i \in \sigma_0\}$.

□

Lemma 6.9 *Let $\delta \in (0, 1)$ and $\gamma > 0$. Let $x_1, \ldots, x_n$ in $\mathbb{R}^n$ satisfy*

$$\mathrm{dist}\,(x_j, \mathrm{span}\,\{x_i \mid i \neq j\}) \geq \gamma,$$

for $j = 1, \ldots, n$. Then there exists a subset $\sigma \subset \{1, \ldots, n\}$ with $|\sigma| \geq (1 - \delta)n$ and such that for all scalars $t_1, \ldots, t_n$

$$\left| \sum_{j \in \sigma} t_j x_j \right| \geq c\gamma\delta \left(\sum_{j \in \sigma} |t_j|^2 \right)^{1/2},$$

where $c > 0$ is an absolute constant.

This was proved in [ST], as a combination of Proposition 4 and the argument from the proof of Corollary 5. The dependence of the right hand side on γ and δ was shown in the Proposition and Theorem 2 of [Gi].

The final lemma is taken from [BS] (Lemma C), and for the sake of completeness we provide a short proof.

Lemma 6.10 *Let $x_1, \ldots, x_m \in \mathbb{R}^n$ satisfy $|\sum t_j x_j| \geq (\sum |t_j|^2)^{1/2}$ for all scalars (t_j). Let $\alpha \in (0, 1)$ and let P be an orthogonal projection in $\mathbb{R}^n$ with $\mathrm{corank}\, P \leq \alpha m$. Then for every $\alpha < \delta < 1$ there exists a subset $\sigma \subset \{1, \ldots, m\}$, with $|\sigma| \geq (1 - \delta)m$ such that for all scalars (t_i),*

$$\left| \sum_{j \in \sigma} t_j P x_j \right| \geq c(\delta - \alpha)^{3/2} \left(\sum_{j \in \sigma} |t_j|^2 \right)^{1/2},$$

where $c > 0$ is an absolute constant.

Sketch of the proof. Similarly as in [BS], without loss of generality we may assume that $n = m$ and the vectors (x_i) are the standard unit vectors (e_i) in $\mathbb{R}^m$. Thus we obtain the resolution of the identity in the space $E = P(\mathbb{R}^m)$, $x = \sum (Pe_i, x)Pe_i$ for $x \in E$. We have $k = \dim E \geq (1 - \alpha)m$. Let $\delta' = (\alpha + \delta)/2$. By the proof of Lemma 6.8, there exists a subset σ' with $\ell = |\sigma'| \geq k - (\delta' - \alpha)k \geq (1 - \delta')m$ such that for all $j \in \sigma'$ we have $\mathrm{dist}\,(Pe_j, \mathrm{span}\,\{Pe_i \mid i \neq j, i \in \sigma'\}) \geq \sqrt{\delta' - \alpha}$.

Therefore by Lemma 6.9, there exists a subset $\sigma \subset \sigma'$, with $|\sigma| \geq \ell - (\delta - \delta')\ell \geq (1 - \delta)m$ such that

$$\left| \sum_{j \in \sigma} t_j P e_j \right| \geq c(\delta' - \alpha)^{1/2}(\delta - \delta') \left(\sum_{j \in \sigma} |t_j|^2 \right)^{1/2},$$

which immediately concludes the proof.   □

Remark 6.11 The conclusion of Lemma 6.10 holds as well if the assumption on the vectors (x_i) is replaced by the condition that the vectors give the John's resolution of the identity: Let $m \geq n$ and let $x_i \in \mathbb{R}^n$, with $|x_i| = 1$, and $c_i > 0$, for $i = 1, \ldots, m$ satisfy $x = \sum_i c_i(x_i, x)x_i$ for all $x \in \mathbb{R}^n$ and $\sum_i c_i = n$. This follows from a similar proof as above, preceded with a slightly modified argument as in the proof of Lemma 6.8.

Proof of Theorem 6.1. Let D be the ellipsoid of minimal volume for K. By Lemma 6.8 there exist $n \geq k \geq (1 - \varepsilon/3)n$ and contact points $x_1, \ldots, x_k$ of K and D that satisfy the conclusion of the lemma with the lower estimate $\sqrt{\varepsilon/3}$. Thus, by Lemma 6.9, there exists a subset $\sigma' \subset \{1, \ldots, k\}$ with $m = |\sigma'| \geq k - (\varepsilon/3)k \geq (1 - 2\varepsilon/3)n$ that the lower ℓ_2-estimate of the lemma is satisfied with the function $c'\varepsilon^{3/2}$, where $c' > 0$ is an absolute constant.

It is not difficult to construct, for an arbitrary $\beta \in (0, 1)$, an orthogonal projection P with corank $P = [\beta m]$ such that $-Px_j \in \beta^{-1}PK$ for every $j \in \sigma'$. Indeed, partition the set σ' into $[\beta m]$ disjoint subsets A_s, with $|A_s| \leq [1/\beta] + 1$ for all s. Let $z_s = \sum_{i \in A_s} x_i$, and let P be the orthogonal projection with ker $P = \text{span}\{z_s\}$. Since $Pz_s = 0$, then $-Px_j \in [1/\beta] \text{conv}\{Px_i : i \neq j, i \in A_s\}$, for all $j \in A_s$ and all s. Thus P is an orthogonal projection of corank $[\beta m]$ such that $-Px_j \in [1/\beta]PK$, for all $j \in \sigma'$.

Let $\beta = \varepsilon/6$ and let P be the corresponding orthogonal projection. Then, by Lemma 6.10, with $\alpha = \varepsilon/6$ and $\delta = \varepsilon/3$ we get a set $\sigma \subset \sigma'$, $|\sigma| \geq m - (\varepsilon/3)m \geq (1 - \varepsilon)n$ such that for all scalars (t_i),

$$\left| \sum_{j \in \sigma} t_j P x_j \right| \geq c\varepsilon^3 \left(\sum_{j \in \sigma} |t_j|^2 \right)^{1/2},$$

where $c > 0$ is a universal constant. Since $|\sigma| \geq k$, by relabelling vectors x_j we get the left hand side of the required inequality:

$$\left\| \sum_{j=1}^k t_j P x_j \right\|_{PK} \geq \left| \sum_{j=1}^k t_j Q x_j \right| \geq c\varepsilon^3 \left(\sum_{j=1}^k |t_j|^2 \right)^{1/2}.$$

The right hand side is obvious from the triangle inequality, since letting $B = PK \cap (-PK)$ we get a convex centrally symmetric body in $P(\mathbb{R}^n)$, and $\|Px_j\|_{PK} \leq \|Px_j\|_B \leq 6/\varepsilon$, for all $j \in \sigma$.   □

References

[BLPS] Banaszczyk W., Litvak A.E., Pajor A., Szarek S.J. (1999) The flatness theorem for non-symmetric convex bodies via the local theory of Banach spaces. Math. Oper. Res. 24(3):728–750

[Ba] Barthe F. (1998) An extremal property of the mean width of the simplex. Math. Ann. 310(4):685–693

[BG] Benyamini Y., Gordon Y. (1981) Random factorization of operators between Banach spaces. J. Anal. Math. 39:45–74

[BM] Bourgain J., Milman V.D. (1986) Distances between normed spaces, their subspaces and quotient spaces. Integral Equations Operator Theory 9:31–46

[BS] Bourgain J., Szarek S.J. (1988) The Banach-Mazur distance to the cube and the Dvoretzky-Rogers factorization. Israel J. Math. 62(2):169–180

[Gi] Giannopoulos A.A. (1995) A note on the Banach-Mazur distance to the cube. In: Geometric Aspects of Functional Analysis (Israel, 1992–1994), Oper. Theory Adv. Appl., 77, Birkhäuser, Basel, 67–73

[Go] Gordon Y. (1988) On Milman's inequality and random subspaces which escape through a mesh in $\mathbb{R}^n$. In: Geometric Aspects of Functional Analysis (1986/87), Lecture Notes in Math., 1317, Springer, Berlin-New York, 84–106

[GGM] Gordon Y., Guédon O., Meyer M. (1998) An isomorphic Dvoretzky's theorem for convex bodies. Studia Math. 127(2):191–200

[Gr] Grünbaum B. (1963) Measures of symmetry for convex sets. Proc. Sympos. Pure Math., Vol. VII, Amer. Math. Soc., Providence, R.I., 233–270

[Gu] Guédon O. (1998) Sections euclidiennes des corps convexes et inégalités de concentration volumique. Thèse de doctorat de mathématiques, Université de Marne-la-Vallée

[JL] Johnson W.B., Lindenstrauss J. Extensions of Lipschitz mappings into a Hilbert space. In: Conference in Modern Analysis and Probability (New Haven, Conn., 1982), 189–206

[LMP] Litvak A.E., Milman V.D., Pajor A. (1999) The covering numbers and "low M^*-estimate" for quasi-convex bodies. Proc. Amer. Math. Soc. 127:1499–1507

[MaT] Mankiewicz P., Tomczak-Jaegermann N. (1990) Pathological properties and dichotomies for random quotients of finite-dimensional Banach spaces. In: Geometry of Banach Spaces (Strobl, 1989), London Math. Soc. Lecture Note Ser., 158, Cambridge Univ. Press, Cambridge, 199–217

[M1] Milman V.D. (1985) Almost Euclidean quotient spaces of subspaces of a finite dimensional normed space. Proc. Amer. Math. Soc. 94:445–449

[M2] Milman V.D. (1985) Random subspaces of proportional dimension of finite dimensional normed spaces: approach through the isoperimetric inequality. In: Banach Spaces (Columbia, Mo., 1984), Lecture Notes in Math., 1166, Springer, Berlin-New York, 106–115

[M3] Milman V.D. (1990) A note on a low M^*-estimate. In: Geometry of Banach Spaces (Strobl, 1989), London Math. Soc. Lecture Note Ser., 158, Cambridge Univ. Press, Cambridge, 219–229

[M4] Milman V.D. (1991) Some applications of duality relations. Geometric Aspects of Functional Analysis (1989–90), Lecture Notes in Math., 1469, Springer, Berlin, 13–40

[MP]   Milman V.D., Pajor A. Entropy and asymptotic geometry of non-symmetric convex bodies. Advances in Math., to appear; see also (1999) Entropy methods in asymptotic convex geometry. C.R. Acad. Sci. Paris, S.I. Math., 329(4):303–308

[MS1]   Milman V.D., Schechtman G. (1985) Asymptotic theory of finite-dimensional normed spaces. Lecture Notes in Math., 1200, Springer, Berlin-New York

[MS2]   Milman V.D., Schechtman G. (1995) An "isomorphic" version of Dvoret-zky's theorem. C. R. Acad. Sci. Paris Sr. I Math. 321:541–544

[MS3]   Milman V.D., Schechtman G. (1999) An "isomorphic" version of Dvoret-zky's theorem, II. In: Convex Geometric Analysis (Berkeley, CA, 1996), Math. Sci. Res. Inst. Publ., 34, Cambridge Univ. Press, Cambridge, 159–164

[PT1]   Pajor A., Tomczak-Jaegermann N. (1985) Remarques sur les nombres d'entropie d'un opérateur et de son transposé. C. R. Acad. Sci. Paris Sr. I Math. 301(15):743–746

[PT2]   Pajor A., Tomczak-Jaegermann N. (1986), Subspaces of small codimension of finite-dimensional Banach spaces. Proc. Amer. Math. Soc. 97(4):637–642

[PT3]   Pajor A., Tomczak-Jaegermann N. (1989) Volume ratio and other s-numbers of operators related to local properties of Banach spaces. J. Funct. Anal. 87(2):273–293

[Pa]   Palmon O. (1992) The only convex body with extremal distance from the ball is the simplex. Israel J. Math. 80(3):337–349

[Pi]   Pisier G. (1989) The Volume of Convex Bodies and Banach Space Geom-etry. Cambridge University Press, Cambridge

[R1]   Rudelson M. Sections of the difference body. Discrete and Computational Geometry, to appear

[R2]   Rudelson M. Distances between non–symmetric convex bodies and the MM^*-estimate. Positivity, to appear

[ST]   Szarek S.J., Talagrand M. (1989) An "isomorphic" version of the Sauer-Shelah lemma and the Banach-Mazur distance to the cube. In: Geometric Aspects of Functional Analysis (1987–88), Lecture Notes in Math., 1376, Springer, Berlin-New York, 105–112

[T]   Tomczak-Jaegermann N. (1989) Banach-Mazur distances and finite-dimensional operator ideals. Pitman Monographs and Surveys in Pure and Applied Mathematics, 38. Longman Scientific & Technical, Harlow; John Wiley & Sons, Inc., New York

A Geometric Lemma and Duality of Entropy Numbers

V.D. Milman[1] and S.J. Szarek[2]

[1] School of Mathematical Sciences, Tel Aviv University, Tel Aviv 69978, Israel
[2] Department of Mathematics, Case Western Reserve University, Cleveland, OH 44106-7058, USA and Université Pierre et Marie Curie, Equipe d'Analyse, Bte 186, 4, Place Jussieu, 75252 Paris, France

1 Introduction

We shall study in this note the following conjecture, to which we shall refer as the "Geometric Lemma"; we state it first in a somewhat imprecise form.

Let n, N be positive integers with $k := \log N \ll n$. If $S \subset \mathbb{R}^n$ is a finite set whose cardinality doesn't exceed N and such that its convex hull $K := \operatorname{conv} S$ admits an equally small Euclidean 1-net (i.e., K can be covered by no more than N translates of the unit Euclidean ball D), then $\frac{1}{2}D \not\subset K$.

At first glance a statement of that nature may appear "trivial". Indeed, on some meta-mathematical level, we are asking whether the "complexity" of an n-dimensional Euclidean ball is smaller than exponential in n (by $\ll$ we mean above "much smaller than"). And our intuition says "NO", no matter what exactly is meant by "complexity". However, the more precise formulation brings together two very different assumptions: small number of vertices of the polytope K and small cardinality of the Euclidean "net" for K. As we shall see in this note, these data are not easy to "combine"; that creates difficulties and, at the same time, contributes to the appeal of the question.

Our interest in the "Lemma" came as an outgrowth of the study of the following problem, which was originally promoted by Carl and Pietsch and is well known in Geometric Functional Analysis (and in the Geometric Operator Theory), usually referred to as "the duality conjecture for entropy numbers of operators". Recall that if X, Y are Banach spaces, $u : X \to Y$ a compact operator and $\varepsilon > 0$, we denote by $N(u, \varepsilon)$ the minimal cardinality of an ε-net of the image $u(B_X)$ of the unit ball B_X of X (in the metric of Y; in other words, $N(u, \varepsilon)$ is the smallest number of balls in Y of radius ε that together cover $u(B_X)$). The conjecture referred to above asserts that there exist universal constants $a, b > 0$ such that

$$a^{-1} \log N(u, b\varepsilon) \leq \log N(u^*, \varepsilon) \leq a \log N(u, b^{-1}\varepsilon) \tag{1}$$

The research of both authors was partially supported by grants from the US-Israel BSF. The second author's research was partially supported by grants from the NSF (U.S.A.).

for any compact operator u and $\varepsilon > 0$. (Here and in what follows, all logarithms are to the base 2.) In terms of the so-called *entropy numbers* (for basic results concerning them and related concepts see [C-S] and [Pie]), defined for an operator u by

$$e_k(u) := \inf\{\varepsilon > 0 : N(u, \varepsilon) \le 2^k\},$$

the assertion of the conjecture (roughly) becomes

$$b^{-1}e_{ak}(u) \le e_k(u^*) \le be_{k/a}(u), \qquad (2)$$

which is its more standard formulation. (For the most part we shall be working with the form (1).) Note that even though ak and k/a are not necessarily integers, (2) makes sense as the definition of $e_k(u)$ works also for noninteger k. Additionally, due to the asymptotic nature of the questions we investigate, the difference between the number and its integer part (or the nearest integer, or, usually, even its double) is immaterial, and so we shall pretend that all numbers are integers as necessary; see also Remark 8.3.

We point out here that, for an operator u, the sequence $(e_k(u))$ tends to 0 as $k \to \infty$ if and only if $(e_k(u^*))$ does (and if and only if u is compact). In such a weak sense the qualitative character of the two sequences is the same. The "Duality Conjecture", if true, would imply that the two sequences are equivalent "distributionally", i.e., in the strongest *quantitative* sense that we may reasonably expect. To the best of our knowledge, it is still not known whether (1) or (2) has a chance to hold with $a = 1$, as inquired originally by Pietsch (for sure not with $a = b = 1$); nevertheless, the above formulation seems to be the most natural one. See also Remark 4.2.

Up to now, the Duality Conjecture in the form stated here has been verified only under very strong assumptions on *both* spaces (see, e.g. [GKS], [PT2]). Our conjectured Geometric Lemma is relevant to the case when one of the spaces, say Y, is a Hilbert space, and X is arbitrary; more specifically, it implies then the second inequality in (1). However, we believe that once a proper argument (assuming there is any) is found, a sort of a "dual" version for the other inequality will be figured out and, possibly, one will also be able to relax the assumptions on Y (paralleling the developments of ideas in [TJ1] and [BPST], where "weaker" – but quantitative – equivalences were established).

There has been a substantial body of work on the Duality Conjecture. Rather than present the current state of the knowledge on the matter, we refer the reader to [BPST], [PT2], [Pi2], their references and the survey [GGP] (in preparation).

As we indicated, the Geometric Lemma remains a conjecture. We did check it for various (classes of) convex bodies, including ℓ_p^n-balls of various radii (cf. one of the comments preceding Proposition 7.2) and some "random bodies" (specifically, "generic" projections of ℓ_1^N-balls – a "canonical" counterexample to many problems in high dimensional convexity, cf. [MTJ]). In the general setting, we were able to prove it "up to a logarithmic factor". In particular, we did show in Theorem 9.3 :

The assertion of the Geometric Lemma holds if we replace the assumption $k \ll n$ by $k \leq c(1 + \log n)^{-6}n$, where $c > 0$ is some universal numerical constant.

The corresponding "entropy duality" result is Theorem 9.4 :

There are numerical constants $a, C > 0$ such that if u is a compact Hilbert space valued operator and $k \in \mathbb{N}$, then $e_{ak}(u^*) \leq C(1 + \log k)^3 e_k(u)$.

Thus, even though the principal appeal of the Geometric Lemma originates in its potential to resolve the Duality Conjecture in the relevant case, the partial solution indicated above already has nontrivial consequences. To the best of our knowledge, in absence of assumptions on *both* spaces, no result of the above type (i.e., with a factor, which is a function of k) appears in the literature, the best estimates to date involving factors of power type in $n := \operatorname{rank} u$ or n/k (see [GKS], [PT2] and the comments following Theorem 9.4).

Finally, let us mention that Theorem 9.3 and, overall, an important part of the discussion deal with formally stronger statements: upper estimates on the mean width of the part of the body in question that is inside the Euclidean ball D. Theorem 9.3 asserts, in particular :

If K is as in the Geometric Lemma, then the mean width of $K \cap D$ does not exceed $C(1 + \log k)^3 \sqrt{k/n}$, where C is a universal numerical constant.

2   Geometric Lemma – Precise Statements

In this section we shall state a geometric version of (the case of) the Duality Conjecture that is implied by the Geometric Lemma. We shall also give precise formulation(s) of the lemma and indicate the relations between various statements. Let us start with the Duality Conjecture, the relevant case of which is

If X is a Banach space, H – a Hilbert space, and $u : X \to H$ a compact operator, then, for any $\varepsilon > 0$, $\log N(u^*, b\varepsilon) \leq a \log N(u, \varepsilon)$, where $a, b > 0$ are universal constants.

Note that one can restrict oneself to the case $\varepsilon = 1$ in the above (rescaling if necessary). Additionally, one may assume (by approximation) that the spaces in question are finite dimensional, and the operator u is one-to-one.

In this context it is convenient to define the *covering number* $N(U, V)$ for U, V-subsets of $\mathbb{R}^n$ (say, with V a closed convex body) by

$$N(U, V) := \min\{N : U \text{ may be covered by } N \text{ translates of } V\}.$$

If we set $U = uB_X$, $V = \varepsilon B_Y$, we get $N(U, V) = N(u, \varepsilon)$. Similarly $N(u^*, \varepsilon) = N(V^\circ, U^\circ)$, where K° denotes the polar body of K (say, with respect to the canonical Euclidean structure). Accordingly, denoting again by $D = D_n$ the standard Euclidean ball (we shall use this convention throughout this note), we see that the above assertion is equivalent to

Conjecture 2.1 *If $U \subset \mathbb{R}^n$ is a symmetric compact convex body, then* $\log N(D, bU^\circ) \leq a \log N(U, D)$ *where, $a, b > 0$ are universal constants.*

For future reference we point out that if V is a Euclidean ball (or, more generally, an ellipsoid), the translates of V in the definition of $N(U, V)$ may be further assumed to have centers contained in U (as required – in the general case – by some authors). This makes covering numbers less sensitive to enlarging the ambient space, while the present definition ensures that they are *always* increasing in the first argument.

Conjecture 2.1 is the one we are *really* going to work with. In the next several sections we shall show how it is implied by the following version of the Geometric Lemma.

Conjecture 2.2 *For any $\gamma > 0$ there exists $c_1 = c_1(\gamma) > 0$ such that if $K \subset \mathbb{R}^n$ verifies*

  (i) $K = \text{conv}\, S$, $\log |S| \leq c_1 n$ *and*

  (ii) $\log N(K, D) \leq c_1 n$,
then $\gamma D \not\subset K$.

Remark 2.3 In fact, Conjecture 2.1 (and Conjecture 2.4 below, see Proposition 5.1 and the comments preceding it) can be derived from the validity of Conjecture 2.2 for just one $\gamma < 1$, the constants involved (a, b, resp. C) depending on that γ and the corresponding c_1. □

It is possible to derive from Conjecture 2.2 a formally stronger statement, for which we need to introduce some notation. First, if $U \subset \mathbb{R}^n$ is a compact symmetric convex body containing the origin in the interior, one denotes by $\| \cdot \|_U$ its Minkowski functional, i.e., the norm, for which U is the unit ball. We shall use the same notation for gauges of nonsymmetric sets. We set also

$$M^*(U) := \int_{S^{n-1}} \sup_{y \in U} \langle x, y \rangle \, d\mu_n(x) = \int_{S^{n-1}} \|x\|_{U^\circ} \, d\mu_n(x)$$

where μ_n is the normalized (i.e., probability) Lebesgue measure on S^{n-1}. The second equality – without the integration signs – is in fact a definition of the polar U°. Both quantities on the right define $M^*(\cdot)$ also for nonsymmetric sets, and the first one even for nonconvex and not containing the origin (we will need that degree of generality later). We mention in passing that $M^*(U)$ equals the mean half-width of U, a well known geometric parameter. For future reference, we also set

$$M(U) := \int_{S^{n-1}} \|x\|_U d\mu_n(x).$$

In the sequel we will seldom formulate explicitly the hypotheses (convexity, symmetry or origin in the interior) on the arguments of $M^*(\cdot)$ and $M(\cdot)$; unless *some* hypotheses are stated, we shall implicitly assume the bare minimum for the formulae to make sense.

We are now ready to state the stronger version of Conjecture 2.2.

Conjecture 2.4 *Let* $K = \operatorname{conv} S \subset \mathbb{R}^n$ *and set* $k = \max\{\log |S|, \log N(K, D)\}$. *Then*

$$M^*(K \cap D) \le C\sqrt{\frac{k}{n}}, \tag{3}$$

where C is a universal constant.

Clearly Conjecture 2.4 $\Rightarrow$ Conjecture 2.2: if $\gamma D \subset K$, then $M^*(K \cap D) \ge M^*(\gamma D) = \gamma$, which is inconsistent with (3) if k/n is small.

In the next section we introduce more notation and recall some (more or less) known results about $M^*(\cdot)$, a few of them quite deep. In section 4 we prove the implications Conjecture 2.4 $\Rightarrow$ Conjecture 2.1 and Conjecture 2.2 $\Rightarrow$ Conjecture 2.4. (We do not know whether Conjecture 2.1 implies *formally* Conjecture 2.2 – or, equivalently, Conjecture 2.4. We devoted that matter some, but not too much, thought; the implication in question could be imaginably useful for constructing a possible counterexample to the Duality Conjecture.) Then, in section 5, we present some "almost isometric" refinements of arguments from section 4, in particular the one suggested in Remark 2.3. In sections 6 through 8 we develop the ideas and tools needed for our study of the conjectures. That study culminates in section 9, where we formulate and prove the results sketched at the end of the introduction.

3   More Notation, Known Results

As general references for notation and basic results of local theory of Banach spaces we suggest the books [M-S], [Pi3] and [TJ2] or the survey paper [G-M]; a handy source for "probability in Banach spaces" is the monograph [L-T].

For a subspace $E \subset \mathbb{R}^n$, we shall denote by P_E the orthogonal projection onto E. Given positive integer $k < n$, $G_{k,n}$ is the Grassmann manifold (of k-dimensional subspaces of R^n) endowed with the canonical rotationally invariant probability measure $\mathcal{P} = \mathcal{P}_{k,n}$. We shall say that a "generic k-dimensional subspace" E of $\mathbb{R}^n$ has certain property if the measure of E's with that property is close to 1 provided n is large (or k, n are large, depending on the context). Similarly, we shall talk about "generic rank k orthogonal projections".

The first auxiliary result is the following "relative" of the Dvoretzky Theorem (see, e.g., [Mi2], Lemma 2.1 and its proof). Such results are usually stated for the symmetric case, but the arguments, based on the "concentration of measure phenomenon", do not *really* require that hypothesis (nor even, in our formulation, that the body in question contains 0, the assertion being insensitive to a "small" translation of the body, cf. Fact 3.2). Let us also remark here that even though in the context of the Duality Conjecture, as usually stated, only the case of symmetric sets is relevant, in the present note we drop the symmetry assumption whenever feasible. We believe that the setting of *general* convex bodies is the natural one, part of the motivation for their study coming from geometry and optimization, where the symmetry hypothesis is often artificial. Additionally, in our arguments nonsymmetric sets actually *have to* appear (e.g., as intersections of two symmetric sets with different centers). On the other hand, lack of symmetry is seldom a source of difficulties, and often – but not always – the passage to the more general framework is merely "formal". Such is, for example, the case of our Conjectures 2.2 and 2.4 from the preceding section: for a possibly nonsymmetric K, apply the statement to $(K - K)/2$ (noting that, e.g., $\log N((K - K)/2, D) \leq 2 \log N(K, D)$ while $M^*(K \cap D) \leq M^*((K - K)/2 \cap D)$; there is also an inequality sort of inverse to the latter for the sets we are interested in, see Remark 8.3). Similarly, if K is symmetric and S is its net, one can always assume that S is also symmetric and contains 0: replace S by $S \cup (-S) \cup \{0\}$, the resulting roughly two-fold increase in the cardinality of S is nonessential in our setting (see Remark 8.3).

Fact 3.1 *Let $V \subset D \subset \mathbb{R}^n$ be a convex set and let $m \leq n$. Let P be a generic orthogonal projection of rank m. Then*

(a) *if $\sqrt{\frac{m}{n}} \geq \varepsilon_0 M^*(V)$, then $PV \subset C\sqrt{\frac{m}{n}}PD$*

(b) *if $\sqrt{\frac{m}{n}} \leq \varepsilon_0 M^*(V)$, then $cM^*(V)PD \subset PV \subset CM^*(V)PD$*

(c) *moreover, if for some $\varepsilon \leq \varepsilon_0$ we have $\sqrt{\frac{m}{n}} \leq \varepsilon M^*(V)$, then*
$$(1 - C\varepsilon)M^*(V)PD \subset PV \subset (1 + C\varepsilon)M^*(V)PD.$$

Above, c, C and ε_0 are universal positive constants, independent of n, V and ε. In all cases "generic" means "except on a set (of projections) of measure $\leq \exp(-c'm)$", where $c' > 0$ is a universal numerical constant.

The next result is closely related to the fact that, for a fixed rank m projection P on $\mathbb{R}^n$, the Euclidean norm of Px is "strongly concentrated" around the value $\sqrt{\frac{m}{n}}$ as x varies over S^{n-1} (clearly, the *average* of $|Px|^2$

equals m/n; here and throughout the paper we use the notation $|\cdot|$ for the Euclidean norm $\|\cdot\|_D$). This well-known phenomenon has been often derived from the isoperimetric inequality for the sphere, but it can be approached also via a direct calculation. Here we choose an equivalent point of view : the point x stays fixed, while the projection P varies over $G_{m,n}$ endowed with the probability measure $\mathcal{P} = \mathcal{P}_{m,n}$.

Fact 3.2 *Let $x \in S^{n-1}$, let $m \le n$ and let P be a generic orthogonal projection of rank m (i.e., considered as an element of $(G_{m,n}, \mathcal{P})$). Then $|Px|$ is strongly concentrated around the value $\sqrt{\frac{m}{n}}$. More precisely,*

(a) *if $\varepsilon > 0$, then*

$$\mathcal{P}\left(|\,|Px| - \sigma_{m,n}| > \varepsilon\right) < \exp\left(-\varepsilon^2 n/2\right),$$

where $\sigma_{m,n}$ is the median of $|Px|$ and $\left|\sigma_{m,n} - \sqrt{\frac{m}{n}}\right| \le \frac{C}{\sqrt{n}}$

(b) *consequently, if $\lambda > 1$, then*

$$\mathcal{P}\left(|Px| > \lambda\sqrt{\frac{m}{n}}\right) < \exp\left(-c(\lambda - 1)^2 m\right)$$

(c) *and, additionally, for $\alpha > 0$,*

$$\mathcal{P}\left(|Px| < \alpha\sqrt{\frac{m}{n}}\right) < (\sqrt{e}\alpha)^m.$$

Above, C and c are universal positive constants.

Part (a) of Fact 3.2 is just the isoperimetric inequality applied to the function $x \to |Px|$ (cf. [J-L], where it was employed in the spirit close to that of our paper). Part (c), better known in the case of the Gaussian measure (see, for example, [Sza]), can be recovered, e.g., from Lemma 6 in [M-P].

From Facts 3.1 and 3.2 we derive the following Milman-Pajor-Tomczak-Talagrand (cf. [Mi1], [PT1], [Ta3]) type result.

Proposition 3.3 *Let $k \le n$, let $A > 0$ and let $K \subset \mathbb{R}^n$ be a symmetric convex body with $\log N(K, D) \le Ak$. Set $\omega := \max\{M^*(K \cap D), \sqrt{k/n}\}$. Then, for a generic rank k orthogonal projection P, we have*

$$c_0\sqrt{\frac{k}{n}}\,|x| \le \max\{\omega\|x\|_K, |Px|\} \quad \text{for all} \quad x \in \mathbb{R}^n, \tag{4}$$

where $c_0 > 0$ is a constant depending only on A. The assertion holds also for not-necessarily-symmetric bodies $K \ni 0$ after one replaces $M^*(K \cap D)$ by $\max_{x \in \mathbb{R}^n} M^*((K - x) \cap D)$ in the definition of ω.

Proof. For a smoother exposition we provide first a detailed proof in the (central) symmetric case and then sketch modifications needed to handle the general setting.

Noting that $\max\{\|\cdot\|_U, \|\cdot\|_V\} = \|\cdot\|_{U\cap V}$ and rescaling, we see that (4) is equivalent to

$$K \cap \omega P^{-1} D \subset c_0^{-1}\omega \sqrt{\frac{n}{k}} D$$

or

$$x \in K \setminus c_0^{-1}\omega \sqrt{\frac{n}{k}} D \Rightarrow |Px| > \omega. \tag{5}$$

Let S be a set with $|S| \leq 2^{Ak}$ such that $K \subset S + D$. A standard argument shows that then in fact $K \subset S + (D \cap 2K)$ (it is here that the symmetry is used; in general we would have $K - K$ in place of $2K$). Moreover, if $S_1 = S \setminus (c_0^{-1}\omega \sqrt{\frac{n}{k}} - 1)D$, then

$$K \setminus c_0^{-1}\omega \sqrt{\frac{n}{k}} D \subset S_1 + (D \cap 2K).$$

Accordingly, to prove (5), hence (4), it suffices to show that, for a generic P, $|P(s+y)| > \omega$ *simultaneously* for all $s \in S_1$ and all $y \in D \cap 2K$. To that end observe that, first, by Fact 3.1(a) or (b) applied with $V = D \cap 2K$, one has $|Py| \leq 2C\omega$ for a generic P and *all* $y \in D \cap 2K$. On the other hand, by Fact 3.2(c), for any fixed $x \in \mathbb{R}^n$,

$$P\left(|Px| \leq \delta \sqrt{\frac{k}{n}} |x|\right) < (\sqrt{e}\delta)^k$$

for any $\delta > 0$. Choosing δ small enough (say, $\delta = (\sqrt{e}2^{A+1})^{-1}$) we get that, for a generic projection P, *all* $s \in S_1$ (note $|S_1| \leq 2^{Ak}$) verify $|Ps| \geq \delta(c_0^{-1}\omega \sqrt{n/k} - 1)\sqrt{k/n} > \delta(c_0^{-1}\omega - 1)$, and so, for $x \in S_1 + (D \cap 2K)$,

$$|Px| \geq (\delta(c_0^{-1}\omega - 1) - 2C\omega)$$

which yields (5) if c_0 is chosen small enough.

If K is not symmetric, a more careful look shows that, in fact, one needs to control *simultaneously* (generic) projections of $K \cap (D + s)$ for all $s \in S_1$ or, equivalently, the projection of $W := \bigcup_{s \in S_1} (K - s) \cap D$. The argument used in the symmetric case carries over directly if A is small (specifically, if $A < c'$, where c' is the constant from Fact 3.1; cf. the proof of Proposition 5.1). For general A, it is more efficient to estimate $M^*(W)$ by $\max_{x \in \mathbb{R}^n} M^*((K - x) \cap D) + C_0\sqrt{k/n}$ via Lemma 8.1 and then "pipe in" conv W in place of $D \cap 2K$ in the argument above (the reader will readily verify that Lemma 8.1 depends only on Fact 3.8 and, moreover, if all y_j's are 0 –

the case which is relevant here – is independent from the rest of the paper). $\square$

We will need an estimate of "covering numbers" known as "Sudakov's inequality".

Fact 3.4 ([L-T], Theorem 3.18) *If $U \subset \mathbb{R}^n$ and $\varepsilon > 0$, then*

$$\log N(U, \varepsilon D) \leq C(\frac{M^*(U)}{\varepsilon})^2 n$$

where C is a universal constant.

Recall also that the problem of duality of entropy numbers (say, in the form (2)) is solved for $k \geq \text{rank } u$ ([K-M], see also [Pi2]). We have

Fact 3.5 *Let $U, V \subset \mathbb{R}^n$ be convex bodies such that $U \ni 0$ and V is 0-symmetric. If $k \geq n$, then*

$$\log N(U, V) \leq k \Rightarrow \log N(V^\circ, \beta U^\circ) \leq \alpha k$$

for some universal constants $\alpha, \beta \geq 1$ (resp. $\alpha = \alpha(\tau)$ if we just assume that $k \geq \tau n$ for some $\tau \in (0, 1)$). Moreover, the above inequality holds – at least if U is also symmetric – with $\beta = \beta(\alpha)$, for any $\alpha > 1$ (resp. $\beta = \beta(\alpha, \tau)$, for $\alpha > 1$ and $\tau \in (0, 1)$).

The Fact was stated in [K-M] just in the case when *both* U and V are symmetric, but the present variant follows formally: just apply the symmetric version to $(U-U)/2$ (and $2k$ in place of k) and note that $((U-U)/2)^\circ \subset 2U^\circ$. We do not know whether the symmetry of U is needed in the last statement; in absence of that hypothesis the present argument yields $\alpha > 2$ in place of $\alpha > 1$. Let us also note that, at least for the first statement and with proper care, one may dispose of the symmetry assumptions altogether (see [M-P]).

We shall need a few more properties of the functional $M^*(\cdot)$.

Fact 3.6 *If $U, V \subset \mathbb{R}^n$ are convex sets, then the function defined on $\mathbb{R}^n$ by*

$$\phi(x) = M^*((x + U) \cap V)$$

is concave. In particular, if both U and V are 0-symmetric, then $\phi(x) \leq \phi(0)$ for $x \in \mathbb{R}^n$.

This follows from the facts that, under Minkowski addition, the set-valued map $x \to (x + U) \cap V$ is concave, while $M^*(\cdot)$ is additive and positively homogeneous.

Let γ_n be the standard Gaussian measure on $\mathbb{R}^n$ (i.e., the one with density $(2\pi)^{-n/2} e^{-|x|^2/2}$). The next result describes the very well known relationship between spherical averages and those with respect to γ_n, and is easily established by integrating the latter in spherical coordinates.

Fact 3.7 *For $U \subset \mathbb{R}^n$, the Gaussian average*

$$\ell_1(U) := \int_{\mathbb{R}^n} \|x\|_U \, d\gamma_n(x)$$

is "essentially the same as" $n^{1/2}M(U)$. More precisely, there are constants $\sigma_n < 1$ with $\sigma_n \to 1$ as $n \to \infty$ such that, for all U as above, $\ell_1(U) = \sigma_n n^{1/2} M(U)$. The same is true if we replace $\ell_1(U)$ by $\ell_1(U^\circ)$ and $M(U)$ by $M^*(U)$.

For the record, $\sigma_n = \sqrt{2}\Gamma(\frac{n}{2})/\Gamma(\frac{n}{2}+1) \in (\sqrt{1-\frac{1}{n}}, 1)$. We could have dispensed with σ_n's in Fact 3.7 if we had defined *both* the Gaussian average and $M(U)$ via second moments of $\|\cdot\|_U$, a rather insignificant modification by the Kahane-Khinchine inequality (see [BLPS], Lemma 3.3, for the nonsymmetric case). However, that would not conform to the standard terminology.

Finally, we mention the following well known

Fact 3.8 *If $S \subset D \subset \mathbb{R}^n$ is a finite set, then*

$$M^*(\text{conv } S) = M^*(S) \leq C\sqrt{\frac{\log|S|}{n}},$$

where C is a numerical constant.

Fact 3.8 is proved most easily by passing to the Gaussian average (Fact 3.7) and a direct computation using tail estimates for the Gaussian density; in the Gaussian setting it is a special case of a much more general phenomenon (see [L-T], (3.6) or our Lemma 8.1). Alternatively, it is implicit in our Fact 3.1. By Fact 3.4, the estimate is exact if the set S is uniformly separated. We recall that a set is called δ-separated if each two its different members are more than δ apart; this leads to the concept of *packing numbers* – equivalent, up to a factor of 2 (in the argument), to that of covering numbers.

In the sequel we shall occasionally write $\Phi \lesssim \Psi$ meaning that there exists a universal numerical constant C such that, for all values of the parameters involved in the definitions of (normally nonnegative) quantities Φ and Ψ, one has $\Phi \leq C\Psi$. E.g., the assertion of Fact 3.8 can be written as $M^*(S) \lesssim \sqrt{\frac{\log|S|}{n}}$. (We point out that this convention differs from the one employed often in, e.g., combinatorics, and using for that concept the symbol $\ll$, reserved in this note for "much smaller than" or "sufficiently smaller than".) Similarly, $\Phi \simeq \Psi$ will indicate two-sided estimates $C^{-1}\Psi \leq \Phi \leq C\Psi$. We will not use that convention when we want to make the dependence on other constants or parameters explicit. Unless stated otherwise, C, c, C_1, c' etc. will stand for positive numerical constants independent of the dimension or any other parameters, whose exact values *may* vary between occurrences.

4   The Implications 2.4 $\Rightarrow$ 2.1 & 2.2 $\Rightarrow$ 2.4

The implication Conjecture 2.4 $\Rightarrow$ Conjecture 2.1. We shall assume the validity of Conjecture 2.4 and show how the results of the preceding section imply then Conjecture 2.1. More precisely, we prove

Proposition 4.1 *Let $k \in \mathbb{N}$. Let $w \geq 1$ be such that, for all n, S, K verifying the assumptions of Conjecture 2.4 for that particular k, one has $M^*(K \cap D) \leq w\sqrt{k/n}$. Then, for all n and for all convex sets $U \subset \mathbb{R}^n$,*

$$\log N(U, D) \leq k \;\Rightarrow\; \log N(D, CwU^\circ) \leq ak, \tag{6}$$

where $a, C > 0$ are universal constants.

Proof. Let $S \subset U$ with $|S| = N(U, D) \leq 2^k$ be such that $S + D \supset U$. Denote $K = \operatorname{conv} S$. We first observe that, for $\rho > 0$,

$$N(D, (\rho + 2)U^\circ) \leq N(D, \rho K^\circ)$$

In fact, any ρ-net of $D(= D^\circ)$ with respect to $\| \cdot \|_{K^\circ}$ is a $(\rho + 2)$-net with respect to $\| \cdot \|_{U^\circ}$. To see that, observe that if $x, y \in D$ and $\|x - y\|_{K^\circ} \leq \rho$, then

$$\|x - y\|_{U^\circ} = \max_{u \in U} \langle x - y, u \rangle$$
$$\leq \max_{s \in S, z \in D} \langle x - y, s + z \rangle$$
$$\leq \|x - y\|_{K^\circ} + |x - y| \leq \rho + 2.$$

(We are being slightly careless here as, in principle, it is possible that $0 \notin K$ and so one can not really speak about $\|\cdot\|_{K^\circ}$. However, this is easily remedied by adding to S a single point, cf. Remark 8.3. Another potential difficulty, K being degenerate, is handled by passing to a lower dimension.)

To derive Conjecture 2.1, it is now enough to appropriately estimate $N(D, \rho K^\circ)$ by $N(K, D)$ for some $\rho \lesssim w$; notice that $N(K, D) \leq N(U, D)$ (cf. the comment following Conjecture 2.1). To that end, apply Proposition 3.3 with the present choice of k, n, K (hence $A = 1$). Let $P = P_F$ be a (generic rank k) projection such that

$$c_0 \sqrt{\frac{k}{n}} |\cdot| \leq \max\{w \sqrt{\frac{k}{n}} \|\cdot\|_K, |P(\cdot)|\}.$$

After dividing out by $\sqrt{k/n}$, this dualizes to

$$c_0 D \subset \operatorname{conv}\{w K^\circ \cup \sqrt{\frac{n}{k}}(D \cap F)\} \; \left(\subset w K^\circ + \sqrt{\frac{n}{k}}(D \cap F)\right).$$

Hence

$$N(D, \rho K^\circ) = N(c_0 D, c_0 \rho K^\circ)$$

$$\leq N(wK^\circ + \sqrt{\frac{n}{k}}(D \cap F), c_0 \rho K^\circ) \tag{7}$$

$$\leq N(\sqrt{\frac{n}{k}}(D \cap F), (c_0 \rho - w)K^\circ)$$

(where we tacitly assumed $c_0\rho - w > 0$). Observe that the polar of $K^\circ \cap F$ (inside the k-dimensional space F) is $P_F K$. Accordingly, if we knew that

$$\log N(P_F K, \beta^{-1}(c_0\rho - w)\sqrt{\frac{k}{n}} P_F D) \leq k, \tag{8}$$

we could conclude from Fact 3.5 that the last member of (7) is bounded by $N(K, D)^\alpha \leq N(U, D)^\alpha \leq 2^{\alpha k}$, as required (above α, β are the constants from Fact 3.5). We now argue as in the proof of Proposition 3.3. If K is symmetric, $K \subset S + D$ implies $K \subset S + (2K \cap D)$. Accordingly, $P_F K \subset P_F S + P_F(2K \cap D)$ (for any $P = P_F$), while, for a generic P_F, $P_F(2K \cap D) \subset 2Cw\sqrt{k/n} P_F D$ by Fact 3.1(a) or (b) and so we get (8) as long as $2Cw \leq \beta^{-1}(c_0\rho - w)$. In particular, $\rho = c_0^{-1}(2C\beta + 1)w \simeq w$ works, as required. In the general case (i.e., K not necessarily symmetric), $(2K \cap D)$ has to be replaced by $\bigcup_{s \in S}(K - s) \cap D$; cf. the end of the proof of Proposition 3.3. $\square$

Remark 4.2 Assuming Conjecture 2.4 (or, by what follows, just Conjecture 2.2), the argument above yields (6), hence Conjecture 2.1, with $a = \alpha$, where α comes from Fact 3.5. In particular, we would obtain then the validity of the case of the Duality Conjecture stated at the beginning of section 2 for any $a > 1$, the price being paid in the magnitude of $b = b(a)$. We also emphasize the that the symmetry hypothesis in Conjecture 2.1 is not used (at least if one doesn't worry about the exact value of the constant a), we leave it there just "for historical reasons." (In any case, that hypothesis can be "disposed of" formally, see the comments following Fact 3.5.) $\square$

The implication Conjecture 2.2 $\Rightarrow$ Conjecture 2.4. Let n, k, S, K be as in Conjecture 2.4. Assuming the validity of Conjecture 2.2, we must show that $\sqrt{\frac{n}{k}}M^*(K \cap D)$ can not be arbitrarily large. Accordingly, throughout the argument we may assume that that quantity is larger than an arbitrary pre-assigned numerical constant (as otherwise we would have been done). Let us denote $n_1 = (\varepsilon_0 M^*(K \cap D))^2 n$ (where ε_0 comes from Fact 3.1; as usual, we pretend that n_1 is an integer), then $\sqrt{\frac{n}{k}}M^*(K \cap D) = \varepsilon_0^{-1}\sqrt{\frac{n_1}{k}}$; clearly we may assume that n_1/k is "large". Apply Fact 3.1 with $m = n_1$. This yields $K_0 = PK$, of which we may think to be contained in $\mathbb{R}^{n_1}$, such that (by the part (b) of the Fact) $K_0 \supset P(K \cap D_n) \supset c\sqrt{\frac{n_1}{n}}D_{n_1}$, while at the same time, by the part (a) of the Fact, $\log N(K_0, 2C\sqrt{\frac{n_1}{n}}D_{n_1}) \leq k$ (as in earlier arguments, we use here the equality $K = \bigcup_{s \in S} s + (K - s \cap D)$ and, in the symmetric case, the inclusion $(K - s) \cap D \subset 2K \cap D$, with appropriate modifications if K is not symmetric; see the end of the proof of Proposition 3.3. Now applying Conjecture 2.2 to $K_1 = (2C\sqrt{\frac{n_1}{n}})^{-1}K_0$ and $\gamma = c(2C)^{-1}$ (this can be done since the cardinality of the set of extreme points of K_1 doesn't exceed that of K) we see that we must have $k > c_1(\gamma)n_1$ or $\sqrt{\frac{n}{k}}M^*(K \cap D) = \varepsilon_0^{-1}\sqrt{\frac{n_1}{k}} < \varepsilon_0^{-1}c_1(\gamma)^{-1/2}$, as required. $\qquad\square$

Remark 4.3 As was the case with the prior implication, the above argument is done "for fixed k", i.e., the validity of Conjecture 2.4 for given k, n is derived from the validity of Conjecture 2.2 for the same k and some other n. The same is (more explicitly) true for Proposition 5.1 from the next section. $\quad\square$

5   The "Almost Isometric" Variants

In this section we shall present some refinements of arguments from the preceding section allowing to prove stronger versions of the implication Conjecture 2.2 $\Rightarrow$ Conjecture 2.4, in particular the one announced in Remark 2.3, i.e. requiring the validity of the former for just *one* $\gamma < 1$.

We note first that in the preceding section we did not use the validity of Conjecture 2.2 for *all* $\gamma > 0$, but just for *some* specific (possibly rather small) $\gamma > 0$, depending on the absolute constants c, C from Fact 3.1(a), (b). Moreover, if we use Fact 3.1(c) instead of (b), an easy modification of the argument shows that we may derive Conjecture 2.4 from Conjecture 2.2 being valid for *some* fixed $\gamma < \frac{1}{2}$.

To get the "almost isometric" variant (any *fixed* $\gamma < 1$) we must work slightly harder; let us state it here for future reference.

Proposition 5.1 *Suppose that there exist constants* $\gamma, \tau \in (0, 1)$ *such that, for every* $n \in \mathbb{N}$ *and* $K = \mathrm{conv}\, S \subset \mathbb{R}^n$ *verifying* $\max\{\log |S|, \log N(K, D)\} \leq \tau n$ *one has* $\gamma D \not\subset K$. *Then, for all* $n \in \mathbb{N}$ *and* $K = \mathrm{conv}\, S \subset \mathbb{R}^n$ *we have*

$$M^*(K \cap D) \leq w\sqrt{\frac{k}{n}}, \tag{9}$$

where $k = \max\{\log |S|, \log N(K, D)\}$ *and* w *is a constant depending only on* γ *and* τ.

More precisely, if for some $\gamma, \tau \in (0, 1)$, *some* $n_0 \in \mathbb{N}$, *and all* $K = \mathrm{conv}\, S \subset \mathbb{R}^{n_0}$ *(resp., for all* $K = \mathrm{conv}\, S \subset RD \subset \mathbb{R}^{n_0}$; *for some* $R > 0$) *the inequality* $k_0 := \max\{\log |S|, \log N(K, D)\} \leq \tau n_0$ *implies* $\gamma D \not\subset K$, *then, for all* $n \in \mathbb{N}$ *and all* $K = \mathrm{conv}\, S \subset \mathbb{R}^n$ *(resp., for all* $K = \mathrm{conv}\, S \subset RD \subset \mathbb{R}^n$; *same* R) *such that* $\max\{\log |S|, \log N(K, D)\} \leq k_0$, *one has* $M^*(K \cap D) \leq w\sqrt{k_0/n}$, *with* $w \lesssim \tau^{-1/2}(1 - \gamma)^{-1}$.

Proof. It is enough to prove the second statement. Observe first that if, for all $K = \text{conv}\, S \subset \mathbb{R}^{n_0}$ with $\max\{\log |S|,\ \log N(K, D)\} \leq k_0$, we have $\gamma D \not\subset K$, then the same is true with n_0 replaced by any $n \geq n_0$: any counterexample $K \subset \mathbb{R}^n$ can be projected back on an n_0-dimensional subspace.

Now choose $\varepsilon > 0$ so that $\gamma \leq (1 - C\varepsilon)/(1 + C\varepsilon)$ and $\varepsilon \leq \min\{\varepsilon_0, 1/2\}$, where C and ε_0 are as in Fact 3.1(c). (Note that the first restriction translates into $\varepsilon \leq C^{-1}\frac{1-\gamma}{1+\gamma} \simeq 1 - \gamma$.) Let $n_1 = \varepsilon^2 M^*(K \cap D)^2 n$ and apply Fact 3.1(c) with $m = n_1$ and $V = K \cap D$ to obtain, for a generic projection P of rank n_1,

$$PD \supset \frac{P(K \cap D)}{(1 + C\varepsilon)M^*(K)} \supset \gamma PD.$$

Without loss of generality we may assume that $\max_{x \in \mathbb{R}^n} M^*((x + D) \cap K)$ is attained at 0 (by Fact 3.6, this is automatically true if K is 0-symmetric) and so, again, generically $P((x + D) \cap K)$ is contained in a ball of radius $(1 + C\varepsilon)M^*(D \cap K)$. This follows from Fact 3.1(c) if $\varepsilon M^*((x+D) \cap K) \geq \sqrt{\frac{n_1}{n}}$ (observe that, by the definition of ε, we have equality if $x = 0$) and holds *a fortiori* if the reverse inequality holds: just enlarge $(x + D) \cap K$ to a convex set – still contained in $x + D$ – for which one has the equality.

We now claim that we must have $k_0 > \min\{c'/2, \tau\}n_1$ (where c' is as in Fact 3.1), from which – in combination with the definition of n_1 – the inequality in the assertion immediately follows. Indeed, if that was not the case, i.e., if $k_0 \leq c'n_1/2$ and $k_0 \leq \tau n_1$, the first of these inequalities would imply that, for a generic projection P (of rank n_1), $P((x+D) \cap K)$ was contained in a ball of radius $(1+C\varepsilon)M^*(D \cap K)$ *simultaneously for all x* in a D-net of K. Setting $K_1 = \frac{PK}{(1+C\varepsilon)M^*(K)} \subset \mathbb{R}^{n_1}$ we see that then (generically) K_1 is contained in an union of less than 2^{k_0} balls of radius 1, hence $\log N(K, D) \leq k_0$, while, on the other hand, K_1 contains (again generically) a ball of radius γ centered at the origin. (Alternatively, we could have applied P to $W = \bigcup_{s \in S}((K - s) \cap D)$, $M^*(W)$ having been estimated using Lemma 8.1; cf. the end of the proof of Proposition 3.3.) Since the number of extreme points of K_1 *never* exceeds that of K (and hence is $\leq 2^{k_0}$), and since $k_0 \leq \tau n_1$ and $\tau n_0 \equiv k_0$ imply $n_0 \leq n_1$, the hypothesis of our statement applies to K_1 (cf. the remark at the beginning of the proof) yielding $K_1 \not\supset \gamma PD$, a contradiction.

To obtain the version of the statement involving R, we observe that, by Fact 3.2(b), $|Ps| \leq 2\sqrt{n_1/n}\,|s|$ *simultaneously for all $s \in S$* provided that n_1/k is larger than some numerical constant C_1 (to ensure the latter, we replace the condition $k > \min\{c'/2, \tau\}n_1$ above by $k > \min\{c'/2, \tau, C_1^{-1}\}n_1$). The radius R_1 of K_1 is then generically less than $\frac{2\sqrt{n_1/n}\,R}{(1+C\varepsilon)M^*(K)} = \frac{2\varepsilon R}{(1+C\varepsilon)} < R$ and so the hypothesis applies to K_1. (In fact we do have a "gain" in the radius as $R_1 \simeq \varepsilon R$, but since we are going to apply the Proposition for a "fixed" γ anyway, this is not going to be exploited.) $\square$

Remark 5.2 The Proposition above states, in essence, that in order to prove the inequality (3) for a fixed k, it is enough to verify whether it holds for the smallest n for which it is non-trivial, i.e., for which the right hand side is less than, say, $\frac{1}{2}$ (or even whether in that case $K \not\supset \frac{1}{2}D$, a weaker condition; same with $\frac{1}{2}$ replaced by any $\gamma < 1$). Going in the opposite direction, from smaller to larger n, is easy: any counterexample in $\mathbb{R}^n$ can be considered as a subset of $\mathbb{R}^m$ for $m > n$; the geometric parameters stay the same (see the comment following Conjecture 2.1), while the functionals $M^*(\cdot)$ in dimensions m and n differ (essentially) by a factor $\sqrt{n/m}$ (this can be seen most easily by replacing, via Fact 3.7, spherical averages with Gaussian means and noting that the latter do not change if we increase dimension). Thus, for a fixed k, the statements of type (3) for various n's are equivalent (in the range of n where the right hand side is *uniformly* non-trivial). $\square$

Remark 5.3 By applying a procedure similar to the proof of Proposition 5.1 for sufficiently small ε, one can show that to deduce Conjecture 2.2 or Conjecture 2.4 it is enough to have a "$\gamma = 1 - \delta$ version" of Conjecture 2.2, where δ is an "arbitrarily good" function of (say) $\frac{k}{n}$. A sample form:

Conjecture 2.2 (or Conjecture 2.4) *is equivalent to the following:*

If k, n, S, K, *are as in* Conjecture 2.4, *then* $(1 - (k/n)^2)D \not\subset K$.

$$(10)$$

Indeed, suppose the above holds and we have a configuration which violates Conjecture 2.2 for some fixed $\gamma < 1$. Apply the previous argument with $\varepsilon = \alpha(\frac{k}{n})^{2/5}$, where α is a small constant. This leads to a $K_0 \subset \mathbb{R}^{n_1}, n_1 \simeq \varepsilon^2 n$ which admits a D-net of cardinality $\leq 2^k$, (and is spanned by $\leq 2^k$ points) with $(1 - C\varepsilon)D_{n_1} \subset K_0$. One routinely verifies that $C\varepsilon < (\frac{k}{n_1})^2$ if α is properly chosen (n_1 is now the "new" n).

Replacing $\alpha(\frac{k}{n})^{2/5}$ by an appropriate expression, one can obtain an analogue of (10) with $(\frac{k}{n})^2$ replaced by an arbitrary preassigned function of $(\frac{k}{n})$.

Let us note here that, on the other hand, K cannot contain a ball of radius substantially larger than 1. Indeed, a simple volume comparison argument shows that if $\gamma D \subset K$, then $\gamma^n < 2^k$ and so $\gamma \leq 1 + \frac{k}{n}$. $\square$

6   Preliminary Estimates for $M^*(K \cap D)$

Our setup is as in Conjecture 2.4, i.e. $K = \operatorname{conv} S \subset \mathbb{R}^n$, $k = \log N(K, D)$, $k_1 = \log |S|$; we shall normally assume that $k_1 \leq k$. We recall that, when needed, we may always assume that n/k is "large".

The first estimate is just a rewording of Fact 3.8.

Proposition 6.1 *If $K \subset RD$, then*

$$M^*(K) \leq CR\sqrt{\frac{k_1}{n}}$$

where C is a numerical constant.

The next estimate is much harder, even though the improvement seems rather minor. Moreover, it is here where the two assumptions, small number of vertices of K and smallness of $N(K, D)$, are being put together.

Proposition 6.2 *If $K \subset RD$, then*

$$M^*(K \cap D) \le C_0 \left(R \sqrt{\frac{k}{n}} \sqrt{\frac{k_1}{n}} \right)^{1/2},$$

where C_0 is a universal constant.

Proof. The conclusion of the Proposition can be rewritten as

$$R < c_0 \delta^2 \frac{n}{\sqrt{k_1 k}} \Rightarrow M^*(K \cap D) < \delta$$

for all $\delta \in (0, 1)$, where $c_0 = C_0^{-2}$. We show first that, in fact, the Proposition is implied by a formally weaker statement

$$R \le c_1 \delta^2 \frac{n}{\sqrt{k_1 k}} \Rightarrow \delta D \not\subset K. \tag{11}$$

(some $c_1 > 0$, with C_0 depending on c_1) and that, moreover, it suffices to obtain (11) just for some *fixed* $\delta \in (0, 1)$, for example for $\delta = 1/6$. To that end, set $M^*(K \cap D) = \eta$. Let $\varepsilon = (2C)^{-1}$, where C comes from Fact 3.1(c), and set $n_1 = \varepsilon^2 \eta^2 n$. We may assume that $n_1 \ge k, k_1$ (as otherwise Proposition 6.2 clearly holds). Consider, as in prior arguments, a generic n_1-dimensional projection PK of K and identify its range with $\mathbb{R}^{n_1}$. We get, by Fact 3.1(c),

$$\frac{\eta}{2} D_{n_1} \subset P(K \cap D_n) \subset \frac{3\eta}{2} D_{n_1}.$$

Rescaling PK by a factor 3η we get an n_1-dimensional body $K_1 \supset D_{n_1}/6$, for which the respective parameters k, k_1 could only decrease. Now, if (11) held for $\delta = 1/6$, it would follow that the radius R_1 of K_1 would have to verify

$$R_1 \ge \frac{c_1}{36} \frac{n_1}{\sqrt{k_1 k}} = c_2 \frac{\eta^2 n}{\sqrt{k_1 k}}. \tag{12}$$

Now *a priori* we know only that $R_1 = (3\eta)^{-1} \cdot \text{radius}(PK) \le (3\eta)^{-1} R$. However, for a generic rank m projection P and for any fixed set Σ with $\log |\Sigma| \le m$, one has $|Px| \simeq \sqrt{\frac{m}{n}} |x|$ *simultaneously* for all $x \in \Sigma$ (by Fact 3.2; this can be made "almost isometric" if $\log |\Sigma|/m$ is "small"). Since the radius of PK is witnessed by $|Ps|$, $s \in S$, and since $\log |S| = k \le n_1$, it would follow that in fact generically

$$\text{radius}(PK) \simeq \sqrt{\frac{n_1}{n}} R \simeq \eta R$$

and so $R_1 \simeq R$, which combined with (12) and the definition of η yields the assertion of Proposition 6.2.

It thus remains to show (11) (in fact just for $\delta = 1/6$, but since that doesn't really simplify the proof, we shall argue the general case). To that end, we need the following special case of "Maurey's Lemma" (see [Pi1]).

Lemma 6.3 *If $S \subset RD$ and $K = \mathrm{conv}\, S$, then, for every $\varepsilon > 0$, setting $s = \lceil (R/\varepsilon)^2 \rceil$, we get that the set*

$$\{ \frac{x_1 + x_2 + \ldots + x_s}{s} : x_j \in S, \ j = 1, \ldots, s \}$$

is an ε-net for K. In particular, if $k_1 = \log |S|$, then $\log N(K, \varepsilon D) \leq 4k_1(\frac{R}{\varepsilon} + 1)^2$.

Now, to prove (11), assume that $\delta D \subset K$. Set $k_2 = 2k$; we shall show that K contains a 2-separated set of cardinality $\geq 2^{k_2}$, which will contradict $\log N(K, D) \leq k$.

Consider a generic k_2-dimensional projection PK of K. Since we are assuming that $\delta D \subset K$, we also have $\delta D_{k_2} \subset PK$. Let Λ be a $\delta/4$-net of PK consisting (for appropriate s) of points of the form $s^{-1}(Px_1 + Px_2 + \ldots + Px_s)$, where $x_j \in S$, $j = 1, \ldots, s$. Since, by the same argument as in the paragraph following (12) and based on Fact 3.2, $\mathrm{radius}(PK) \lesssim \sqrt{k_2/n}R$ in the generic case, Lemma 6.3 implies that it is enough to take $s \approx 4(\sqrt{k_2/n}R/(\delta/4))^2 \simeq \frac{k_2}{n}\frac{R^2}{\delta^2}$, hence $\log |\Lambda| \lesssim \frac{R^2}{\delta^2}\frac{k_2 k_1}{n}$. Now, let Δ be a maximal $\delta/4$-separated subset of Λ; noticing that Δ is a $\frac{\delta}{2}$-net for $PK \supset \delta D_{k_2}$ we infer that $|\Delta| \geq 2^{k_2}$. Set $\tilde{\Lambda} = \{s^{-1}(x_1 + \ldots + x_s) : x_j \in S, j = 1, \ldots, s\} \subset K$, in particular

$$\log |\tilde{\Lambda}| \lesssim \frac{R^2}{\delta^2}\frac{k_2 k_1}{n} \tag{13}$$

and let $\tilde{\Delta}$ be the subset of $\tilde{\Lambda}$ corresponding to elements of Δ. We shall show that the elements of $\tilde{\Delta}$ are generically 2-separated; as $|\tilde{\Delta}| = |\Delta|$, this will yield the desired contradiction.

By Fact 3.2, a generic P shortens a given distance by the factor $\sqrt{k_2/n}$ and so, typically, the distance between two elements of $\tilde{\Delta}$ will be $\gtrsim \sqrt{n/k_2}\frac{\delta}{4} \gg 2$. Accordingly, we can afford to settle for a factor *smaller* than $\sqrt{n/k_2}$, but we need to control *all* distances between elements of $\tilde{\Delta}$. To this end, observe that, by Fact 3.2(b), for a fixed $x \in \mathbb{R}^n \setminus \{0\}$ and for $\lambda > 2$,

$$\mathcal{P}\left(|Px| > \lambda\sqrt{\frac{k_2}{n}}|x| \right) < \exp\left(-c'\lambda^2 k_2\right). \tag{14}$$

Choose $\lambda = \sqrt{\frac{n}{k_2}} \cdot \frac{\delta}{8}$ (we may assume $\lambda > 2$). If we knew that, for all $x \in (\tilde{\Delta} - \tilde{\Delta}) \setminus \{0\}$,

$$|Px| \le \lambda \sqrt{\frac{k_2}{n}} |x| \tag{15}$$

we could infer that, for all such x, one has $|x| > 2$, as required (recall that $|Px| > \delta/4$, the elements of Δ being $\delta/4$-separated). However, we do not know *a priori* which elements of $\tilde{\Lambda}$ will end up in $\tilde{\Delta}$, and so we need to require (15) for a generic P and for *all* $x \in (\tilde{\Lambda} - \tilde{\Lambda})$. By (13) and (14), this can be assured provided that

$$|\tilde{\Lambda} - \tilde{\Lambda}| \cdot \exp\left(-c'\lambda^2 k_2\right) \le \exp\left(C\frac{R^2}{\delta^2}\frac{k_2 k_1}{n}\right) \cdot \exp\left(-c'\lambda^2 k_2\right) \ll 1$$

or, say,

$$C\frac{R^2}{\delta^2}\frac{k_2 k_1}{n} \le \frac{c'}{2}\lambda^2 k_2 = c''n\delta^2 .$$

Considering that $k_2 = 2k$, the above is equivalent, for a properly chosen $c > 0$, to the estimate on R assumed in (11). This concludes the proof of Proposition 6.2.   □

7 "Boxing in" the Set K.

In the preceding section we did obtain some estimates for $M^*(K \cap D)$ provided the set K was "nicely" bounded. Observe that, e.g., the estimate from Proposition 6.2 is nontrivial (i.e., $\ll 1$) if $K \subset RD$ with $R \ll \frac{n}{k}$ ($k_1 \le k$ is tacitly assumed). However, *a priori* no *reasonable* bound on the radius of K is given (one only has, clearly, $R \le 2^{k+}$). We shall show now that in fact it is enough to prove Conjecture 2.2 or Conjecture 2.4 in the case when K is "reasonably" bounded. The approach rests again on considering projections of K, this time *deterministic* ones. For simplicity, *in this section* we shall restrict our analysis to the 0-symmetric case.

Recall that, as explained in the last paragraph of section 5, no K verifying our assumptions can contain $(1 + \frac{k}{n})D$, i.e., there exist $u_1, |u_1| = 1$ such that

$$K \subset \{|\langle \cdot, u_1 \rangle| \le 1 + \frac{k}{n}\}.$$

Let $K' := P_{\{u_1\}^\perp} K$ (the projection onto orthogonal complement of u_1). Clearly K' verifies our standard assumptions in $\{u_1\}^\perp$ and so we can find $u_2 \perp u_1, |u_2| = 1$, such that

$$K_1 \subset \left\{|\langle \cdot, u_2 \rangle| \le 1 + \frac{k}{n-1}\right\} \cap \{u_1\}^\perp$$

and hence

$$K \subset \left\{ |\langle \cdot, u_2 \rangle| \leq 1 + \frac{k}{n-1} \right\}.$$

Continuing in this way we get an orthonormal sequence $u_1, \ldots, u_{n_1}, n_1 = n/2$, such that if $E = [u_1, \ldots, u_{n_1}]$ (where $[\cdot]$ denotes the linear span), then

$$K_1 := P_E K \subset \left\{ |\langle \cdot, u_j \rangle| \leq 1 + \frac{k}{n_1}, j = 1, 2, \ldots, n_1 \right\}.$$

We did thus show

Proposition 7.1 *If, for some $n, k_1, k \in \mathbb{N}$ and $\alpha > 0$ there exists a symmetric set $K = \operatorname{conv} S \subset \mathbb{R}^n$ such that*

$$\log N(K, D) \leq k \text{ and } \log |S| \leq k_1, \text{ while } \alpha D \subset K,$$

then, for $n_1 = n/2$, there exists (a symmetric set) $K_1 \subset \mathbb{R}^{n_1}$ satisfying

$$\log N(K_1, D) \leq k, \ K_1 = \operatorname{conv} S_1, \ \log |S_1| \leq k_1,$$

$$\alpha D_{n_1} \subset K_1 \text{ and } K_1 \subset 2B_\infty^{n_1} \subset 2\sqrt{n_1} D_{n_1},$$

where $B_\infty^{n_1} = [-1, 1]^{n_1}$ is the $\ell_\infty^{n_1}$ ball.

It follows that for our purposes (i.e. proving Conjecture 2.2 or Conjecture 2.4, or (10)) it is enough to consider sets $K \subset RD$, where $R \leq 2\sqrt{n}$, or even $K \subset 2B_\infty^n$.

A more precise analysis yields a slightly better bound on R; we do not *really* use it in the sequel but present here as the argument seems to be of some interest, in particular it can be adapted to show that Conjecture 2.4 holds for multiples of the unit ball of ℓ_1^n (and similar sets). Again, only the estimate on $\log N(K, D)$ is used, and, again, it is enough to produce an $n/2$-dimensional projection of K which is contained in RD. The starting point is a well known formula for the asymptotic order of "covering numbers" of the ℓ_1^n-ball $B_1^n \subset \mathbb{R}^n$. We have, for $R \in [1, \sqrt{n}]$ (see [Sch])

$$\log N(RB_1^n, D) \simeq R^2 \left(\log \frac{n}{R^2} + 1 \right) \tag{16}$$

It follows from (16) by a direct calculation that if $\log N(RB_1^n, D) \leq k$, we must have $R \lesssim \sqrt{\frac{k}{\log \frac{n}{k}}}$. This example is representative for the general case, we have

Proposition 7.2 *In the notation and under the assumptions of Proposition 7.1, we have*

$$K_1 \subset C_1 \sqrt{\frac{k}{\log \frac{n}{k}}} D_{n_1}$$

where C_1 is a universal constant.

Proof. Let $R = C_1 \sqrt{\frac{k}{\log \frac{n}{k}}}$, the constant $C_1 > 0$ to be determined later. By a reasoning analogous to the one which led to Proposition 7.1, we see that either there is an $n_1 = n/2$-dimensional projection of K contained in RD (in which case we are done), or there exists a sequence $v_1, v_2, \ldots, v_{n_1}$ of elements of K such that

$$\text{dist}(v_j, [v_i : i < j]) > R, \ j = 1, \ldots, n_1.$$

For simplicity, let us assume (as we may) that the Gramm-Schmidt orthonormalization applied to (v_j) yields the standard basis (e_j), and so

$$\langle v_j, e_i \rangle = 0 \ \text{ and } \ \langle v_i, e_i \rangle > R \ \text{ if } \ 1 \leq j < i \leq n_1. \tag{17}$$

Set $T := \text{conv}\{\pm v_j\} \subset K$. We shall show that the covering numbers of T are "roughly" at least as large as those of $RB_1^{n_1}$: first for the ℓ_∞-norm and, as a consequence, for the Euclidean norm. We start by recalling an estimate "dual" to (16)

$$\log N(D, rB_\infty^n) \simeq \frac{\log(r^2 n + 1)}{r^2},$$

valid for $r \in [\frac{1}{\sqrt{n}}, 1]$, and a related one

$$\log N(tB_1^n, B_\infty^n) \simeq t \left(\log \frac{n}{t} + 1 \right)$$

for $t \in [1, n]$, both obtained by, roughly speaking, counting the lattice points contained in respective bodies (in a quite general setting, an essentially equivalent problem to that of calculating the covering numbers $N(\cdot, B_\infty^n)$), cf. [Sch]. Let $t \in [1, n]$; it is elementary to show that (17) implies that the linear map u defined by $ue_i = t^{-1}v_i$ sends the integer lattice $\mathbb{Z}^{n_1}$ to a set which is R/t-separated in the ℓ_∞-norm. At the same time, $u(tB_1^{n_1}) \subset T$ and so

$$\log N \left(T, \frac{R}{2t} B_\infty^{n_1} \right) \geq \log N(tB_1^n, B_\infty^n) \geq ct \log \left(\frac{n}{t} + 1 \right),$$

where $c > 0$ is a numerical constant. On the other hand, denoting $r = \frac{R}{2t}$, one has

$$\log N(T, rB_\infty^{n_1}) \leq \log N(T, D) + \log N(D, rB_\infty^{n_1})$$
$$\leq \log N(T, D) + C_2 \frac{\log(r^2 n + 1)}{r^2},$$

whence

$$\log N(T, D) \geq ct \log \left(\frac{n}{t} + 1 \right) - C_2 \frac{\log(r^2 n + 1)}{r^2}.$$

Choosing t so that the first term on the right is twice bigger than the second (in particular $t \simeq R^2$), we get an estimate (16) with RB_1^n replaced by T. As before, this can be reconciled with $\log N(T, D) \leq \log N(K, D) \leq k$ only if $R \leq C_1 \sqrt{\frac{k}{\log \frac{n}{k}}}$, C_1 depending only on c and C_2. $\qquad \square$

Remark 7.3 The argument above is based on the fact that

$$\log N(K, D) \leq k \Rightarrow \log N(K, rB_\infty^n) \lesssim k$$

if $r \simeq \sqrt{\log \frac{n}{k}/k}$. Accordingly, if we were able to obtain from K, verifying the assumptions of Proposition 7.1 (say, by projections), a body K_1, for which $\log N(K_1, rB_\infty^n) \geq Ak$ for large A, this would yield a contradiction and imply the Geometric Lemma.   □

The last result of this section tells us that, for our purposes, we may additionally assume that $M^*(K)$ is "fairly small" (at least, temporarily, in the symmetric case). We have

Proposition 7.4 *Let* $k_1, k \in \mathbb{N}$ *and* $\alpha > 0$. *Suppose that, for some* $n \in \mathbb{N}$, *there exists a symmetric set* $K = \text{conv } S \subset \mathbb{R}^n$ *(resp. additionally* $K \subset RD$ *for some* $R > 0$*) such that*

$$\log N(K, D) \leq k \text{ and } \log |S| \leq k_1, \text{ while } \alpha D \subset K,$$

then, for $n_1 = n/2$, *there exists (a symmetric set)* $K_1 \subset \mathbb{R}^{n_1}$ *satisfying*

$$\log N(K_1, D_{n_1}) \leq k, K = \text{conv } S_1, \log |S_1| \leq k_1, \alpha D_{n_1} \subset K_1$$

and

$$M^*(K_1) \leq C(1 + \log n),$$

(resp. $M^*(K_1) \leq C(1 + \log \frac{R}{\alpha})$ *and* $K_1 \subset RD_{n_1}$*), where* C *is a universal numerical constant.*

If one replaces the hypothesis $\alpha D \subset K$ *by a weaker one,* $M^*(K \cap D) \geq \alpha$, *one gets a similar conclusion, the only changes being that in the new setting* $n_1 \simeq \alpha^2 n$, $\frac{1}{2}D_{n_1} \subset K_1$ *and* $M^*(K_1) \leq C(1 + \log(\alpha^2 n))$ *.(resp.* $M^*(K_1) \leq C(1 + \log R)$*).*

Proof. By [F-T] and [Pi3],Theorem 2.5, there exists $u \in GL(n)$ such that

$$M(uK) \cdot M^*(uK) \leq C(1 + \log n), \tag{18}$$

where C is a universal numerical constant or, more precisely, such that $M(uK) \cdot M^*(uK)$ does not exceed the so-called K-convexity constant of $(\mathbb{R}^n, \|x\|_K)$; uK is often referred to as the ℓ-*position* of K. It is well-known and easily seen that if $E \subset \mathbb{R}^n$ is an m-dimensional subspace, then $M(B \cap E)$ exceeds $M(B)$ by at most (asymptotically) $\sqrt{\frac{n}{m}}$. (Indeed, for Gaussian averages we have, identifying E with $\mathbb{R}^m$, $\int_E \|x\|_B \, d\gamma_m(x) \leq \int_{\mathbb{R}^n} \|x\|_B \, d\gamma_n(x)$ – essentially by the triangle inequality – and it remains to apply Fact 3.7.) A *fortiori*, the same is true with $B \cap E$ replaced by $P_E(B)$ and, by duality, for $M^*(\cdot)$. Let us choose E, $\dim E = n_1 \geq n/2$, such that $P_E(uD)$ is a ball, say $P_E(uD) = \lambda D_{n_1}$ (we identify E with $\mathbb{R}^{n_1}$). We then have

$$\alpha \lambda D_{n_1} \subset P_E(uK)$$
$$\log N(P_E(uK), \lambda D_{n_1})\} \leq k$$

and so, if we set $K_1 = \lambda^{-1} P_E(uK)$ (again considered as a subset of $\mathbb{R}^{n_1}$, we also drop the subscript n_1 in D_{n_1} in what follows), then

$$\alpha D \subset K_1$$
$$\log N(K_1, D) \leq k \tag{19}$$
$$M(K_1) \cdot M^*(K_1) \leq C_0(1 + \log n).$$

The entropy estimate in (19) implies that the volume of K_1 does not exceed 2^k times the volume of D. As a consequence,

$$M(K_1) \geq 2^{-\frac{k}{n_1}} \geq \frac{1}{2}$$

(this follows just from the Hölder inequality) and so

$$M^*(K_1) \leq C_1(1 + \log n).$$

as required. To settle the variant involving the condition $K \subset RD$ we observe that in that case we obtain (additionally) first $P_E(uK) \subset R\lambda D$ and then, after rescaling, $K_1 \subset RD$, as required.

To get the assertion when just $M^*(K \cap D) \geq \alpha$ is assumed, we argue as in the proof of the implication 2.2 $\Rightarrow$ 2.4 or, more precisely, the proof of Proposition 5.1, cf. Remark 5.2): we first apply to our configuration a generic projection of rank $n_0 \simeq \alpha^2 n$ to obtain K_0, $\frac{1}{2} D_{n_0} \subset K_0 \subset \mathbb{R}^{n_0}$, and then repeat the procedure described above.   □

Remark 7.5 Proposition 7.4 is the only point where symmetry intervenes in a significant way (the arguments of Propositions 7.1 and 7.2 can be routinely modified to yield nonsymmetric variants). Indeed, it is not known whether (18) can be achieved for a general convex body K (via an *affine* map u; see [BLPS] and [Rud] for results to date). We could have approached the issue by using [LTJ] to pass to an $n/2$-dimensional projection of K verifying (18). However, as mentioned already in the paragraph preceding Fact 3.1, our *final* estimates can be formally derived from the symmetric case, and so we decided to take the easy way here. See also Remark 8.3.   □

8   "Combining" the Sets.

We start with the following lemma, which is a variant of Theorem 2 of [Ta2] (cf. [L-T], (3.6)).

Lemma 8.1 Let $(y_j), (A_j), j = 1, \ldots, N$, and $R > 0$, be such that $y_j \in RD$ and $A_j \subset RD$ for all $j \leq N$. Then

$$M^*\left(\bigcup_{j \leq N} (y_j + A_j)\right) \leq \max_{j \leq N} M^*(A_j) + C_0 R \sqrt{\frac{\log N}{n}}. \tag{20}$$

Proof. Rescaling reduces the Lemma to the case when $R = 1$, which we shall assume from now on. We have

$$M^* \left(\bigcup_j (y_j + A_j) \right) \leq \max_{j \leq N} M^*(\{y_j\}) + M^* \left(\bigcup_j A_j \right).$$

Since the first term on the right does not exceed $C\sqrt{\frac{\log N}{n}}$ by Proposition 6.1, it is enough to prove (20) when all y_j's are 0. This in turn follows from the isoperimetric inequality (see [M-S]) : as $A_j \subset D$, the function $\| \cdot \|_{A_j^\circ}$ is 1-Lipschitz and so, for $t > 0$,

$$\mu_n(\|x\|_{A_j^\circ} - M^*(A_j) > t) \leq e^{-nt^2/2},$$

where μ_n is the normalized Lebesgue measure on S^{n-1} (note a slight abuse of notation: we write $\|x\|_{A^\circ} = \max\{\langle x, y \rangle : y \in A\}$ even though this is not necessarily a norm, or even a seminorm; it would be more proper to employ the term "the support function of A" used in geometry). Hence

$$\mu_n(\max_j \|x\|_{A_j^\circ} - \max_j M^*(A_j) > t)$$

$$\leq \mu_n(\max_j (\|x\|_{A_j^\circ} - M^*(A_j)) > t) \leq \min\{N \cdot e^{-nt^2/2}, 1\} \qquad (21)$$

and so

$$M^*(\bigcup_j A_j) = \int \max_j \|x\|_{A_j^\circ} d\mu_n(x)$$

$$\leq \max_j M^*(A_j) + \int_0^\infty \mu_n(\max_j \|x\|_{A_j^\circ} - \max_j M^*(A_j) > t) \, dt$$

$$\leq \max_j M^*(A_j) + C_2 \sqrt{\frac{\log N}{n}},$$

where the last inequality follows easily from (21). $\qquad \square$

From the Lemma we derive the following

Proposition 8.2 *Let $K_1, K_2 \subset \mathbb{R}^n$ be convex sets such that $\log N(K_j, D) \leq k$ for $j = 1, 2$. If K_1, K_2 are symmetric, then*

$$M^*((K_1 + K_2) \cap D) \leq M^*(K_1 \cap D) + M^*(K_2 \cap D) + C\sqrt{\frac{k}{n}}.$$

*In the general case, the functional $M^*(\cdot \cap D)$ needs to be replaced **everywhere** by $\max_{x \in \mathbb{R}^n} M^*((\cdot - x) \cap D)$.*

Proof. Let (x_i) and (y_j) be D-nets of K_1 and K_2 respectively. Then

$$K_1 + K_2 = \bigcup_{i,j}((x_i + D) \cap K_1) + ((y_j + D) \cap K_2)$$

$$= \bigcup_{i,j} x_i + y_j + ((K_1 - x_i) \cap D + ((K_2 - y_j) \cap D).$$

In particular, $(K_1 + K_2) \cap D$ is contained in the "subunion" restricted to $x_i + y_j \subset 3D$. Hence, if K_1, K_2 are symmetric, then, by Lemma 8.1 and Fact 3.6,

$$M^*((K_1 + K_2) \cap D)$$

$$\leq 3C_0\sqrt{\frac{2k}{n}} + \max_{i,j} M^*(((K_1 - x_i) \cap D) + ((K_2 - y_j) \cap D))$$

$$= 3C_0\sqrt{\frac{2k}{n}} + \max_i M^*((K_1 - x_i) \cap D) + \max_j M^*((K_2 - y_j) \cap D)$$

$$\leq C\sqrt{\frac{k}{n}} + M^*(K_1 \cap D) + M^*(K_2 \cap D),$$

as required. The not-necessarily-symmetric case is proved the same way. $\square$

Remark 8.3 The special case of the Proposition with $K_1 = K, K_2 = -K$ shows that if $\log N(K, D) \leq k$, then

$$M^*((K - K) \cap D) \leq 2 \max_{x \in \mathbb{R}^n} M^*((K - x) \cap D) + C\sqrt{\frac{k}{n}}$$

(a variant with $K - K$ replaced by $(K - K)/2$ and without factor 2 on the right hand side also holds, e.g., by the argument of Proposition 8.5 below). As already mentioned in the paragraph preceding Fact 3.1, inequalities going in the opposite direction are even easier. Consequently, when estimating $M^*(K \cap D)$, there is no major difference between a symmetric and a non-symmetric setting. More generally, Propositions 8.2 and 8.5 show that the functional in question is stable with respect to doubling the set S or the cardinality of the 1-net, and justify our occasional lack of rigor when adding a few points to S or not differentiating between k and $k + 1$. $\square$

Clearly, there is a lot of flexibility in applying Lemma 8.1, e.g. for "combining" more than two sets. For example, by iteration one gets (for the sake of brevity, we state this and the next result just in the symmetric case)

Corollary 8.4 *Let $K_1, K_2, \ldots, K_s \subset \mathbb{R}^n$ be symmetric convex sets verifying the assumptions from Proposition 8.2. Then*

$$M^*((K_1 + K_2 + \cdots + K_s) \cap D) \leq M^*(K_1 \cap D) + \cdots + M^*(K_s \cap D) + Cs\sqrt{\frac{k}{n}}.$$

For completeness, we also state a variant of Proposition 8.2 for convex hulls (rather than Minkowski sums), which we do not need for the direct purposes of this paper. Its appeal lies in the fact that the multiplicative constant on the right hand side is 1, a feature that is important in some contexts.

Proposition 8.5 *Let $K_1, K_2 \subset \mathbb{R}^n$ be symmetric convex sets such that $\log N(K_j, D) \leq k$ for $j = 1, 2$. Then*

$$M^*(\text{conv}(K_1 \cup K_2) \cap D) \leq \max\{M^*(K_1 \cap D), M^*(K_2 \cap D)\} + C\sqrt{\frac{k}{n}} + C\sqrt{\frac{\log n}{n}}.$$

Proof. Set $M = \max\{M^*(K_1 \cap D), M^*(K_2 \cap D)\}$. Arguing as in the proof of Proposition 8.2, we get

$$\text{conv}(K_1 \cup K_2) = \bigcup_{i,j,t \in [0,1]} (1 - t)((x_i + D) \cap K_1) + t((y_j + D) \cap K_2)$$

$$= \bigcup_{i,j,t \in [0,1]} (1 - t)x_i + ty_j + ((1 - t)((K_1 - x_i) \cap D) + t((K_2 - y_j) \cap D))$$

Similarly as in Proposition 8.2, to analyze $\text{conv}(K_1 \cup K_2) \cap D$ it is enough to consider only the subsegments of the segments $(1 - t)x_i + ty_j$ that lie in $2D$. Given $\varepsilon > 0$, let S' be an ε-net for the union of such subsegments with $|S'| \leq (1 + 4/\varepsilon)2^{2k}$. Applying Lemma 8.1 gives

$$M^*(\text{conv}(K_1 \cup K_2) \cap D) \leq M + \varepsilon + C_0\sqrt{\frac{\log(1 + 4/\varepsilon) + 2k}{n}}.$$

Optimizing over $\varepsilon > 0$ we get the assertion. $\qquad\qquad\qquad\square$

9   Further Estimates for $M^*(K \cap D)$

Similarly as in section 6, the setup is as in Conjecture 2.4, i.e.

$$K = \text{conv}\, S, \quad k = \max\{\log |S|, \log N(K, D)\} \qquad (22)$$

(we did suppress above the dimension n of the ambient space as it may vary from point to point; cf. Remark 5.2). We recall that the objective is to show that $M^*(K \cap D)$ is "small" if n/k is "large". Thus far we did prove (in section 6, Proposition 6.2) that this holds provided $K \subset RD$ with $R \ll \frac{n}{k}$, while (in section 7, Proposition 7.2) it is shown that one may assume, without loss of generality, that $R \leq C\sqrt{\frac{k}{\log \frac{n}{k}}}$, where C is a numerical constant. Admittedly, the gap between the two estimates is significant. Still, they do allow to deduce our "objective" if $k \ll \frac{n^{2/3}}{(\log n)^{1/3}}$. In this section we shall narrow the gap substantially by proving

Proposition 9.1 *There exists a constant $c > 0$ such that whenever $S, K \subset \mathbb{R}^n$ and k verify (22) and $K \subset RD$, with $R \leq \exp\left(c(\frac{n}{k})^{1/6}\right)$, then $\frac{1}{2}D \not\subset K$.*

Corollary 9.2 *If n, S, K and k are as in Proposition 9.1 and, for some $R \geq 1$, $K \subset RD$, then*

$$M^*(K \cap D) \leq C(1 + \log R)^3 \sqrt{\frac{k}{n}},$$

where C is a universal constant.

Proof. More generally, any condition of the type $R \leq \psi(\sqrt{\frac{n}{k}})$ (for $\psi : \mathbb{R}^+ \to \mathbb{R}^+$, $\psi \nearrow +\infty$) in the theorem translates into an estimate $M^*(K \cap D) \leq C_1 \psi^{-1}(R) \sqrt{k/n}$ (here of course $\psi(x) = \exp(cx^{1/3})$). This follows from the second statement of Proposition 5.1 (with $\gamma = \frac{1}{2}$): the condition $R \leq \psi(\sqrt{\frac{n}{k}})$ (which assures $\frac{1}{2}D \not\subset K$) translates into $k \leq (\psi^{-1}(R))^{-2}n$ and so the hypothesis of that statement is satisfied with $\tau = (\psi^{-1}(R))^{-2}$, which yields $w \leq C_1 \psi^{-1}(R)$ in (9), as required. $\square$

The next two corollaries summarize the progress obtained in this note towards the Geometric Lemma and the Duality Conjecture, and so we state them as theorems.

Theorem 9.3 *There exists a constant $C > 0$ such that if $S, K \subset \mathbb{R}^n$ and k are as in (22), then $M^*(K \cap D) \leq C(1 + \log k)^3 \sqrt{k/n}$. In particular, there exists a constant $c > 0$ such that if $k \leq c(1 + \log n)^{-6}n$, then $M^*(K \cap D) < \frac{1}{2}$ (and, consequently, $\frac{1}{2}D \not\subset K$).*

Proof. Consider first the case when K is symmetric. Let $w = w(k)$ be the smallest constant such that the inequality $M^*(K \cap D) \leq w\sqrt{k/n}$ holds for all $n \geq k$ and for all S, K verifying the hypothesis (22) with K symmetric. Consider K, for which $M^*(K \cap D) = w\sqrt{k/n}$ (by compactness, the supremum of $M^*(K \cap D)/\sqrt{k/n}$ is achieved; one could of course devise an argument not using that fact). By the last part of Proposition 7.4, this yields a set $K_1 \subset \mathbb{R}^{n_1}$ with $n_1 \simeq w^2 k$, verifying (22) for the same value of k, and such that $\frac{1}{2}D_{n_1} \subset K_1$. After further halving the dimension (via Proposition 7.1) we may additionally attain $R \leq 2\sqrt{n_1} \simeq w\sqrt{k}$; Corollary 9.2 yields then

$$\frac{1}{2} \leq C(1 + \log(w\sqrt{k}))^3 \sqrt{\frac{k}{n_1}} \simeq (1 + \log(wk))^3 \frac{1}{w},$$

hence $w \lesssim (1 + \log(wk))^3$, which is only possible if $w \lesssim (1 + \log k)^3$, as required. The not-necessarily-symmetric case follows formally, see the comments preceding Fact 3.1. $\square$

Finally, let us restate Theorem 9.3 in terms of covering numbers and entropy numbers (the restating requires only the definitions and a direct application of Proposition 4.1).

Theorem 9.4 *There exist numerical constants $a, C > 0$ such that, for all n, all convex sets $K \subset \mathbb{R}^n$ and all k,*

$$\log N(K, D) \leq k \;\Rightarrow\; \log N(D, CwK^\circ) \leq ak.$$

where $w = w(k) := (1 + \log k)^3$. Similarly, for a compact operator u, whose range is a Hilbert space,

$$e_{ak}(u^*) \leq Cwe_k(u),$$

for all k and with the same $w = w(k)$. Moreover, the second statement (and the first in the symmetric case) holds for any given $a > 1$, with the price being paid then in the magnitude of $C = C(a)$.

This should be compared with the "best to date" duality results for operators of rank $\leq n$ (see Corollary 2.4 of [Pi2]), where an analogous estimate with $w = (1 + (\frac{n}{k})^2)(\log(2 + \frac{n}{k}))^2$ is obtained (our estimate is superior for $k \ll n(\log n)^{-3/2} \log \log n$).

Proof of Proposition 9.1. Let $n, k \in \mathbb{N}$, $R \in [1, \infty)$ and assume that $K = \operatorname{conv} S \subset \mathbb{R}^n$ is such that $\max\{\log|S|, \log N(K, D)\} \leq k$ and $\frac{1}{2}D \subset K \subset RD$. Since the symmetric set $(K - K)/2$ verifies the same hypotheses with k replaced by $2k$, we may and shall assume that K and S were symmetric to begin with, and that $S \ni 0$ (cf. Remark 8.3 and the comments in the paragraph preceding Fact 3.1). By Proposition 7.4, at the price of halving the dimension we may further assume that

$$M^*(K) \leq C'(1 + \log R). \tag{23}$$

To take advantage of various estimates we obtained for $M^*(\cdot \cap D)$ we will, roughly speaking, decompose the set K into a Minkowski sum of "more easily manageable" sets. Let us first demonstrate a single step of such a decomposition. Let $t \in [1, R)$; by Fact 3.4 (Sudakov's inequality) combined with (23) we have

$$k_1 := \log N(K, tD) \lesssim \left(\frac{1 + \log R}{t}\right)^2 n. \tag{24}$$

Consider the corresponding t-net of K, i.e. the set $\mathcal{N}_1$ verifying

$$S_1 + tD \supset K, \log|\mathcal{N}_1| \leq k_1.$$

Assign to each $s \in S$ an $s' \in \mathcal{N}_1$ such that $s \in s' + tD$ and let S_2 consist of all the differences $s - s'$, then $\log|S_2| \leq k$. (K and S being symmetric with $S \ni 0$, we may arrange that the same is true for $\mathcal{N}_1$ and S_2; these conditions are not indispensable for the argument, but they do clarify the picture.) Set $K_1 = \operatorname{conv} \mathcal{N}_1$ and $K_2 = \operatorname{conv} S_2$, then

$$K \subset K_1 + K_2 \tag{25}$$

and

$$K_2 \subset tD. \tag{26}$$

Now, by (25), Proposition 8.2 and the estimates for cardinalities of $\mathcal{N}_1$ and S_2,

$$\frac{1}{2} \le M^*(K \cap D) \le M^*(K_1 \cap D) + M^*(K_2 \cap D) + C_1\sqrt{\frac{k}{n}}.$$

The first term on the right can be now efficiently handled via Proposition 6.2 (k_1 being rather small if t is "large"), while the second term is more susceptible to majorizing even via Proposition 6.1: the radius of K_2 being, by (26), significantly smaller than that of K if t is not "too large". To be absolutely precise, in the process we lost control of $N(K, D)$; we only know that $\log N(K, 2D) \le k$ (which follows trivially from $K_2 \subset K - K = 2K$). This is readily remedied by considering instead the chain of inequalities

$$\frac{1}{4} \le M^*\left(\frac{1}{2}K \cap D\right) \le M^*\left(\frac{1}{2}K_1 \cap D\right) + M^*\left(\frac{1}{2}K_2 \cap D\right) + C_1\sqrt{\frac{k}{n}}. \tag{27}$$

Now, by Proposition 6.2 and (24),

$$M^*\left(\frac{1}{2}K_1 \cap D\right) \lesssim \left(\frac{R}{2}\sqrt{\frac{k}{n}}\sqrt{\frac{k_1}{n}}\right)^{1/2} \lesssim \left(R\sqrt{\frac{k}{n}}\frac{1+\log R}{t}\right)^{1/2}$$

and similarly (as $\log N(\frac{1}{2}K_2, D) \le k$)

$$M^*(\frac{1}{2}K_2 \cap D) \lesssim \left(t\sqrt{\frac{k}{n}}\sqrt{\frac{k}{n}}\right)^{1/2}.$$

Combining these with (27) and optimizing over $t \in [1, R)$ we obtain

$$\frac{1}{4} \lesssim \left(R(1 + \log R)(\frac{k}{n})^{3/2}\right)^{1/4}.$$

This inequality (obtained assuming that $\frac{1}{2}D \subset K \subset RD$) is impossible if $k \le c_1 n/(R(1 + \log R))^{2/3}$ for sufficiently small $c_1 > 0$, leading (cf. the proof of Corollary 9.2) to an estimate

$$M^*(K \cap D) \lesssim (R(1 + \log R))^{1/3}\sqrt{\frac{k}{n}},$$

which already is an improvement over Proposition 6.2 and (cf. the remarks at the beginning of this section) allows to deduce that $M^*(K \cap D)$ is "small" provided $R \ll (\frac{n}{k})^{3/2}/\log\frac{n}{k}$ or if $k \ll \frac{n^{3/4}}{(\log n)^{1/4}}$.

To obtain a better estimate, we – roughly speaking – "decompose" K into a Minkowski sum of $\log R$ sets. Let us return to the setup described in (23) and the paragraph preceding it. To simplify the notation assume that $R = 2^m$ for some $m \in \mathbb{N}$. For $j = 1, 2, \ldots, m$ set $R_j = 2^{m-j}$ and let $\mathcal{N}_j$ be an R_j-net of K; by (23) and Fact 3.4 one may assume that

$$\log |\mathcal{N}_j| \leq C_2 \left(\frac{1 + \log R}{R_j} \right)^2 n.$$

This estimate is clearly not optimal for the last few j's, we improve it by setting $\mathcal{N}_j = S$ when the right hand side exceeds k. In particular we get $\mathcal{N}_m = S$ and

$$\log N(K, R_j D) \leq k_j := \log |\mathcal{N}_j| \leq \min \left\{ C_2 \left(\frac{1 + \log R}{R_j} \right)^2 n, \ k \right\}. \tag{28}$$

As in the "two term decomposition", we set $S_1 = \mathcal{N}_1$, while for $j > 1$ we assign to each $s \in \mathcal{N}_j$ an $s' \in \mathcal{N}_{j-1}$ such that $s \in s' + R_{j-1}D$ and let S_j consist of all the differences $s - s'$; then

$$\log |S_j| \leq k_j. \tag{29}$$

(For a more transparent argument, we may again arrange that all S_j's are symmetric and contain 0, and that $\mathcal{N}_j \subset \mathcal{N}_{j+1}$.) Set $K_j = \text{conv } S_j$, then

$$\frac{K}{2} \subset \frac{K_1}{2} + \frac{K_2}{2} + \cdots + \frac{K_m}{2} \tag{30}$$

and

$$\frac{K_j}{2} \subset \frac{1}{2} R_{j-1}D = R_j D, \quad \log N\left(\frac{K_j}{2}, D\right) \leq k. \tag{31}$$

Similarly as before, by (30), Proposition 8.4 and (28),

$$\frac{1}{4} \leq \sum_{j=1}^{m} M^* \left(\frac{K_j}{2} \cap D \right) + C_3 m \sqrt{\frac{k}{n}}.$$

On the other hand, by Proposition 6.2, (29), (31) and (28),

$$M^* \left(\frac{K_j}{2} \cap D \right) \lesssim \left(R_j \sqrt{\frac{k}{n}} \sqrt{\frac{k_j}{n}} \right)^{1/2}$$

$$\lesssim \left(R_j \sqrt{\frac{k}{n}} \frac{1 + \log R}{R_j} \right)^{1/2} = \left(\sqrt{\frac{k}{n}} (1 + \log R) \right)^{1/2} \tag{32}$$

for $j = 1, 2, \ldots, m$. Combining the last two inequalities gives

$$\frac{1}{4} \lesssim m\sqrt{\frac{k}{n}} + m\left(\sqrt{\frac{k}{n}}(1 + \log R)\right)^{1/2} \lesssim (\log R)^{3/2}\left(\frac{k}{n}\right)^{1/4}, \tag{33}$$

which is impossible if $\frac{k}{n} < c_2(\log R)^{-6}$ (for properly chosen $c_2 > 0$ or, equivalently, $R < \exp\left(c(\frac{n}{k})^{1/6}\right)$ with $c = c_2/\log e$, as required. This completes the proof of Proposition 9.1.   □

Remark 9.5 A significant step in the proof of Proposition 9.1 involved reducing the argument – via Proposition 7.4 – to the case when $M^*(K)$ is "controlled". We wish to point out that even if $M^*(K)$ is bounded by a universal constant, our argument doesn't give estimates substantially better than those contained in Proposition 9.1 (and Corollaries 9.2, 9.3) for the general case. The only improvement is that the exponents $1/6, 3$ and -6 in the respective statements are then replaced by $1/4, 2$ and -4.   □

Remark 9.6 Another reason for the logarithmic factor in, say, Corollary 9.2, is that we use a Sudakov type inequality (Fact 3.4) to estimate the cardinality of nets of K for different "degrees of resolution" and then put these estimates together to majorize $M^*(K \cap D)$. This has an inherent error as it doesn't capture the possible difference between the "Dudley majoration" and the "Sudakov minoration" (cf. [L-T], (12.2) and (12.3)) for the expectation of a supremum of a Gaussian process. The "obvious" way to (attempt to) remedy this problem would be to try to use the majorizing measures ([Ta1]) as the basis for calculation. However, even if we were successful in implementing this program, it appears that we couldn't remove *all* logarithmic factors: the quantities k_j in (32) appear with the exponent $1/4$ as opposed to $1/2$ in the standard "entropy integral" and so the most improvement we could hope for would be replacing m by $m^{1/2}$ in the term $C_9m(\sqrt{k/n}(1 + \log R))^{1/2}$ in (33), resulting in the same "gain" in the exponents as in the previous Remark. Moreover, even if we were able to simultaneously "force" the boundedness of $M^*(K)$, avoid the "Sudakov-Dudley discrepancy" and somehow handle better the term $C_3m\sqrt{k/n}$ in (33) (coming from Proposition 8.4), we would still be left with a $m^{1/2} \simeq (\log R)^{1/2}$ factor at the right end of (33), leading to exponents $1/2, 1$ and -2 in Proposition 9.1 and Corollaries 9.2, 9.3.   □

Remark 9.7 The procedure of decomposing the set K into a Minkowski sum of "manageable" sets is actually somewhat noncanonical. Let us explain that point in the simpler case of "splitting" into a sum of just two terms (by demonstrating which we started our proof of Proposition 9.1). What happens is that the construction of the set K_2 is based on a kind of " retraction" of S to S_2 given by the correspondence $s \to s - s'$, which *a priori* can be a rather irregular map. The following approach is more natural. For a closed convex body $B \subset \mathbb{R}^n$ let $\mathcal{R}_B$ be the metric projection of $\mathbb{R}^n$ onto B (i.e., the "nearest point" map); then $\mathcal{R}_B$ and $\mathcal{Q}_B := I - \mathcal{R}_B$ are *contractions* (all operations being considered with respect to the Euclidean metric). Now redefine S_2 as $\mathcal{Q}_{K_1}(S)$. The prior argument carries over to this setting, in fact we do even have $N(S_2, D) \leq |S| \leq 2^k$, S_2 being a contraction of S. However, later in the process we use Proposition 6.2 to estimate $M^*(K_2 \cap D)$ and for that we need to control $N(K_2, D)$, which is not easily attainable : the maps $\mathcal{Q}_B$ being nonlinear, there is no reason why the set $\mathcal{Q}_{K_1}(K) = \mathcal{Q}_{K_1}(\operatorname{conv} S)$ would contain $K_2 := \operatorname{conv} S_2 = \operatorname{conv} \mathcal{Q}_{K_1}(S)$. Accordingly, this modification of the argument does not improve the estimates obtained in any substantial way. □

References

[BLPS] Banaszczyk W., Litvak A., Pajor A., Szarek S.J. (1999) The flatness theorem for non-symmetric convex bodies via the local theory of Banach spaces. Math. of Operation Research 24(3):728–750

[BPST] Bourgain J., Pajor A., Szarek S.J., Tomczak-Jaegermann N. (1989) On the duality problem for entropy numbers of operators. In: Lindenstrauss J., Milman V.D. (Eds.) Geometric Aspects of Functional Analysis (1987–88), Lecture Notes in Math., 1376, Springer, Berlin-New York, 50–63

[C-S] Carl B., Stephani I. (1990) Entropy, Compactness and the Approximation of Operators. Cambridge Tracts in Mathematics, 98, Cambridge University Press, Cambridge

[F-T] Figiel T., Tomczak-Jaegermann N. (1979) Projections onto Hilbertian subspaces of Banach spaces. Israel J. Math. 33:155–171

[G-M] Giannopoulos A.A., Milman V. Euclidean structure in Banach spaces. In: Johnson W.B., Lindenstrauss J. (Eds.) Handbook on the Geometry of Banach Spaces, Elsevier Science, to appear

[GGP] Gluskin E.D., Gordon Y., Pajor A. Entropy, approximation numbers and other parameters. In: Johnson W.B., Lindenstrauss J. (Eds.) Handbook on the Geometry of Banach Spaces, Elsevier Science, to appear

[GKS] Gordon Y., König H., Schütt C. (1987) Geometric and probabilistic estimates for entropy and approximation numbers of operators. J. Approx. Theory 49(3):219–239

[J-L] Johnson W.B., Lindenstrauss J. (1984) Extensions of Lipschitz mappings into a Hilbert space. In: Conference in Modern Analysis and Probability (New Haven, Conn., 1982), Contemp. Math. 26, Amer. Math. Soc., Providence, R.I., 189–206

[K-M] König H., Milman V. (1987) On the covering numbers of convex bodies. In: Geometrical Aspects of Functional Analysis (1985/86), Lecture Notes in Math., 1267, Springer, Berlin-New York, 82–95

[L-T] Ledoux M., Talagrand M. (1991) Probability in Banach Spaces. Isoperimetry and processes. In: Ergebnisse der Mathematik und ihrer Grenzgebiete (3), 23, Springer-Verlag, Berlin

[LTJ] Litvak A., Tomczak-Jaegermann N. Random aspects of high-dimensional convex bodies. Preprint

[MTJ]   Mankiewicz P., Tomczak-Jaegermann N. Random Banach spaces. In: Johnson W.B., Lindenstrauss J. (Eds.) Handbook on the Geometry of Banach Spaces, Elsevier Science, to appear

[Mi1]   Milman V. (1985) Random subspaces of proportional dimension of finite-dimensional normed spaces: approach through the isoperimetric inequality. Banach Spaces (Columbia, Mo., 1984), Lecture Notes in Math., 1166, Springer, Berlin-New York, 106–115

[Mi2]   Milman V. (1990) A note on a low M^*-estimate. Geometry of Banach Spaces (Strobl, 1989), London Math. Soc. Lecture Note Ser., 158, Cambridge Univ. Press, Cambridge, 219–229

[M-P]   Milman V., Pajor A. Entropy and asymptotic geometry of nonsymmetric convex bodies. Advances in Math., to appear

[M-S]   Milman V., Schechtman G. (1986) Asymptotic theory of finite-dimensional normed spaces. With an appendix by M. Gromov. Lecture Notes in Mathematics, 1200, Springer-Verlag, Berlin-New York

[PT1]   Pajor A., Tomczak-Jaegermann N. (1986) Subspaces of small codimension of finite-dimensional Banach spaces. Proc. Amer. Math. Soc. 97(4):637–642

[PT2]   Pajor A., Tomczak-Jaegermann N. (1989) Volume ratio and other s-numbers of operators related to local properties of Banach spaces. J. Funct. Anal. 87(2):273–293

[Pie]   Pietsch A. (1987) Eigenvalues and s-numbers. Cambridge Studies in Advanced Mathematics, 13, Cambridge University Press, Cambridge-New York

[Pi1]   Pisier G. (1981) Remarques sur un résultat non publié de B. Maurey. (French) Seminar on Functional Analysis, 1980–1981, Exp. No. V, 13 pp., Ecole Polytech., Palaiseau

[Pi2]   Pisier G. (1989) A new approach to several results of V. Milman. J. Reine Angew. Math. 393:115–131

[Pi3]   Pisier G. (1989) The Volume of Convex Bodies and Banach Space Geometry. Cambridge Tracts in Mathematics, 94, Cambridge University Press, Cambridge

[Rud]   Rudelson M. Distances between non-symmetric convex bodies and the MM^*-estimate. Available at http://xxx.lanl.gov/abs/math/9812010, preprint

[Sch]   Schütt C. (1984) Entropy numbers of diagonal operators between symmetric Banach spaces. J. Approx. Theory 40(2):121–128

[Sza]   Szarek S.J. (1990) Spaces with large distance to ℓ_∞^n and random matrices. Amer. J. Math. 112:899–942

[Ta1]   Talagrand M. (1987) Regularity of Gaussian processes. Acta Math. 159(1-2): 99–149

[Ta2]   Talagrand M. (1992) A simple proof of the majorizing measure theorem. Geom. Funct. Anal. 2(1):118–125

[Ta3]   Talagrand M. (1992) Sudakov-type minoration for Gaussian chaos processes. Israel J. Math. 79(2-3):207–224

[TJ1]   Tomczak-Jaegermann N. (1987) Dualité des nombres d'entropie pour des opérateurs à valeurs dans un espace de Hilbert. (French) C. R. Acad. Sci. Paris Sér. I Math. 305(7):299–301

[TJ2]   Tomczak-Jaegermann N. (1989) Banach-Mazur distances and finite-dimensional operator ideals. Pitman Monographs and Surveys in Pure and Applied Mathematics, 38, Longman Scientific & Technical, Harlow

Stabilized Asymptotic Structures and Envelopes in Banach Spaces

V.D. Milman[1] and N. Tomczak-Jaegermann[2]

[1] School of Mathematical Sciences, Tel Aviv University, Tel Aviv 69978, Israel
[2] Department of Mathematical Sciences, University of Alberta, Edmonton, Alberta, Canada T6G 2G1

1 Introduction

In this note we make a few remarks concerning asymptotic structures of infinite-dimensional Banach spaces. These structures carry information on some linear topological and geometric properties that appear "everywhere" "far away" in a Banach space X, and depend upon stabilization at infinity of finite-dimensional subspaces of X. Nevertheless, the phenomena they describe may happen to be purely infinite-dimensional, without finite-dimensional analogues. This approach was first introduced in [Mi1] and [Mi2], and gained importance due to the solutions of the distortion problem: first by a deep construction in [Ts] of a Banach space "custom made" for this purpose, and then in [OS1], [OS2], for "classical" Banach spaces, including Hilbert space and other ℓ_p spaces, $p \neq 1, \infty$.

The asymptotic infinite-dimensional structure was introduced in [MMT] (continuing the line of [MiT]), to help to classify Banach spaces by their asymptotic properties "at infinity". On the other end of the spectrum is the study of the family of *all* finite-dimensional subspaces of a given infinite dimensional Banach space. Investigation of asymptotic behavior when the dimensions tend to infinity reveals regularities behind the increasing diversity of the structure that appears "at the beginning part" of the space. This is the so-called Local Theory, or Asymptotic Finite Dimensional Theory. Although the two theories are not directly connected they complement and influence each other. The last two sections of this note provide further examples of this effect.

Most of the results of this note were known to us shortly after the paper [MMT] was written, however, their usefulness was not clear at that time. The further development of this direction, including the deep and rich survey article [O] and recent papers quoted there, shows an interest and possible usefulness of such results. So we decided to present this note, including short and self-contained proofs, in spite of the fact that some parts of it can be also deduced from the theory developed in the meantime in [KOS] and [OS3].

Supported in part by a BSF grant, an NSF grant and an NSERC grant.

2   Notation and Preliminaries

We use standard Banach space notation found in [LT], and use [MMT] for
the notation of asymptotic structures.

2.1   Asymptotic structures of a Banach space X are defined with respect
to a fixed family $\mathcal{B} = \mathcal{B}(X)$ of infinite-dimensional subspaces of X, which
satisfies two conditions. The filtration condition says that

> For every $X_1, X_2 \in \mathcal{B}$ there exists $X_3 \in \mathcal{B}$ such that $X_3 \subset X_1 \cap X_2$.

The norming condition says that there exists $C < \infty$ such that

$$\|x\| \leq C \sup \|x\|_{X/X_0} \qquad \text{for all } x \in X,$$

where the supremum is taken over all subspaces $X_0 \in \mathcal{B}$ and $\|\cdot\|_{X/X_0}$ denotes
the norm in the quotient space X/X_0.

2.1.1   Natural examples of such families are the family $\mathcal{B}^0(X)$ of all subspaces
of X of finite-codimension, and the families of all tail subspaces with respect
to a fixed basis, a fixed FDD, or a fixed minimal system in X. These families
will be denoted by $\mathcal{B}^t(X)$, if the reference system is clear in the context.
    Recall that $\{u_i\}$ is called a *minimal* system in X, if there exists a se-
quence $\{u_i^*\}$ in X^* such that $\{u_i, u_i^*\}$ is a biorthogonal system. Unless otherwise stated, we shall assume that $\{u_i\}$ is *fundamental* (i.e., $\overline{\text{span}}\,[u_i]_{i \geq 1} =
X$) and that $\{u_i^*\}$ is *norming* (i.e., there exists $C < \infty$ such that $\|x\| \leq
C \sup\{|x^*(x)| \mid \|x^*\| \leq 1, x^* \in \overline{\text{span}}\,[u_i^*]_{i \geq 1}\}$, for all $x \in X$) (we sometimes
say C-norming, to emphasize the constant C). It is well known that every
Banach space contains a minimal (and fundamental and 1-norming) minimal
system. If $\{u_i\}$ is such a minimal system in X, a tail subspace is a subspace
of the form $X^n = \overline{\text{span}}\,[u_i]_{i > n}$, for some $n \in \mathbb{N}$.

2.1.2   If $\mathcal{B}$ is a family satisfying 2.1, we may introduce an equivalent norm
on X in such a way that $\mathcal{B}$ is 1-norming in the new norm. Therefore unless
otherwise stated, throughout this note we shall assume that family $\mathcal{B}$ is 1-
norming. It is then clear that the following condition holds:

> for every finite-dimensional subspace $W \subset X$ and every $\varepsilon > 0$ there is $Z \in \mathcal{B}$
> such that $\|x\| \leq (1+\varepsilon)\|x+z\|$, for all $x \in W$ and $z \in Z$.

2.2   For $n \in \mathbb{N}$, by $\mathcal{M}_n$ we denote the space of all n-dimensional Banach
spaces with fixed normalized monotone bases, and the distance given by (the
logarithm of) the equivalence constant of the bases. Recall that two basic
sequences $\{e_i\}$ and $\{f_i\}$ are C-equivalent, for some $C > 0$, (and we write

$\{e_i\} \overset{C}{\sim} \{f_i\}$) whenever there exist $a, b > 0$ with $ab \leq C$ and such that for all finite sequences of scalars $\{a_i\}$ we have

$$(1/a)\|\sum_i a_i f_i\| \leq \|\sum_i a_i e_i\| \leq b\|\sum_i a_i f_i\|.$$

If $F = \overline{\mathrm{span}}\,[f_i]$ we may write $\{e_i\} \overset{C}{\sim} F$ instead of $\{e_i\} \overset{C}{\sim} \{f_i\}$. Recall that $\mathcal{M}_n$ is a compact metric space.

2.3 Let us recall the language of *asymptotic games* ([MMT]) that is convenient for describing asymptotic structures. In such a game (with respect to a fixed family $\mathcal{B}$ satisfying 2.1) there are two players **S** and **V**. Rules of moves are the same for all games. Set $Y_0 = X$. For $k \geq 1$, in the kth move, player **S** chooses a subspace $Y_k \in \mathcal{B}(X)$, $Y_k \subset Y_{k-1}$, and then player **V** chooses a vector $x_k \in S(Y_k)$ in such a way that the vectors $x_1, \ldots, x_k$ form a basic sequence with the basis constant smaller than or equal to 2. Further rules will ensure that the games considered here will stop after a fixed finite number of steps.

2.4 A space $E \in \mathcal{M}_n$ with the basis $\{e_i\}$ is called an *asymptotic space* for X (with respect to $\mathcal{B}$) if for every $\varepsilon > 0$ we have

$$\forall Y_1 \in \mathcal{B} \ \exists y_1 \in S(Y_1) \ \ \forall Y_2 \in \mathcal{B}, Y_2 \subset Y_1 \ \exists y_2 \in S(Y_2) \ \ \cdots$$
$$\ldots \forall Y_n \in \mathcal{B}, Y_n \subset Y_{n-1} \ \exists y_n \in S(Y_n)$$
$$\{y_1, \ldots, y_n\} \overset{1 \pm \varepsilon}{\sim} E.$$

2.4.1 Any n-tuple $(y_1, \ldots, y_n)$ obtained as above is called *permissible*. The vector y_i is called an ith *winning* move of **V** in a *vector game* for E and ε.

　　The set of all n-dimensional asymptotic spaces for X is denoted by $\{X\}_n$.

2.4.2 For a Banach space X with an FDD, asymptotic spaces with respect to this FDD were studied in [KOS], and they play an important role in [OS3]. They correspond to the family of the tail spaces with respect to this FDD. If $\{F_j\}$ is a FDD for X, i.e., $X = \overline{\mathrm{span}}\,[F_j]_j$ and $\dim F_j < \infty$ for all j, and if $\{u_i\}$ is a minimal system such that there exist integers $m_1 = 0 < m_2 < m_3 < \cdots$ such that $F_j = \mathrm{span}\,[u_i]_{m_j < i \leq m_{j+1}}$ for all j, then it can be easily seen that asymptotic spaces with respect to the FDD are the same as defined by all tail subspaces with respect to $\{u_i\}$.

2.5 It is easy to see that the set $\{X\}_n$ is closed in $\mathcal{M}_n$. It was proved in [MMT], 1.4 and 1.5, that this set can be also characterized in terms of a different asymptotic game called a *subspace game*. Namely, $\{X\}_n$ coincides with the smallest subset $\mathcal{F} \subset \mathcal{M}_n$ such that for every $\varepsilon > 0$ the following moves of an asymptotic game have the property:

$$\exists Y_1 \in \mathcal{B} \;\; \forall y_1 \in S(Y_1) \;\; \exists Y_2 \in \mathcal{B}, Y_2 \subset Y_1 \;\; \forall y_2 \in S(Y_2) \;\; \cdots$$

$$\cdots \exists Y_n \in \mathcal{B}, Y_n \subset Y_{n-1} \;\; \forall y_n \in S(Y_n)$$

$$\exists F \in \mathcal{F} \;\; \{y_1, \ldots, y_n\} \overset{1+\varepsilon}{\sim} F.$$

2.5.1 We refer to any such subspace Y_i as an ith *winning* move of **S** in a *subspace game* for $\{X\}_n$ and ε, and to vectors $\{y_1, \ldots, y_i\}$ (with $1 \le i \le n$) as to the first i moves of **V** in the same subspace game. Note that the basis constant of $\{y_1, \ldots, y_i\}$ is less than or equal to 2.

In particular, an i-tuple $(x_1, \ldots, x_i)$ forms the first i moves of **V** for a subspace game for $\{X\}_n$ and $\varepsilon > 0$ if and only if

$$\exists Y_{i+1} \in \mathcal{B} \;\; \forall y_{i+1} \in S(Y_{i+1}) \;\; \cdots \;\; \exists Y_n \in \mathcal{B}, Y_n \subset Y_{n-1} \;\; \forall y_n \in S(Y_n)$$

$$\exists E \in \{X\}_n \;\; \{x_1, \ldots, x_i, y_{i+1}, \ldots, y_n\} \overset{1+\varepsilon}{\sim} E.$$

2.6 Asymptotic spaces can be described in terms of countably branching trees (*cf.* [MiSh], [KOS], [OS3]). We shall use two kinds of such trees to describe moves of player **V** in asymptotic games: one for a vector game and another for a subspace game.

2.6.1 For $n \in \mathbb{N}$ let T_n be a countably branching tree of length n on $\mathbb{N}$. This means that $T_n = \{(s_1, \ldots, s_j) \mid s_i \in \mathbb{N} \text{ for } 1 \le i \le j, \; 1 \le j \le n\}$, ordered by the relation $(s_1, \ldots, s_j) \prec (t_1, \ldots, t_k)$ if $j \le k$ and $s_i = t_i$ for all $1 \le i \le j$.

2.6.2 For each $E \in \{X\}_n$ and $\varepsilon > 0$ we can build an asymptotic tree $\mathcal{T}(E, \varepsilon)$ on $S(X)$ indexed by T_n and consisting of winning moves of **V** in a vector game for E and ε. That is, $\mathcal{T}(E, \varepsilon) = \{x(s_1, \ldots, s_i) \in S(X) \mid (s_1, \ldots, s_i) \in T_n\}$, with the order on $\mathcal{T}$ induced by T_n.

Denoting by $\{e_i\}$ the natural basis in E we get that any branch $x(s_1) \prec x(s_1, s_2) \prec \ldots \prec x(s_1, \ldots, s_i)$ of $\mathcal{T}(E, \varepsilon)$ is $(1 + \varepsilon)$-quivalent to $\{e_1, \ldots, e_j\}$, for $1 \le j \le n$. Moreover, for any node $\sigma = x(s_1, \ldots, s_i) \in \mathcal{T}(E, \varepsilon)$ with $1 \le i < n$ and any subspace $Y \in \mathcal{B}$ there is a successor $\sigma' = x(s_1, \ldots, s_i, s'_{i+1}) \in \mathcal{T}(E, \varepsilon)$ with $\sigma \prec \sigma'$ and $\sigma' \in Y$.

If $\tau \in T_n$ and $x(\tau) \in \mathcal{T}(E, \varepsilon)$, we refer to $x(\tau)$ as the τ'th *winning* move of **V** in a vector game determined by $\mathcal{T}(E, \varepsilon)$.

2.6.3 For $n \in \mathbb{N}$ and $\varepsilon > 0$ by $\mathcal{T}(n, \varepsilon)$ we denote a tree on $S(X)$ indexed by T_n and consisting of moves of **V** in a subspace game for $\{X\}_n$ and ε. That is, $\mathcal{T}(n, \varepsilon) = \{x(s_1, \ldots, s_i) \in S(X) \mid (s_1, \ldots, s_i) \in T_n\}$, where, in the notation from 2.5, we have $x(s_1) = y_1$, $x(s_1, s_2) = y_2$, $\ldots$ $x(s_1, \ldots, s_n) = y_n$. The order on $\mathcal{T}$ is induced by T_n. We call such a tree a *permissible* tree.

3   Stabilizing General Asymptotic Structure

3.1   Let X be a Banach space. It is natural to ask whether it is possible to achieve a higher level of regularity, or stabilization, for the asymptotic structure of X, while keeping the same set of all asymptotic spaces? It turns out that for a general family $\mathcal{B}(X)$, this can be done by passing to an appropriate subspace. We shall show in Theorem 3.2 below that there exists a subspace $Z \subset X$ with a basis $\{z_i\}$ such that for every $\varepsilon > 0$, any n successive blocks of $\{z_i\}$ far enough are $(1 + \varepsilon)$-equivalent to some asymptotic space, while all asymptotic spaces of X appear also as asymptotic spaces of Z, in fact, they are represented in an asymptotic way on a subsequence of $\{z_i\}$.

For the asymptotic structure determined by a (shrinking or boundedly complete) minimal system it is possible to achieve such a high level of stabilization simultaneously in the space and in the dual. This will be shown in Theorem 4.2 in the next Section.

Results also leading to interesting more regular asymptotic structures were proved in [KOS] and [OS3]. For example for spaces with FDD, it is shown in [KOS] that by a suitable blocking of the given FDD one can obtain a coarser FDD satisfying condition analogous to (i). Paper [OS3] studies the asymptotic structure determined by the family $\mathcal{B}^0(X)$. In both cases, Theorem 3.2 can be deduced from the theory developed in these papers.

3.2   Theorem. *Let X be a Banach space and let $\mathcal{B}$ be a family satisfying conditions 2.1 and 2.1.2. There exists a subspace $Z \subset X$ with a basis $\{z_i\}$ such that*

(i) *for all $n \in \mathbb{N}$ and $\varepsilon > 0$ there exists $N = N(n, \varepsilon)$ such that for any normalized successive blocks $z_N < w_1 < \cdots < w_n$ of $\{z_i\}$ there exists $E \in \{X\}_n$ such that $\{w_1, \ldots, w_n\} \overset{1+\varepsilon}{\sim} E$.*

(ii) *for every $n \in \mathbb{N}$ and every space $E \in \{X\}_n$ the following is true for every $\varepsilon > 0$:*

$$\forall M_1 \in \mathbb{N} \; \exists m_1 > M_1 \; \forall M_2 > M_1 \; \exists m_2 > M_2 \; \cdots$$

$$\cdots \forall M_n > M_{n-1} \; \exists m_n > M_n \; \{z_{m_1}, \ldots, z_{m_n}\} \overset{1+\varepsilon}{\sim} E.$$

Moreover, for $k \in \mathbb{N}$, the basis constant of $\{z_i\}_{i \geq k}$ is less than or equal to $1 + \varepsilon_k$, for some sequence $\varepsilon_k \downarrow 0$.

3.2.1   Let us say, for future reference, that a subspace $Z = \overline{\operatorname{span}}[z_i] \subset X$ is *stabilizing* in X if Z satisfies conditions (i) and (ii) above.

3.3   *Proof.* The proof of the theorem is based on two stabilization procedures. The first describes a construction of a subspace of X for a given sequence of vectors satisfying (i), which ensures that for any choice of a next vector in this subspace, condition (i) will be automatically satisfied. The second procedure describes an inductive choice of vectors to satisfy condition (ii). Fix $\varepsilon_k \downarrow 0$.

3.3.1 Assume that we have constructed for $n \geq 1$ a normalized basic sequence $z_1, \ldots, z_n$ and subspaces $Y_1 \supset Y_2 \supset \ldots \supset Y_n$ from $\mathcal{B}(X)$ such that for every $1 \leq j \leq n$ we have $z_j \in S(Y_j)$ and the following condition holds:

(*) for all $2 \leq k \leq j$ and $1 \leq i \leq \min(j - k + 1, k)$ and all normalized successive blocks $y_1 < \ldots < y_{i-1} \in \operatorname{span}[z_k, \ldots, z_{j-1}]$ and $u \in S(Y_j)$, the i-tuple $(y_1, \ldots, y_{i-1}, u)$ forms the first i moves of **V** for a subspace game for $\{X\}_k$ and ε_k.

Now fix $2 \leq k \leq n + 1$ and $1 \leq i \leq \min(n - k + 2, k)$ and consider any normalized successive blocks $y_1 < \ldots < y_{i-1}$ in $\operatorname{span}[z_k, \ldots, z_n]$. By (*), these blocks form the first $i - 1$ moves of **V** for a subspace game for $\{X\}_k$ and ε_k, and let $Y'_{n+1} \in \mathcal{B}$ be an ith winning move for **S** in this game. Then for every $u \in S(Y'_{n+1})$, $(y_1, \ldots, y_{i-1}, u)$ forms the first i moves of **V** for a subspace game for $\{X\}_k$ and ε_k. We may additionally require, by 2.1.2, that $\|z\| \leq (1 + \varepsilon_k)\|z + u\|$, for any $z \in \operatorname{span}[z_k, \ldots, z_n]$ and $u \in Y'_{n+1}$.

Repeating this procedure for fixed k and i as above and a finite ε'-net of $(i-1)$-tuples $(y_1, \ldots, y_{i-1})$ (with an appropriate ε'), then varying k and i over finite sets, and using the filtration property of $\mathcal{B}$, we get a subspace $Y_{n+1} \in \mathcal{B}$, $Y_{n+1} \subset Y_n$ such that for all k and i as above and all normalized $y_1 < \ldots < y_{i-1}$ in $\operatorname{span}[z_k, \ldots, z_n]$ and $u \in S(Y_{n+1})$, the i-tuple $(y_1, \ldots, y_{i-1}, u)$ forms first i moves of **V** for a subspace game for $\{X\}_k$ and ε_k.

It is clear that with an arbitrary choice of $z'_{n+1} \in S(Y_{n+1})$, the sequence $z_1, \ldots, z_n, z'_{n+1}$ satisfies condition (*) for $j = n + 1$, and hence for all $1 \leq j \leq n + 1$, by the inductive hypothesis. The actual choice of vectors z_j will be made in such a way as to satisfy condition (ii) of the theorem, and will be described below.

3.3.2 Condition (ii) can be formally expressed by countably many conditions in terms of trees representing a "dense" set of asymptotic spaces of X. Then a sequence of vectors $\{z_i\}$ is constructed in such a way that each such tree is represented (preserving the tree structure) on a certain subsequence of $\{z_i\}$. Since all trees are finite, there are no technical problems with approximations and convergence, and we can avoid extensive formalism in describing these procedures.

Since this type of argument will be also used later, let us describe in more detail the set of trees we shall deal with.

For every $n \in \mathbb{N}$ let T_n be the tree of length n defined in 2.6.1. Let $\{\nu_\ell^{(n)}\}_\ell$ be an order preserving enumeration of the nodes of T_n, i.e., for any nodes $\tau = \nu_\ell^{(n)}$ and $\tau' = \nu_{\ell'}^{(n)}$ in T_n, whenever $\tau \prec \tau'$ then $\ell \leq \ell'$. Additionally, assume that if τ and τ' are successors of a common node, that is, $\tau = (s_1, \ldots, s_j, k)$

and $\tau' = (s_1, \ldots, s_j, k')$, for some $(s_1, \ldots, s_j) \in T_n$, then $k \leq k'$ implies $\ell \leq \ell'$. It is not difficult to see that such an enumeration exists.

For every $m, n \in \mathbb{N}$ with $2 \leq n \leq m$, let $\mathcal{A}_{m,n}$ be a finite ε_m-net in $\{X\}_n$, and let $\{E_{m,n,j}\}_j$ be an enumeration of $\mathcal{A}_{m,n}$. Consider the enumeration of $\bigcup \mathcal{A}_{m,n}$ given by $\{E_{m,n,j}\}_{m,n,j}$ taken in the lexicographic order.

Relabel the sequence $\{E_{m,n,j}\}_{m,n,j}$ by $\{F_\mu\}_\mu$. Each $F_\mu = E_{m,n,j}$ has the corresponding asymptotic tree $\mathcal{T}(F_\mu) = \mathcal{T}(E_{m,n,j}, \varepsilon_m)$ (as in 2.6.1 and 2.6.2). The tree $\mathcal{T}(F_\mu)$ is indexed by T_n and denote the associated enumeration $\{\nu_\ell^{(n)}\}_\ell$ of T_n by $\{\nu_\ell(F_\mu)\}_\ell$.

Note that although for different values $\mu \neq \mu'$ the spaces F_μ and $F_{\mu'}$ may happen to the same, but they appear in a context of different approximations, and hence their corresponding trees $\mathcal{T}(F_\mu)$ and $\mathcal{T}(F_{\mu'})$ are different.

3.3.3 We can now describe the inductive construction of the sequence $\{z_j\}$. and is done in Let $Y_1 \in \mathcal{B}$ and $z_1 \in S(Y_1)$ be arbitrary. Vectors $z_2, z_3, \ldots$ will be defined as winning moves of $\mathbf{V}$ in vector games determined by the trees $\mathcal{T}(F_\mu)$'s, for appropriate μ's.

This is done in the following manner. Vector z_2 uses the tree $\mathcal{T}(F_1)$; then vectors z_3 and z_4 use trees $\mathcal{T}(F_1)$ and $\mathcal{T}(F_2)$, respectively; vectors z_5, z_6 and z_7 use trees $\mathcal{T}(F_1)$, $\mathcal{T}(F_2)$ and $\mathcal{T}(F_3)$, respectively, and so on. Furthermore, when a tree $\mathcal{T}(F_\mu)$ is used for the first time in this construction, the corresponding vector is defined as the $\nu_1(F_\mu)$'th winning move of $\mathbf{V}$ in a vector game determined by $\mathcal{T}(F_\mu)$ (see 3.3.2). When $\mathcal{T}(F_\mu)$ is used for the second time, the corresponding vector is a $\nu_2(F_\mu)$'th winning move for $\mathbf{V}$ in the same game, and so on.

It is not then difficult to put the above description into a formula. For $p > 1$ write $p - 1 = s(s+1)/2 + \mu$, for some $1 \leq \mu \leq s+1$ and $s \geq 0$. Then assuming $z_1, \ldots, z_{p-1}$ have been defined, we first define a subspace $Y_p \in \mathcal{B}$, as at the end of 3.3.1. Then a vector $z_p \in S(Y_p)$ is chosen as the $\nu_{s+1}(F_\mu)$'th winning move of $\mathbf{V}$ in a vector game determined by $\mathcal{T}(F_\mu)$.

The comment at the end of 3.3.1 shows that the condition (i) is satisfied. On the other hand, since for every F_μ, the asymptotic tree is represented on a subsequence of $\{z_p\}_p$, then condition (ii) is satisfied as well. $\qquad \square$

3.3.4 Remark. Given an infinite subsequence $M = \{m_p\} \subset \mathbb{N}$, the construction in 3.3.3 could have been done in such a way as to satisfy condition (ii) of the theorem only on M, while the only condition imposed outside of M would be that (i) is valid. For this, in the nth step if $n = m_p$, for some $p \in \mathbb{N}$, we define the vector z_n the same way as in 3.3.3, otherwise, for $n \notin M$, we put $z_n \in Y_n$ arbitrarily.

4   Stabilizing Asymptotic Structure in X and X^*

To tackle the asymptotic structures simultaneously in the space and its dual we shall assume that X has a shrinking minimal system $\{u_i, u_i^*\}$, and the asymptotic structures in X and in X^* are determined by the tail families $\mathcal{B}^t(X)$ and $\mathcal{B}^t(X^*)$ with respect to $\{u_i\}$ and $\{u_i^*\}$, respectively. However the first definition is general.

4.1   Let X be a Banach space with a family $\mathcal{B}(X)$ satisfying 2.1. Let $Z \subset X$ and $V \subset X^*$. We say that V *asymptotically norms* Z if there exists $C < \infty$ such that for every $n \in \mathbb{N}$, $E \in \{Z\}_n$ and $\varepsilon > 0$ the following two conditions are satisfied:

(i) there exist a tree $\mathcal{T}(E, \varepsilon) = \{x(\tau)\}_{\tau \in T_n}$ on $S(Z)$ representing E and a permissible tree $\mathcal{T}_*(n, \varepsilon) = \{w^*(\tau)\}_{\tau \in T_n}$ on $S(V)$ such that for all $\tau, \tau' \in T_n$ we have $|w^*(\tau)(x(\tau))| \geq 1/(C + \varepsilon)$ and $w^*(\tau)(x(\tau')) = 0$ if $\tau \neq \tau'$.

(ii) for all scalar sequences (a_i) there exist a tree $\mathcal{T}(E, \varepsilon) = \{x(s_1, \ldots, s_i)\} \subset S(Z)$ and a permissible tree $\mathcal{T}_*(n, \varepsilon) = \{w^*(s_1, \ldots, s_i)\} \subset S(V)$ and a scalar sequence (b_i) such that for any branch $\gamma = (s_1, \ldots, s_n)$ of T_n, letting $x = \sum a_i x(s_1, \ldots, s_i)$ and $w^* = \sum b_i w^*(s_1, \ldots, s_i)$ we have $\|w^*\|_{X^*} = 1$ and $\|x\| \leq (C + \varepsilon)|w^*(x)|$.

4.2   **Theorem.** *Let X be a Banach space with a shrinking, fundamental and norming minimal system $\{u_i, u_i^*\}$. There exist block subspaces $Z = \overline{\operatorname{span}}[z_i] \subset X$ and $V = \overline{\operatorname{span}}[z_i^*] \subset X^*$ which are stabilizing in X and X^* respectively and such that V asymptotically norms Z. Moreover, if X is reflexive, Z asymptotically norms V as well.*

4.3   We require two known facts.

4.3.1 If X has a shrinking minimal system, there exists a norm on X, 2-equivalent to the original norm, with the property that for every $\delta > 0$ and every tail subspace $\widetilde{V} \in \mathcal{B}^t(X^*)$ there exists a tail subspace $\widetilde{X} \in \mathcal{B}^t(X)$ such that for every $x \in S(\widetilde{X})$ there is $f \in S(\widetilde{V})$ with $f(x) \geq 1 - \delta$. For the proof see [MiSh] (*cf.* also [MMT], 4.1.1). Notice that if X has a basis, this fact is trivial without the shrinking assumption, with constant 2 replaced by the basis constant of X. Without loss of generality we may assume that the original norm on X already has this property.

4.3.2 If X has a minimal system and the norm satisfies 4.3.1, then for every $E \in \{X\}_n$ and $\varepsilon > 0$ the following is true: for every scalar sequence (a_i) there exists a tree $\mathcal{T}(E, \varepsilon) = \{x(s_1, \ldots, s_i)\} \subset S(X)$ representing E and a tree $\mathcal{T}_*(n, \varepsilon) = \{w^*(s_1, \ldots, s_i)\} \subset S(X^*)$, as in 2.6.3, and a scalar sequence (b_i) such that $w^*(\tau)(x(\tau')) = 0$ for any $\tau \neq \tau' \in T_n$. Moreover, for any branch $\gamma = (s_1, \ldots, s_n)$ of T_n, letting $x = \sum a_i x(s_1, \ldots, s_i)$ and $w^* = \sum b_i w^*(s_1, \ldots, s_i)$ we have $\|w^*\|_{X^*} = 1$ and $\|x\| \leq (1 + \varepsilon)|w^*(x)|$.

This is a reformulation of [MiSh] Theorem 2.2, also used in [MMT], 4.5. A similar fact is proved in 5.2 and 5.3 below.

4.3.3 *Proof.* For the proof of the theorem consider the family $\{T(F_\mu)\}$ of trees in $S(X)$ representing asymptotic spaces from $\{X\}_n$, for $n \in \mathbb{N}$, defined in 3.3.2. Let $\{T_*(H_\mu)\}$ be an analogous family of trees in $S(X^*)$ representing $\{X^*\}_n$, for $n \in \mathbb{N}$.

Finally, consider the family $\{(F_\mu, (a_i))\}$, where $\{F_\mu\}$ is a sequence dense in $\bigcup\{X\}_n$, defined in 3.3.2, and, with a fixed F_μ, (a_i) runs over some ε_m-net in the unit ball $B(\ell_\infty^m)$. (Here m and n are determined by F_μ.) We may also assume that each such net contains all standard unit vectors $e_j \in B(\ell_\infty^m)$.

For each couple $(F_\mu, (a_i))$, where (a_i) is not the unit vector, let $T'(F_\mu, (a_i)) \subset S(X)$ and $T'_*(F_\mu, (a_i)) \subset S(X^*)$ be arbitrary trees determined by 4.3.2.

If $(a_i) = e_j$ for some j we define the trees differently to make sure that condition (i) in 4.1 holds. First we use a subspace game in X^* and transport the resulting subspaces to X by 4.3.1. Then combine this choice of subspaces in X with a winning strategy in a vector game in X for F_μ to get a tree $T'(F_\mu, e_j)$ in $S(X)$ and a tree $T'_*(F_\mu, e_j)$ in $S(X^*)$ such that each node $x^*(s_1, \ldots, s_i) \in T'_*(F_\mu, e_j)$ norms (up to an appropriate constant) the corresponding node $x(s_1, \ldots, s_i) \in T'(F_\mu, e_j)$; and nodes in both trees are successive blocks. Moreover, $T'_*(F_\mu, e_j)$ satisfies 2.6.3, so it is a permissible tree. (For a given F_μ, these trees do not depend on the vector e_j and we may consider the same tree for all j.)

4.3.4 Now the end of the proof is very much that same as in 3.3.3 and 3.3.4. We construct subspaces $V_1 \supset V_2 \supset \ldots \supset V_n \supset \ldots$ in X^* and $Y_1 \supset Y_2 \supset \ldots \supset Y_n \supset \ldots$ in X and vectors $z_1, z_1^*, z_2, z_2^*, \ldots$ with $z_n \in S(Y_n)$ and $z_n^* \in S(V_n)$ and such that the sets $\operatorname{supp}(z_n) \cup \operatorname{supp}(z_n^*)$ are successive.

Subspaces are defined by 3.3.1 in X and X^*, and, additionally, Y_n and V_n are related by 4.3.1. The choices of vectors follow two different families of trees. For example, if n is even, z_n and z_n^* are chosen independently of each other to represent trees $T(F_\mu)$ and $T(H_\mu)$, respectively (see 3.3.3 and 3.3.4). For n odd, z_n and z_n^* are chosen to represent the trees $T'(F_m, (a_i))\}$ and $T'_*(F_m, (a_i))\}$, respectively.

It is clear that the subspaces $Z = [z_i]$ and $V = [z_i^*]$ are stabilizing in X and X^*, and V asymptotically norms Z.

The moreover part is obtained by adding in the procedure above the third pair of families of trees determined by $\{(H_\mu, (a_i))\}$, where $\{H_\mu\}$ is a sequence dense in $\bigcup\{X^*\}_n$, and (a_i) runs over appropriate nets in the unit ball $B(\ell_\infty^m)$. □

5   More on Asymptotic Structure

Geometry and structure of the set of all finite-dimensional asymptotic spaces of a given space X is still quite unclear, except for a few rather simple observations in [MMT], 1.8. The following result shows that for two arbitrary asymptotic spaces of X, their direct sum is also an asymptotic space. Under the additional assumption of asymptotic unconditionality this fact is trivial. In the general case, if one of the spaces is ℓ_p^n $(1 \leq p \leq \infty)$, this was proved and used in a fundamental way in [MMT], Section 5. Another interesting application (for the ℓ_p^n case) was given in [HT].

5.1   Proposition. *Let X be a Banach space and let $\mathcal{B}$ be a family satisfying condition 2.1. There exists $C < \infty$ such that the following holds. For $t = 1, 2$, let $E_t \in \{X\}_{n_t}$. Let $\{J_1, J_2\}$ be a partition of $\{1, \ldots, n_1 + n_2\}$ with $|J_t| = n_t$ for $t = 1, 2$, and let $\varepsilon > 0$. There exists an asymptotic space in $\{X\}_{n_1 + n_2}$ with basis $\{f_i\}_i$ such that for $t = 1, 2$ we have $\{f_i\}_{i \in J_t} \overset{1+\varepsilon}{\sim} E_t$ and*

$$\|x_1\| \leq C\|x_1 + x_2\| \qquad \text{for all} \quad x_t \in \mathrm{span}\,[f_i]_{i \in J_t}.$$

5.2   The proof is an asymptotic version of the classical Mazur technique of constructing basic sequences. To emphasize this analogy we formulate its main ingredient in terms of permissible n-tuples, which is sufficient for our purposes.

Lemma. *Let X be a Banach space with a bimonotone basis. Under the assumptions of the proposition there exists a permissible $(n_1 + n_2)$-tuple $(y_i)_i$ such that $\{y_i\}_{i \in J_t} \overset{1+\varepsilon}{\sim} E_t$ for $t = 1, 2$ and such that for every scalar sequence $\{a_i\}_{i \in J_1}$ there exist a permissible n_1-tuple $(g_i)_{i \in J_1}$ in $S(X^*)$ and a sequence of scalars $\{b_i\}_{i \in J_1}$ such that for all $i \in J_1$ we have $g_i(y_j) = 0$ if $1 \leq j \leq n_1 + n_2$, $i \neq j$ and $g_i(y_i) \leq 1$, and $\|\sum b_i g_i\| \leq 1$ and*

$$\left\| \sum_{i \in J_1} a_i y_i \right\| \leq (1 + \varepsilon) \left(\sum_{i \in J_1} a_i y_i \right) \left(\sum_{i \in J_1} b_i g_i \right).$$

5.3   *Proof of Proposition 5.1.* If X has a basis, the proposition follows immediately from the lemma by a standard argument. In general, we prove it first in the stabilizing subspace $Z \subset X$ from Theorem 3.2. Note that this argument does not use the permissibility of functionals (g_i).   □

5.3.1   The proof of the lemma is a modification of the argument from [MiSh] Theorem 2.2, *cf.* also [MMT], 4.5, and therefore we only outline it here. (Let us mention a misprint in the statement of 4.5. in [MMT], where it should say that $g_i(y_i) \leq 1$ for $1 \leq i \leq n$.)

Assume for simplicity that $J_1 = \{1, 3, \ldots, 2n-1\}$ and $J_2 = \{2, 4, \ldots, 2n\}$, for some $n \in \mathbb{N}$, the general case requires only a modification in the notation.

5.3.2   *Proof of Lemma 5.2.* We first sketch the proof without showing the permissibility of (g_i).

Let $Y_1 \in \mathcal{B}^t(X)$ be the first winning move of **S** in the subspace game for $\{X\}_{2n}$ and ε. Consider an asymptotic tree $\mathcal{T}(E_1, \varepsilon) = \{x(s_1, \ldots, s_j)\}$, as in

2.6.2, and let $x(\bar{s}_1) \in S(Y_1)$ be the first winning move of $\mathbf{V}$ in Y_1 in a vector game determined by this tree.

Fix $\delta > 0$ to be defined later. Let $\mathcal{N}$ be a δ-net in the ball in ℓ_∞^m. Fix an arbitrary $\{a_i\}_{i \in J_1} \in \mathcal{N}$, and relabel it by $\{a_1', \dots, a_n'\}$. For every branch $\gamma = (\bar{s}_1, s_2, \dots, s_n)$ of the index tree T_n, let

$$w_\gamma = w(\bar{s}_1, s_2, \dots, s_n) = \sum_j a_j' x(\bar{s}_1, s_2, \dots, s_j).$$

Let $f_\gamma = f(\bar{s}_1, s_2, \dots, s_n) \in S(X^*)$ be a functional almost norming w_γ, $f_\gamma(w_\gamma) \sim \|w_\gamma\|$.

5.3.3 Let $\phi(\bar{s}_1, \dots, s_{n-1})$ be a w^*-cluster point of $\{f(\bar{s}_1, \dots, s_n)\}$ (with $s_n \to \infty$); then let $\phi(\bar{s}_1, \dots, s_{n-2})$ be a w^*-cluster point of $\{\phi(\bar{s}_1, \dots, s_{n-1})\}$ (with $s_{n-1} \to \infty$), and so on, let $\phi(\bar{s}_1)$ be a w^*-cluster point of $\{\phi(\bar{s}_1, s_2)\}$ (with $s_2 \to \infty$). Assume for simplicity that $\phi(\bar{s}_1)$ has a finite support and let $g_1 = \phi(\bar{s}_1)/\|\phi(\bar{s}_1)\|$.

Repeat this construction for all $\{a_i'\} \in \mathcal{N}$ and let $k_1 > \max \operatorname{supp} g_1$ for all g_1's corresponding to $\{a_i'\} \in \mathcal{N}$. We may also assume that $k_1 > \max \operatorname{supp} x(\bar{s}_1)$.

Let $Y_2 \subset Y_1$ be a 2'nd winning move of $\mathbf{S}$ in the subspace game for $\{X\}_{2n}$ and ε. Let $v_1 \in S(Y_2)$ with $k_1 < \min \operatorname{supp} v_1$ be a first winning move for $\mathbf{V}$ in the vector game for E_2 and ε. Clearly, $g_1(v_1) = 0$ for all g_1's constructed above. Assume that v_1 has a finite support and let $k_2 = \max \operatorname{supp} v_1$. This ends the first step of the construction.

5.3.4 For $k \in \mathbb{N}$, by Q_k we denote the natural projection in X^* onto the kth tail subspace. (Note that we shall not use the uniform boundedness of the norms of the Q_k's.)

In the second step, let $Y_3 \subset Y_2$ be a 3'nd winning move of $\mathbf{S}$ in the subspace game for $\{X\}_{2n}$ and ε. Return to an arbitrary $\{a_i'\} \in \mathcal{N}$ and the corresponding functional $\phi(\bar{s}_1)$. Pick $\bar{s}_2$ such that $Q_{k_2}\phi(\bar{s}_1, \bar{s}_2) \sim Q_{k_2}\phi(\bar{s}_1) = \phi(\bar{s}_1)$ (in norm) and such that $x(\bar{s}_1, \bar{s}_2) \in Y_3$ and $k_2 < \min \operatorname{supp} x(\bar{s}_1, \bar{s}_2)$. Pretend that $\phi(\bar{s}_1, \bar{s}_2)$ is finitely supported and let $h_2 = (I - Q_{k_2})\phi(\bar{s}_1, \bar{s}_2)$. Set $g_2 = h_2/\|h_2\|$.

Proceeding the same way as in 5.3.3 let $k_3 > \max \operatorname{supp} \phi(\bar{s}_1, \bar{s}_2)$ for functionals $\phi(\bar{s}_1, \bar{s}_2)$ corresponding to all $\{a_i'\} \in \mathcal{N}$. Finally, pick $Y_4 \subset Y_3$ to be a 4'th winning move of $\mathbf{S}$ in the subspace game for $\{X\}_{2n}$ and ε. Let $v_2 \in S(Y_4)$ be the second winning move for $\mathbf{V}$ in the subspace game for E_2 and ε (with the first move equal to v_1) and such that $k_3 < \min \operatorname{supp} v_2$. Assume that v_2 is finitely supported and let $k_4 = \min \operatorname{supp} v_2$.

Observe that letting $b_1 = \|\phi(\bar{s}_1)\|$ and $b_2 = \|h_2\|$, the functional $b_1 g_1 + b_2 g_2$ is well approximated in norm by $\phi(\bar{s}_1, \bar{s}_2)$.

5.3.5 We repeat the above construction by choosing in each step, for an arbitrary $\{a_i'\} \in \mathcal{N}$, an appropriate branch of the tree $\mathcal{T}(E_1, \varepsilon)$, and choosing the functionals g_i and scalars b_i so that $b_1 g_1 + \ldots + b_j g_j$ is well approximated in norm by $\phi(\bar{s}_1, \ldots, \bar{s}_j)$, for $j = 1, \ldots, n$.

At the end, we let $y_{2j-1} = x(\bar{s}_1, \ldots, \bar{s}_j)$ and $y_{2j} = v_j$, for $j = 1, \ldots, n$ and all conditions of 5.2 easily follow.

To get the permissibility of the functionals (g_i) it is clearly sufficient that we take $Y_1 \in \mathcal{B}^t(X)$ to additionally satisfy that the corresponding tail subspace in X^* is a first winning move of $\mathbf{S}$ in the subspace game for $\{X^*\}_n$ and ε; and then we define $k_1, k_3, \ldots$ large enough determined by subsequent winning moves of $\mathbf{S}$ in this game. $\square$

5.4   The proof of the lemma actually shows a statement about trees representing any asymptotic space $E = E_1 \in \{X\}_n$, without mentioning E_2 (for X with a shrinking minimal system). This is in the spirit of 4.3.2, but stronger: one can choose a tree $\mathcal{T}(E, \varepsilon)$ that works simultaneously for all sequences (a_i) (but the choice of $\mathcal{T}_*(n, \varepsilon)$ still depends on (a_i)). We leave the details to the interested reader.

6   Duality for Envelope Functions

Let us recall the definitions of envelope functions ([MiT] and [MMT], 1.9). The *upper* and the *lower envelopes* for X are functions $r_X(\cdot)$ and $g_X(\cdot)$, respectively, defined on c_{00} (= the space of all scalar sequences eventually equal to 0) by the formulas

$$r_X(a) = \sup \| \sum_i a_i e_i \| \quad \text{and} \quad g_X(a) = \inf \| \sum_i a_i e_i \|,$$

where $a = (a_1, \ldots, a_n, 0 \ldots) \in c_{00}$, and the supremum and the infimum are taken over all natural bases $\{e_i\}$ of asymptotic spaces $E \in \{X\}_n$ and all n.

It immediately follows from the definition of the set $\{X\}_n$ that $r(\cdot)$ and $g(\cdot)$ are unconditional and subsymmetric. It is easy to see that $r(\cdot)$ is a norm on c_{00} and that $g(\cdot)$ satisfies the triangle inequality on disjointly supported vectors. We also let $\hat{g}_X(\cdot)$ be the largest norm on c_{00} such that $\hat{g}_X(a) \leq g_X(a)$ for $a \in c_{00}$.

Also recall that the class of asymptotic ℓ_p spaces is intimately connected with envelopes: X is an asymptotic ℓ_p space for a given $1 \leq p \leq \infty$ (resp., for some $1 \leq p \leq \infty$) if both r_X and g_X (and hence also $\hat{g}_X$) are equivalent to the ℓ_p norm (resp., r_X and $\hat{g}_X$ are equivalent norms) on c_{00} ([MiT], [MMT]).

6.1   The following observation shows that for reflexive Banach spaces, envelopes of X and X^* are in natural duality. This immediately generalizes the duality theorem for reflexive asymptotic ℓ_p spaces ([MMT], 4.3). This is an easy consequence of Theorem 4.2 on stabilizing subspaces.

Recall that the reflexivity assumption is necessary in this context (*cf.*, [MMT], 4.2.3), and, in particular, general asymptotic ℓ_p spaces need not be reflexive, even for $1 < p < \infty$. (Let us also observe that the assumptions of Corollary 5.2 in [MMT] should be exactly the same as in Theorem 4.3.)

Here the asymptotic structures in X and X^* are understood with respect to the families $\mathcal{B}^t(X)$ and $\mathcal{B}^t(X^*)$, respectively, of tail subspaces with respect to a minimal system $\{u_i, u_i^*\}$.

6.1.1 For a Banach space Y, by $r_Y^{\#}$ (resp., $g_Y^{\#}$) we denote the norm on c_{00} dual to r_Y (resp., $\hat{g}_Y$).

Proposition. *If X is a reflexive Banach space with a minimal fundamental and norming system, then $(1/4)r_{X^*}^{\#}(a) \le \hat{g}_X(a) \le 4r_{X^*}^{\#}(a)$ and $(1/4)\hat{g}_{X^*}^{\#}(a) \le r_X(a) \le 4\hat{g}_{X^*}^{\#}(a)$ for $a \in c_{00}$.*

6.2 *Proof.* We first show that if X has a shrinking minimal system and the norm satisfies 4.3.1, then $r_{X^*}^{\#}(a) \le \hat{g}_X(a)$ and $r_X(a) \le g_{X^*}^{\#}(a)$ for $a \in c_{00}$. Let $Z \subset X$ and $V \subset X^*$ be subspaces from Theorem 4.2.

6.2.1 Let $E \in \{X\}_n$ and $\varepsilon > 0$. By condition (i) of 4.1, there exist blocks in $S(Z)$, $\{x_i\} \overset{1+\varepsilon}{\sim} E$ and $\{w_i^*\} \subset S(V)$ representing some asymptotic space from $\{X^*\}_n$ such that $w_i(x_i) \ge 1 - \varepsilon$ and $w_i(x_j) = 0$ if $i \ne j$. Fix $a = (a_i) \in c_{00}$. Thus for any $(b_i) \in c_{00}$ we have

$$(1-\varepsilon)\sum |a_i b_i| \le (\sum b_i w_i^*)(\sum a_i x_i)$$
$$\le \|\sum b_i w_i^*\|_{X^*} \cdot \|\sum a_i x_i\|_X \le r_{X^*}(b)\|\sum a_i x_i\|_X.$$

Taking the supremum over all $b \in c_{00}$ with $r_{X^*}(b) \le 1$ and then the infimum over all $E \in \{X\}_n$ we get $r_{X^*}^{\#}(a) \le g_X(a)$, for all $a \in c_{00}$. Since the left hand side is a convex function, this also implies $r_{X^*}^{\#}(a) \le \hat{g}_X(a)$.

6.2.2 Pick $a = (a_i) \in c_{00}$ and any $E \in \{X\}_n$. By condition (ii) of 4.1, there exist blocks $\{x_i\} \subset S(Z)$ and $\{w_i^*\} \subset S(V)$ representing suitable asymptotic spaces as before, and $b = (b_i) \in c_{00}$ such that $\|\sum b_i w_i^*\|_{X^*} \le 1$ and

$$\|\sum a_i x_i\|_X \le (1+\varepsilon)(\sum b_i w_i^*)(\sum a_i x_i).$$

The expression on the right hand side is less than or equal to

$$(1+\varepsilon)\sum |a_i b_i| \le (1+\varepsilon)g_{X^*}(b)g_X^{\#}(a) \le (1+\varepsilon)g_{X^*}^{\#}(a).$$

Taking the supremum over all $E \in \{X\}_n$ we get $r_X(a) \le g_{X^*}^{\#}(a)$.

6.2.3 The rest of the proof is formal. The space X with the original norm satisfies $r_{X^*}^{\#}(a) \le 4\hat{g}_X(a)$ and $r_X(a) \le 4g_{X^*}^{\#}(a)$. If X is reflexive then we have $r_X^{\#}(a) \le 4\hat{g}_{X^*}(a)$ for $a \in c_{00}$, and hence by duality, $r_X \ge 4\hat{g}_{X^*}^{\#}$, which completes the proof of the second estimates in 6.1.1. The proof of the first ones is similar. □

6.2.4 Note that the proof above does not use the whole strength of Theorem 4.2, but rather only 4.3.1 and 4.3.2.

6.3 It is well known that duality theorems may lead to some complementation results. We give a simple corollary in this direction.

Corollary. *Let X be a reflexive Banach space with a minimal fundamental and norming system. Let $D < \infty$ and assume that there exists a space $E = \text{span}\,[e_i] \in \{X\}_n$ ($n \in \mathbb{N}$) such that $\|\sum a_i e_i\|_E \leq D\hat{g}_X(a)$ for all (a_i). Then for every $\varepsilon > 0$ there exists a permissible n-tuple $\{x_i\} \overset{1+\varepsilon}{\sim} E$ such that $\text{span}\,[x_i]$ is $8D$-complemented in X.*

6.3.1 *Proof.* The argument is standard and we sketch it very briefly. Assume that the norm on X satisfies 4.3.1. Fix $0 < \varepsilon' < \varepsilon$ sufficiently small and let $\{x_i\}$ and $\{w_i^*\}$ be as in 6.2.1. Then the operator $P : X \to \text{span}\,[x_i]$ defined by $P(x) = \sum w_i^*(x)x_i$ is a small perturbation of a projection (depending on ε'). Moreover, using the fact that $r_{X^*} \leq 4g_X^\#$, we get $\|P\| \leq 4(1 + \varepsilon')D$. □

6.3.2 Remark. A similar calculation also shows that under the assumptions of the corollary we get that for all scalars (b_i),

$$(1 - \varepsilon)(4\,D)^{-1}\, r_{X^*}(b) \leq \Big\| \sum b_i w_i^* \Big\|_{X^*} \leq r_{X^*}(b)$$

So we can say a bit imprecisely that if $(\mathbb{R}^n, \hat{g}_X)$ is D-equivalent to an asymptotic space from $\{X\}_n$ then $(\mathbb{R}^n, r_{X^*})$ is $4D$-equivalent to an asymptotic space from $\{X^*\}_n$.

References

[HT]   Habala P., Tomczak-Jaegermann N. Finite representability of ℓ_p in quotients of Banach spaces. Positivity, to appear

[KOS]   Knaust H., Odell E., Schlumprecht Th. On asymptotic structure, the Szlenk index and UKK properties in Banach spaces. Preprint

[LT]   Lindenstrauss J., Tzafriri L. (1977) Classical Banach Spaces I, Sequence Spaces, Springer Verlag

[MMT]   Maurey B., Milman V.D., Tomczak-Jaegermann N. (1995) Asymptotic infinite-dimensional theory of Banach spaces. GAFA Israeli Seminar, Birkhauser Verlag, 149–175

[Mi1]   Milman V.D. (1969) Spectrum of continuous bounded functions on the unit sphere of a Banach space. Funct. Anal. Appl. 3:67–79

[Mi2]   Milman V.D. (1971) The geometric theory of Banach spaces, Part II, Usp. Mat. Nauk 26:73–149 (in Russian), (English translation: Russian Math. Surveys 26:79–163)

[MiSh] Milman V.D., Sharir M. (1979) Shrinking minimal systems and complementation of l_p^n-spaces in reflexive Banach spaces. Proc. London Math. Soc. 39:1–29

[MiT]   Milman V.D., Tomczak-Jaegermann N. (1993) Asymptotic l_p spaces and bounded distortions. In: Banach Spaces, Contemp. Math. 144:173–196

[O]   Odell E. On subspaces, asymptotic structure and distortion of Banach spaces; connections with logic. In: Analysis and Logic, Proc. of Conf. Mons, 1997. LMS Lecture Notes in Mathematics, Cambridge University Press, to appear

[OS1]   Odell E., Schlumprecht T. (1993) The distortion of Hilbert space. Geom. Functional Anal. 3:201–207

[OS2]   Odell E., Schlumprecht T. (1994) The distortion problem. Acta Math. 173:259–281

[OS3]   Odell E., Schlumprecht T. Trees and branches in Banach spaces. Preprint

[Ts]   Tsirelson B.S. (1974) Not every Banach space contains l_p or c_0. Functional Anal. Appl. 8:138–141

On the Isotropic Constant of Non-symmetric Convex Bodies

G. Paouris

Department of Mathematics, University of Crete, Iraklion 714-09, Greece

Abstract. We show that Bourgain's estimate $L_K \leq c\sqrt[4]{n}\log n$ for the isotropic constant holds true for non-symmetric convex bodies as well.

1 Introduction

Let K be a convex body in $\mathbb{R}^n$ with volume $|K| = 1$. Then, K is called *isotropic* if there exists a constant $L_K > 0$ such that

$$\int_K \langle x, \theta \rangle^2 dx = L_K^2 \tag{1}$$

for every $\theta \in S^{n-1}$. It is not hard to check (see [MP] for the origin symmetric case) that every convex body has an isotropic image under $GL(n)$. Moreover, this *isotropic position* is uniquely determined up to orthogonal transformations, hence the *isotropic constant* L_K is an invariant for the class $\{TK : T \in GL(n)\}$.

An important problem asks if there exists an absolute constant $C > 0$ such that $L_K \leq C$ for every isotropic convex body K with centroid at the origin o. This question has many equivalent reformulations: Let us mention the hyperplane problem which asks if every convex body of volume 1 has a hyperplane section through its centroid with "area" greater than an absolute constant.

Bourgain [B] has shown that $L_K \leq c\sqrt[4]{n}\log n$ for every origin symmetric isotropic convex body K in $\mathbb{R}^n$. This is the best known general estimate for the isotropic constant. Dar [D1] proved that $L_K \leq c'\sqrt{n}$ for every convex body with centroid at the origin. The purpose of this note is to extend Bourgain's estimate to non-symmetric isotropic bodies:

Theorem. *If K is an isotropic convex body in $\mathbb{R}^n$, then $L_K \leq c\sqrt[4]{n}\log n$.*

We shall actually follow Bourgain's argument, as presented in [D2]. We will make no assumption about the origin.

2 Proof of the Theorem

In what follows, K is an isotropic, not necessarily symmetric convex body in $\mathbb{R}^n$. The letters c, c', c_1, c_2 etc. will denote absolute positive constants.

Observe that (1) is equivalent to

$$\int_K \langle x, Tx\rangle dx = (\operatorname{tr}T)L_K^2 \tag{2}$$

for every $T \in L(\mathbb{R}^n)$. In particular, if $T \in SL(n)$ is symmetric and positive, the arithmetic-geometric means inequality gives $nL_K^2 \leq (\operatorname{tr}T)L_K^2$, which implies the following:

Lemma 1. *For every symmetric and positive $T \in SL(n)$ we have*

$$nL_K^2 \leq \int_K \langle x, Tx\rangle dx. \quad \Box \tag{3}$$

Lemma 2. *For every $\theta \in S^{n-1}$,*

$$\int_K \exp\left(\frac{|\langle x, \theta\rangle|}{c_1 L_K}\right) dx \leq 2. \tag{4}$$

Proof. This is a consequence of Borell's lemma (see [MS], Appendix III): There exists $c_2 > 0$ such that

$$\left(\int_K |\langle x, \theta\rangle|^p dx\right)^{1/p} \leq c_2 p \int_K |\langle x, \theta\rangle| dx \tag{5}$$

for every $p \geq 1$ and $\theta \in S^{n-1}$. If K is isotropic, then $\int_K |\langle x, \theta\rangle| dx \leq L_K$ for every $\theta \in S^{n-1}$, and the Lemma follows from (5). $\quad\Box$

If V is a convex body in $\mathbb{R}^n$, the *mean width* $w(V)$ of V is the quantity

$$w(V) = \int_{S^{n-1}} \left\{\max_{z \in V}\langle z, \theta\rangle - \min_{z \in V}\langle z, \theta\rangle\right\} \sigma(d\theta),$$

where σ is the rotationally invariant probability measure on S^{n-1}. Well-known results from [L], [FT] and [P] show that for every symmetric convex body V in $\mathbb{R}^n$ there exists $T \in SL(n)$ for which

$$w(TV)w((TV)^\circ) \leq c_3 \log n, \tag{6}$$

where $(TV)^\circ$ is the polar body of TV. We will need the following extension to the non-symmetric case:

Lemma 3. *Let K be a convex body in $\mathbb{R}^n$ with $|K| = 1$. There exists a symmetric and positive $T \in SL(n)$ such that*

$$w(TK) \leq 2c_3 \sqrt{n} \log n. \tag{7}$$

Proof. Consider the difference body $V = K - K$ of K. Then, we can find $T \in SL(n)$ such that $w(TV)w((TV)^\circ) \leq c_3 \log n$. Since mean width is invariant under orthogonal transformations, we may clearly assume that T is symmetric and positive. Now, if $\| \cdot \|$ is the norm induced to $\mathbb{R}^n$ by TV,

$$w((TV)^\circ) = \int_{S^{n-1}} \|\theta\| \sigma(d\theta)$$

$$\geq \left(\int_{S^{n-1}} \|\theta\|^{-n} \sigma(d\theta) \right)^{-1/n} = \left(\frac{|D_n|}{|TV|} \right)^{1/n},$$

where D_n is the Euclidean unit ball. Hence,

$$w(TV) \leq c_3 \left(\frac{|TV|}{|D_n|} \right)^{1/n} \log n \leq c_3 \sqrt{n} |TV|^{1/n} \log n. \tag{8}$$

Observe that $TV = T(K - K) = TK - TK$. From the Rogers-Shephard inequality [RS] we have $|TV| \leq \binom{2n}{n} |TK| \leq 4^n$. Hence,

$$w(TV) \leq 4c_3 \sqrt{n} \log n. \tag{9}$$

Finally,

$$w(TV) = 2 \int_{S^{n-1}} \max_{z \in TK - TK} \langle z, \theta \rangle \sigma(d\theta)$$

$$= 2 \int_{S^{n-1}} \left\{ \max_{z \in TK} \langle z, \theta \rangle - \min_{z \in TK} \langle z, \theta \rangle \right\} \sigma(d\theta)$$

$$= 2w(TK).$$

This shows that $w(TK) \leq 2c_3 \sqrt{n} \log n$. □

The last ingredient of the proof is the Dudley-Fernique decomposition of a convex body A:

Lemma 4. *Let $A \subseteq RD_n$ be a convex body in $\mathbb{R}^n$, where $R > 0$. There exist finite sets $Z_j \subset \mathbb{R}^n$, $j \in \mathbb{N}$ with*

$$\log |Z_j| \leq c_5 n \left(\frac{2^j w(A)}{R} \right)^2,$$

which satisfy the following: For every $x \in A$ and every $m \in \mathbb{N}$ we can find $z_j \in Z_j \cap (3R/2^j) D_n$, $j = 1, \ldots, m$ and $w_m \in (R/2^m) D_n$ such that

$$x = z_1 + \ldots + z_m + w_m.$$

Proof. Recall that the covering number $N(A, tD_n)$ is the smallest integer N for which there exist N translates of tD_n whose union covers A. Using Sudakov's inequality [S] we see that

$$\log N(A, tD_n) \leq \log N(A - A, tD_n) \leq c_4 n \left(\frac{w(A - A)}{t} \right)^2 = 4c_4 n \left(\frac{w(A)}{t} \right)^2.$$

For every $j \in \mathbb{N}$ we find $N_j \subset \mathbb{R}^n$ with $|N_j| = N(A, (R/2^j)D_n)$ such that $A \subset \cup_{y \in N_j}(y + (R/2^j)D_n)$, and set $Z_j = N_j - N_{j-1}$, $j \geq 1$ (and $N_0 = \{o\}$). If $x \in A$ and $m \in \mathbb{N}$, for every $j \leq m$ there exists $y_j \in N_j$ such that $|x - y_j| \leq R/2^j$. We write

$$x = y_1 + (y_2 - y_1) + \ldots + (y_m - y_{m-1}) + (x - y_m),$$

and conclude the proof with $z_j = y_j - y_{j-1}$ and $w_m = x - y_m$. $\square$

Proof of the Theorem. Let K be an isotropic convex body. By Lemma 3, there exists a symmetric and positive $T \in SL(n)$ such that $w(TK) \leq 2c_3\sqrt{n}\log n$. Lemma 1 shows that

$$nL_K^2 \leq \int_K \langle x, Tx \rangle dx \leq \int_K \max_{z \in TK} |\langle z, x \rangle| dx. \tag{10}$$

Let $A = TK$ in Lemma 4, and consider the sets Z_j, $j \in \mathbb{N}$. Then, for every $x \in K$,

$$\max_{z \in TK} |\langle z, x \rangle| \leq \sum_{j=1}^m \max_{z \in Z_j \cap (3R/2^j)D_n} |\langle z, x \rangle| + \max_{w \in (R/2^m)D_n} |\langle w, x \rangle|$$

$$\leq \sum_{j=1}^m \frac{3R}{2^j} \max_{z \in Z_j \cap (3R/2^j)D_n} |\langle \bar{z}, x \rangle| + \frac{R}{2^m}|x|,$$

where $\bar{z}$ is the unit vector parallel to z. Using the above and taking into account the fact that $\int_K |x|dx \leq \sqrt{n}L_K$, we see that

$$nL_K^2 \leq \sum_{j=1}^m \frac{3R}{2^j} \int_K \max_{z \in Z_j} |\langle \bar{z}, x \rangle| dx + \frac{R}{2^m}\sqrt{n}L_K. \tag{11}$$

Now, Lemma 2 shows that for every $t > 0$

$$\mathrm{Prob}\left(x \in K : \max_{z \in Z_j} |\langle \bar{z}, x \rangle| \geq t \right) \leq 2|Z_j| \exp(-t/c_1 L_K),$$

and this implies that

$$\int_K \max_{z \in Z_j} |\langle \bar{z}, x \rangle| dx \leq c_6 L_K \log |Z_j| \leq c_7 n L_K \left(\frac{w(TK)2^j}{R} \right)^2.$$

Inserting this information into (11) we see that

$$nL_K^2 \leq c_8 L_K \left(nw^2(TK)\frac{2^m}{R} + \sqrt{n}\frac{R}{2^m} \right).$$

Choosing $m \in \mathbb{N}$ such that $R/2^m \simeq \sqrt[4]{n}w(TK)$, we get

$$nL_K^2 \leq c_9 n^{\frac{3}{4}} w(TK)L_K,$$

and the estimate $w(TK) \leq 2c_3\sqrt{n}\log n$ completes the proof. $\square$

Remark. If K is isotropic and has its centroid at the origin, then

$$L_K \simeq \int_K |\langle x, \theta \rangle|dx \simeq |K \cap \theta^\perp|^{-1}$$

for every $\theta \in S^{n-1}$ (see [F] for precise estimates). Therefore, in this case, the Theorem implies that all hyperplane sections of K through the origin have "area" greater than $1/c\sqrt[4]{n}\log n$, where $c > 0$ is an absolute constant.

Acknowledgement

The author wants to thank the Erwin Schrödinger International Institute for Mathematical Physics, Vienna, for the hospitality and support.

References

[B]   Bourgain J. (1991) On the distribution of polynomials on high dimensional convex sets. Lecture Notes in Mathematics 1469, Springer, Berlin, 127–137

[D1]   Dar S. (1997) On the isotropic constant of non-symmetric convex bodies. Israel J. Math. 97:151–156

[D2]   Dar S. (1995) Remarks on Bourgain's problem on slicing of convex bodies. In: Geometric Aspects of Functional Analysis, Operator Theory: Advances and Applications 77:61–66

[F]   Fradelizi M. (1999) Hyperplane sections of convex bodies in isotropic position. Beiträge Algebra Geom. 40(1):163–183

[FT]   Figiel T., Tomczak-Jaegermann N. (1979) Projections onto Hilbertian subspaces of Banach spaces. Israel J. Math. 33:155–171

[L]   Lewis D.R. (1979) Ellipsoids defined by Banach ideal norms. Mathematika 26:18–29

[MP]   Milman V.D., Pajor A. (1989) Isotropic position and inertia ellipsoids and zonoids of the unit ball of a normed n-dimensional space. Lecture Notes in Mathematics 1376, Springer, Berlin, 64–104

[MS]   Milman V.D., Schechtman G. (1986) Asymptotic theory of finite dimensional normed spaces. Lecture Notes in Mathematics 1200, Springer, Berlin

[P]   Pisier G. (1982) Holomorphic semi-groups and the geometry of Banach spaces. Ann. of Math. 115:375–392

[RS]   Rogers C.A., Shephard G. (1957) The difference body of a convex body. Arch. Math. 8:220–233

[S]   Sudakov V.N. (1971) Gaussian random processes and measures of solid angles in Hilbert spaces. Soviet Math. Dokl. 12:412–415

Concentration on the ℓ_p^n Ball

G. Schechtman[1] and J. Zinn[2]

[1] Department of Mathematics, Weizmann Institute of Science, Rehovot, Israel
[2] Department of Mathematics, Texas A&M University, College Station, Texas, USA

Abstract. We prove a concentration inequality for functions, Lipschitz with respect to the Euclidean metric, on the ball of ℓ_p^n, $1 \leq p < 2$ equipped with the normalized Lebesgue measure.

1 Introduction

In [SZ] the authors proved an inequality which can be interpreted as giving the right order of the tail distribution of the ℓ_q^n norm on the ℓ_p^n ball equipped with the normalized Lebesgue measure. More precisely and specializing to the case of $q = 2$ and $p = 1$, we proved there that $\mu(\{x : \|x\|_2 > t\})$ is bounded above by $C\exp(-ctn)$ for $t > T/\sqrt{n}$ (where C, c and T are absolute constants) and bounded from below by a similar quantity (with different absolute constants). The measure μ can be either the normalized Lebesgue measure on B_1^n - the ball of ℓ_1^n or the normalized Lebesgue measure on ∂B_1^n - the sphere of ℓ_1^n. (For other p's the relevant measure on the sphere is a different one than the usual surface measure.)

Following a question of M. Gromov, we generalize here this inequality so as to give the right deviation inequality for a general Lipschitz function with respect to the Euclidean metric and for any deviation, i.e., not only for large enough t.

More precisely we prove in Theorem 3.1 that

$$\mu\left(\left\{x : |f(x) - \int f d\mu| > t\right\}\right) \leq C\exp(-ctn)$$

For some absolute positive constants C, c and for all $t > 0$.

We also prove a similar result for μ replaced with the normalized Lebesgue measure on the ℓ_p^n ball, $1 < p < 2$. Here the right hand side of the inequality above takes the form $C\exp(-ct^p n)$. This is treated in Section 4. Somewhat surprisingly, for the function $f(x) = \|x\|_2$ we get better concentration results than for a general Lipschitz function. This is done in Section 5.

The first author was supported in part by BSF. The second author was supported in part by BSF and NSF.

2   Preliminaries

Let λ denote the normalized Lebesgue measure on ∂B_1^n - the unit sphere of ℓ_1^n.

We recall the following known lemma (an equivalent version of which was also used and proved in [SZ]).

Lemma 2.1 *Let $X_1, X_2, \ldots, X_n$ be independent random variables each with density function $\frac{1}{2}e^{-|t|}$ and put $S = (\sum_{i=1}^n |X_i|)$. Then $\left(\frac{X_1}{S}, \frac{X_2}{S}, \ldots, \frac{X_n}{S}\right)$ induces the measure λ on ∂B_1^n. Moreover, $\left(\frac{X_1}{S}, \frac{X_2}{S}, \ldots, \frac{X_n}{S}\right)$ is independent of S.*

We also recall the following fact which is essentially Theorem 3.1 in [SZ] (combine the statement of (1) of Theorem 3 with the last line on page 220).

Theorem 2.2 *([SZ]) There are absolute positive constants T, c such that for all $t > T/\sqrt{n}$, putting $X = (X_1, X_2, \ldots, X_n)$,*

$$\Pr\left(\|X\|_2/S > t\right) \leq \exp(-ctn).$$

We shall also make essential use of a theorem of Talagrand [Tal] giving a fine deviation inequality, with respect to the probability measure induced by X on $\mathbb{R}^n$, for functions which have controlled Lipschitz constants with respect to both the ℓ_2^n and the ℓ_1^n norms. Maurey [Mau] discovered a relatively simple proof of this inequality while Bobkov and Ledoux [BL] found another simple proof and a far reaching generalization of this inequality. We also refer to the lecture notes [Led] by Ledoux which gives a very nice treatment of this and related inequalities.

Theorem 2.3 *([Tal],[Mau],[BL]) Let $F : \mathbb{R}^n \to \mathbb{R}$ be a function satisfying*

$$|F(x) - F(y)| \leq \alpha\|x - y\|_2 \quad and \quad |F(x) - F(y)| \leq \beta\|x - y\|_1.$$

Then

$$\Pr\left(|F(x) - \mathbb{E}F| > r\right) \leq C\exp(-c\min(r/\beta, r^2/\alpha^2))$$

for some absolute positive constants C, c and all $r > 0$. In particular,

$$\Pr\left(\left|\frac{S}{n} - 1\right| > r\right) \leq C\exp(-cn\min(r, r^2)).$$

We refer to [Led], (4.3) on p. 53 from which a similar inequality, with $\mathbb{E}F$ replaced by the median of F follows immediately. Replacing the median with the mean is standard (see e.g. [MS] Prop V.4, page 142). The "In particular" part follows from the fact that the function $F(x_1, \ldots, x_n) = \frac{1}{n}\sum_{i=1}^n |x_i|$ satisfies

$$|F(x) - F(y)| \leq \frac{1}{\sqrt{n}}\|x - y\|_2 \quad and \quad |F(x) - F(y)| \leq \frac{1}{n}\|x - y\|_1.$$

We shall also use the following simple corollary to Theorem 2.3.

Corollary 2.4 *For some absolute positive constants C, c and all $r > 0$,*

$$\Pr\left(\left|\frac{S}{n} - 1\right| > r\right) \leq C\exp(-c\sqrt{n}r).$$

3   The Main Result

Theorem 3.1 *There exist positive constants C, c such that if $f : \partial B_1^n \to \mathbb{R}$ satisfies $|f(x) - f(y)| \leq \|x - y\|_2$ for all $x, y \in \partial B_1^n$ then, for all $t > 0$,*

$$\lambda(\{x : |f(x) - \mathbb{E}(f)| > t\}) \leq C\exp(-ctn).$$

Remark. Considering functions that depend only on the first $n-1$ variables, we easily get a similar statement for the full ball B_1^{n-1}.

The main technical part of the proof of Theorem 3.1 is contained in the following lemma. The probability distribution in its statement is the one introduced in the previous section.

Lemma 3.2 *Put $X = (X_1, X_2, \ldots, X_n)$. Then for some absolute positive constants C, c and for every Lipschitz function with Lipschitz constant 1*

$$\Pr\left(\left|f\left(\frac{X}{S}\right) - f\left(\frac{X}{n}\right)\right| > t\right) \leq Ce^{-ctn}, \text{ for all } 0 < t \leq 2.$$

Proof. Since f has Lipschitz constant 1,

$$\Pr\left(\left|f\left(\frac{X}{S}\right) - f\left(\frac{X}{n}\right)\right| > t\right) \leq \Pr\left(\frac{\|X\|_2}{S}\left|\frac{S}{n} - 1\right| > t\right).$$

We distinguish between two cases.

Case 1. $t \leq \frac{T}{\sqrt{n}}$, where T is from Theorem 2.2.

Using first Corollary 2.4 and then Theorem 2.2

$$\begin{aligned}
\Pr\left(\frac{\|X\|_2}{S}\left|\frac{S}{n} - 1\right| > t\right) &= \mathbb{E}_{\|X\|_2 \atop S}\Pr\left(\left|\frac{S}{n} - 1\right| > \frac{tS}{\|X\|_2}\right)\\
&\leq C\mathbb{E}_{\|X\|_2 \atop S}\exp\left(-\frac{ct\sqrt{n}}{\|X\|_2/S}\right)\\
&\leq C\int_0^{T/\sqrt{n}} \frac{ct\sqrt{n}}{u^2}\exp\left(-\frac{ct\sqrt{n}}{u}\right)du \qquad\qquad (1)\\
&\quad + C\int_{T/\sqrt{n}}^1 \frac{ct\sqrt{n}}{u^2}\exp\left(-\frac{ct\sqrt{n}}{u}\right)\exp(-cun)du. \quad (2)
\end{aligned}$$

Here and elsewhere in this note C and c denote absolute constants, not necessarily the same in each instance.

The first summand, (1), is equal to $C\exp(-\frac{ct\sqrt{n}}{u})\Big|_0^{T/\sqrt{n}} = C\exp(-c'tn)$. For the second summand, (2), observe that the maximum of $\exp(\frac{-ct\sqrt{n}}{u} - cun)$ occurs when $u^2 = t/\sqrt{n}$ and at this point the maximum is: $\exp(-2c\sqrt{t}n^{3/4})$. Hence (2) is dominated by

$$C\exp(-2c\sqrt{t}n^{3/4})\int_{T/\sqrt{n}}^1 \frac{ct\sqrt{n}}{u^2} \le C'tn\exp(-2c\sqrt{t}n^{3/4}).$$

Since, for $t \le \frac{T}{\sqrt{n}}$, $t^{1/2}n^{3/4} \ge T^{-1/2}tn$, the last quantity is bounded by $Ctn\exp(-ctn)$. Since we may assume that $tn \ge 1$, this quantity is bounded by $C\exp(-ctn)$.

Case 2. $\frac{T}{\sqrt{n}} \le t \le 2$.

We write

$$\Pr\left(\frac{\|X\|_2}{S}\left|\frac{S}{n}-1\right| > t\right) = \Pr\left(\frac{\|X\|_2}{S}\left|\frac{S}{n}-1\right| > t \text{ and } \frac{\|X\|_2}{S} \le t\right)$$
$$+ \Pr\left(\frac{\|X\|_2}{S}\left|\frac{S}{n}-1\right| > t \text{ and } \frac{\|X\|_2}{S} > t\right). \quad (3)$$

Using Lemma 2.1 and Theorem 2.3,

$$\Pr\left(\frac{\|X\|_2}{S}\left|\frac{S}{n}-1\right| > t \text{ and } \frac{\|X\|_2}{S} \le t\right)$$
$$\le \mathbb{E}_{\frac{\|X\|_2}{S}}\left(\exp\left(\frac{-cntS}{\|X\|_2}\right)I_{\left(\frac{\|X\|_2}{S}\le t\right)}\right)$$
$$\le C\int_0^t \frac{tn}{u^2}\exp\left(\frac{-ctn}{u}\right)\Pr\left(u \le \frac{\|X\|_2}{S} \le t\right)du$$
$$\le C\int_0^t \frac{tn}{u^2}\exp\left(\frac{-ctn}{u}\right)du.$$

Now,

$$\int_0^{1/\sqrt{n}} \frac{tn}{u^2}\exp(-ctn/u)\,du = C\exp(-ctn^{3/2}).$$

While if T is large enough,

$$\int_{1/\sqrt{n}}^t \frac{tn}{u^2}\exp(-ctn/u)\,du \le \int_{1/\sqrt{n}}^t \frac{tn}{u^2}\exp(-cn)\,du$$
$$= nte^{-cn}[\sqrt{n} - 1/t] \le n^{3/2}te^{-cn} \le Ce^{-c'n}.$$

This takes care of the first summand in (3). For the second summand we use a similar line of inequalities to that of Case 1, using Corollary 2.4 and Theorem 2.2.

$$\Pr\left(\frac{\|X\|_2}{S}\left|\frac{S}{n}-1\right|>t \text{ and } \frac{\|X\|_2}{S}>t\right)$$

$$\leq C\int_t^1 \frac{t\sqrt{n}}{u^2}\exp\left(\frac{-ct\sqrt{n}}{u}-cun\right)du$$

$$\leq C\exp(-ctn)\int_t^1 \frac{t\sqrt{n}}{u^2}\exp\left(\frac{-ct\sqrt{n}}{u}\right)du$$

$$\leq C\exp(-ctn).$$

□

Proof of Theorem 3.1. Given a function on ∂B_1^n which is Lipschitz with constant one, extend it to a function on all of $\mathbb{R}^n$ satisfying the same Lipschitz condition. (There are many ways to do it, for example $\bar{f}(x)=\inf_{y\in\partial B_1^n}(f(y)+\|x-y\|_2)$.) We shall continue to call the extended function f.

Theorem 2.3 implies easily that it Y is an independent copy of $X=(X_1,\ldots,X_n)$ then for all $t>0$

$$\Pr\left(\left|f\left(\frac{X}{n}\right)-f\left(\frac{Y}{n}\right)\right|>t\right)\leq Ce^{-ctn}.$$

Combining this with Lemma 3.2 and Lemma 2.1 we get that

$$\lambda\times\lambda(\{(x,y):\ |f(x)-f(y)|>t)\leq Ce^{-ctn}$$

from which the statement of the theorem follows by standard arguments (see e.g. [MS] Prop V.4, page 142). □

Next we show in Proposition 3.4 below that, for each fixed $t\leq 1/4$, the result of Theorem 3.1 is best possible, except for the choice of the universal constants c,C. Note that the function $f(x)=\|x\|_\infty=\max\{|x_1|,\ldots,|x_n|\}$ is Lipschitz with constant one. We first prove a somewhat weaker result whose proof is much simpler.

Proposition 3.3 *For each $2\log n/n\leq t\leq 1/4$,*

$$\lambda\left(\left\{x:\ \left|\|x\|_\infty-\int\|\cdot\|_\infty d\lambda\right|>t\right\}\right)\geq e^{-4tn}.$$

Proof. Note first that by Lemma 2.1, $\int\|\cdot\|_\infty d\lambda=\mathbb{E}(\max\{|X_1|,\ldots,|X_n|\})/\mathbb{E}S$. It is now easy to deduce that $\int\|\cdot\|_\infty d\lambda<2\log n/n$. Thus,

$$\lambda\left(\left\{x:\ \left|\|x\|_\infty-\int\|\cdot\|_\infty d\lambda\right|>t\right\}\right)$$

$$\geq\Pr\left(\max_{i\leq n}|X_i|\geq(2\log n/n+t)S\right)$$

$$\geq \Pr(|X_1| \geq 2tS) \geq \Pr\left(|X_1| \geq \frac{2t}{1-2t}\sum_{i=2}^{n}|X_i|\right)$$

$$= \mathbb{E}e^{\frac{-2t}{1-2t}\sum_{i=2}^{n}|X_i|} = (1-2t)^{n-1}$$

$$\geq e^{-\frac{2t(n-1)}{1-2t}} \geq e^{-4tn}.$$

$\square$

Proposition 3.4 *For each $t \leq 1/4$,*

$$\lambda\left(\left\{x : \left|\|x\|_\infty - \int \|\cdot\|_\infty d\lambda\right| > t\right\}\right) \geq ce^{-4tn}$$

for some absolute constant $c > 0$.

Proof. By Proposition 3.3 it is enough to assume $t < 2\log n/n$. Let $M = \|X\|_\infty$, $\alpha = t + \frac{\mathbb{E}M}{n} < 1, \beta = \frac{\alpha}{1-\alpha}$ and let $S_k = \sum_{i=1}^{k}|X_i|$.

$$P\left(\frac{M}{S} > \alpha\right) \geq nP(X_n > \alpha S) - \binom{n}{2}P(X_{n-1}, X_n > \alpha S) = I - II.$$

But,

$$I = nP(X_n > \frac{\alpha}{1-\alpha}S_{n-1}) = n\mathbb{E}\exp(-\beta S_{n-1})$$

$$= n(\mathbb{E}\exp(-\beta X))^{n-1} = n\left(\frac{1}{1+\beta}\right)^{n-1}$$

while,

$$II = \binom{n}{2}P(X_{n-1} - \beta X_n > \beta S_{n-2}, X_n - \beta X_{n-1} > \beta S_{n-2})$$

$$\leq \binom{n}{2}P((1-\beta^2)X_{n-1}, (1-\beta^2)X_n > (\beta^2 + \beta)S_{n-2})$$

$$= \binom{n}{2}\left(\mathbb{E}\exp\left(-2\left(\frac{\beta^2 + \beta}{1-\beta^2}\right)X\right)\right)^{n-2} = \binom{n}{2}\left(\frac{1}{1 + \frac{2(\beta^2+\beta)}{1-\beta^2}}\right)^{n-2}$$

$$= \binom{n}{2}\left(\frac{1-\beta}{1+\beta}\right)^{n-2} \leq \frac{n^2}{2}\left(\frac{1}{1+\beta}\right)^{2(n-2)}.$$

Factoring,

$$I - II \geq \frac{n}{(1+\beta)^{n-1}}\left[1 - \frac{n}{2}(1-\alpha)^{n-3}\right]. \tag{4}$$

Notice that

$$\frac{n}{(1+\beta)^{n-1}} = n(1-\alpha)^{n-1} \geq n\exp\left(-(n-1)\frac{\alpha}{1-\alpha}\right)$$

$$= n \exp\left(\frac{\alpha}{1-\alpha}\right) \exp(-(nt/(1-\alpha)) \exp(-\mathbb{E}M/(1-\alpha))$$

$$\geq n^{\frac{-\alpha}{1-\alpha}} \exp\left(\frac{\alpha}{1-\alpha}\right) \exp(-(nt/(1-\alpha)) \exp\left(-\frac{(\mathbb{E}M - \log n)}{1-\alpha}\right).$$

But,

$$\mathbb{E}M = \int_0^\infty P(M > t)\, dt = \int_0^\infty \left[1 - (1 - \exp(-t))^n\right] dt$$

$$= \int_0^1 \frac{1 - u^n}{1 - u}\, du = \int_0^1 \sum_{j=0}^{n-1} u^j\, du = \sum_{j=0}^{n-1} \frac{1}{j+1},$$

so $\mathbb{E}M - \log n$ converges to a positive limit (Euler's constant). Since both $n^{\frac{-\alpha}{1-\alpha}}$ and $\exp(\frac{\alpha}{1-\alpha})$ tend to 1, we get from (4) that

$$I - II \geq c\left(1 - \frac{n}{2}(1 - \alpha)^{n-3}\right) e^{-4tn} \geq c\left(1 - \frac{n}{2}\exp(-\alpha(n-3))\right) e^{-4tn}.$$

Hence, it suffices to have $(n - 3)\alpha \geq \log n$ or $nt \geq \log n - \mathbb{E}M + 3\alpha$. This last quantity is asymptotically negative, since $\log n - \mathbb{E}M$, is asymptotically negative and $\alpha \to 0$. □

4 Concentration on the ℓ_p^n Ball, $1 < p < 2$

Theorem 3.1 and the well known concentration estimate for Lipschitz function on the Euclidean sphere or ball (which is of the form $C\exp(-ct^2 n)$) suggest that a similar result with estimate $C\exp(-ct^p n)$ holds on the ℓ_p^n ball, $1 < p < 2$. As we shall show here this is indeed the case. The proof seems to require another result of Talagrand [Tal2]. Alternatively one can use results from Latała and Oleszkiewicz [LO] in this volume. Actually [LO] was motivated in part by a question of the authors whose motivation was Theorem 4.1 below. Since the proof is very similar to that of the case $p = 1$ we only sketch it.

Theorem 4.1 *There exist positive constants C, c such that if $1 < p < 2$ and $f : \partial B_p^n \to \mathbb{R}$ satisfies $|f(x) - f(y)| \leq \|x - y\|_2$ for all $x, y \in B_p^n$ then, for all $t > 0$,*

$$\lambda(\{x :\ |f(x) - \mathbb{E}(f)| > t\}) \leq C\exp(-ct^p n).$$

Here λ denotes the measure on the ℓ_p^n sphere which assign to each set A the normalized Lebesgue measure, on the ℓ_p^n ball, of the set $\{tA :\ 0 \leq t \leq 1\}$. It is then easy to deduce the same result (with different constants) for the normalized Lebesgue measure on the ℓ_p^n ball.

As we indicated above, the role of Theorem 2.3 will be replaced here with the following result [Tal2] (which is also a combination of Theorems 1 and 2 of [LO]).

Theorem 4.2 ([LO]) *Let $F : \mathbb{R}^n \to \mathbb{R}$ be a function satisfying $|F(x) - F(y)| \le \alpha\|x - y\|_2$, let $\Pr$ denotes the probability distribution on $\mathbb{R}^n$ with density $c_p^n \exp(-|x_1|^p - \cdots - |x_n|^p)$ and denote $S = (|X_1|^p + \cdots + |X_n|^p)^{1/p}$. Then*

$$\Pr(|F(X) - \mathbb{E}F| > r) \le C\exp(-c(r/\alpha)^p)$$

for some absolute positive constants C, c and all $r > 0$. In particular,

$$\Pr\left(\left|\frac{S}{n^{1/p}} - \frac{\mathbb{E}S}{n^{1/p}}\right| > r\right) \le C\exp(-cn^{p/2}r^p).$$

The analogues of Lemma 2.1 and Theorem 2.2 also hold, i.e. with the new interpretation of $\Pr$, λ and S, $\left(\frac{X_1}{S}, \frac{X_2}{S}, \ldots, \frac{X_n}{S}\right)$ induces the measure λ on ∂B_p^n, and $\left(\frac{X_1}{S}, \frac{X_2}{S}, \ldots, \frac{X_n}{S}\right)$ is independent of S. Also

Theorem 4.3 ([SZ]) *There are absolute positive constants T, c such that for all $t > T/n^{1/2-1/p}$, putting $X = (X_1, X_2, \ldots, X_n)$,*

$$\Pr(\|X\|_2/S > t) \le \exp(-ct^p n).$$

Denote $\alpha = \alpha(n, p) = \frac{\mathbb{E}S}{n^{1/p}}$ and note that $\alpha(n, p)$ is bounded away from zero and ∞ by universal constants, for $1 < p < 2$. As for the case $p = 1$, Theorem 4 will follow (using Theorem 4.2) once we establish

Lemma 4.4 *For every Lipschitz function f with constant 1*

$$\Pr\left(\left|f\left(\frac{X}{S}\right) - f\left(\frac{\alpha X}{n^{1/p}}\right)\right| > t\right) \le Ce^{-ct^p n}.$$

For some absolute positive constants C, c and for all $0 < t \le 2$.

Sketch of proof. As in the proof of Lemma 3.2, the proof reduces to estimating

$$\Pr\left(\frac{\|X\|_2}{S}\left|\frac{S}{n^{1/p}} - \alpha\right| > t\right) = \mathbb{E}_{\frac{\|X\|_2}{S}}\Pr\left(\left|\frac{S}{n} - \alpha\right| > \frac{tS}{\|X\|_2}\right)$$

which by Theorem 4.2 is dominated by

$$C\mathbb{E}_{\frac{\|X\|_2}{S}} \exp\left(-\frac{ct^p n^{p/2}}{(\|X\|_2/S)^p}\right)$$

$$= C\int_0^\infty \frac{pct^p n^{p/2}}{u^{p+1}} \exp\left(-\frac{ct^p n^{p/2}}{u^p}\right)\Pr\left(\frac{\|X\|_2}{S} > u\right) du. \qquad (5)$$

Case 1. $t \le Tn^{1/2-1/p}$.

The right hand side of (5) is dominated by

$$\int_0^{Tn^{1/2-1/p}} \frac{pct^p n^{p/2}}{u^{p+1}} \exp\left(-\frac{ct^p n^{p/2}}{u^p}\right) du$$

$$+ \int_{Tn^{1/2-1/p}}^1 \frac{pct^p n^{p/2}}{u^{p+1}} \exp\left(-c\left(\frac{t^p n^{p/2}}{u^p} + u^p n\right)\right) du$$

$$= I + II.$$

I is equal to $\exp(-c't^p n)$ while in II the integrand is dominated by $\frac{ct^p n^{p/2}}{u^{p+1}}$ $\exp(-c't^{p/2}n^{p/4+1/2})$ and thus II is dominated by $t^p n \exp(-ct^{p/2}np/4 + 1/2)$. The bound $t \le Tn^{1/2-1/p}$ implies now that II is also dominated by $\exp(-c't^p n)$.

Case 2. $t > Tn^{1/2-1/p}$.
Write the right hand side of (5) as

$$C\left(\int_0^{Tn^{1/2-1/p}} + \int_{Tn^{1/2-1/p}}^t + \int_t^1\right) \frac{pct^p n^{p/2}}{u^{p+1}} \exp\left(-\frac{ct^p n^{p/2}}{u^p}\right) \Pr\left(\frac{\|X\|_2}{S} > u\right) du$$

$$= I + II + III. \tag{6}$$

Estimating $\Pr\left(\frac{\|X\|_2}{S} > u\right)$ by 1, I is bounded by $\exp(-c't^p n)$. II is bounded by

$$\exp(-cn^{p/2}) \int_{Tn^{1/2-1/p}}^t \frac{pct^p n^{p/2}}{u^{p+1}} du \le ct^p n \exp(-cn^{p/2}) \le \exp(-c'n^{p/2}).$$

Finally III of (6) is dominated by

$$\int_t^1 \frac{pct^p n^{p/2}}{u^{p+1}} \exp\left(-c\left(\frac{t^p n^{p/2}}{u^p} + u^p n\right)\right) du$$

$$\le \exp(-ct^p n) \int_t^1 \frac{pct^p n^{p/2}}{u^{p+1}} \exp\left(-c\frac{t^p n^{p/2}}{u^p}\right) du$$

$$< \exp(-ct^p n).$$

□

5   The Function $f(x) = \|x\|_2$

As we remarked in the introduction, for t larger than an absolute constant divided by $\sqrt{n}$, the conclusion of Theorem 3.1 is best possible for the function $f(x) = \|x\|_2$. It turns out that for smaller values of t a stronger inequality holds.

Proposition 5.1 *Let*

$$\alpha(n, t) = \begin{cases} nt, & \text{for } t > n^{-1/2} \\ n^{3/4}t^{1/2}, & \text{for } n^{-5/6} < t \le n^{-1/2} \\ n^2 t^2, & \text{for } 0 < t \le n^{-5/6}. \end{cases}$$

Then, for some absolute constants $0 < c, C < \infty$,

$$\lambda\left(\left\{x : \left|\|x\|_2 - \int \|\cdot\|_2 d\lambda\right| > t\right\}\right) \leq Ce^{-c\alpha(n,t)}.$$

For $t > T/\sqrt{n}$ the proposition was proved in [SZ]. For the lower values of t, as in the proof of the main result, it is enough to prove

$$\Pr\left(\frac{\|X\|_2}{S}\left|\frac{S}{n} - 1\right| > t\right) \leq Ce^{-c\alpha(n,t)} \tag{7}$$

and, for Y independent of X,

$$\Pr\left(\left|\|X\|_2 - \|Y\|_2\right| > nt\right) \leq Ce^{-c\alpha(n,t)}. \tag{8}$$

Proof of (7). As in Case 1 in the proof of Lemma 3.2 (using Theorem 2.3 instead of Corollary 2.4),

$$\Pr\left(\frac{\|X\|_2}{S}\left|\frac{S}{n} - 1\right| > t\right) = \mathbb{E}_{\frac{\|X\|_2}{S}} \Pr\left(\left|\frac{S}{n} - 1\right| > \frac{tS}{\|X\|_2}\right)$$

$$\leq C\mathbb{E}_{\frac{\|X\|_2}{S}} \exp\left(-cn\left(\left(\frac{tS}{\|X\|_2}\right) \wedge \left(\frac{tS}{\|X\|_2}\right)^2\right)\right).$$

Now,

$$\mathbb{E}_{\frac{\|X\|_2}{S}}\left[\exp\left(-cn\left(\left(\frac{tS}{\|X\|_2}\right) \wedge \left(\frac{tS}{\|X\|_2}\right)^2\right)\right) 1_{\|X\|_2 \leq tS}\right] \leq e^{-cn}$$

while by Theorem 2.2,

$$\mathbb{E}_{\frac{\|X\|_2}{S}}\left[\exp\left(-cn\left(\left(\frac{tS}{\|X\|_2}\right) \wedge \left(\frac{tS}{\|X\|_2}\right)^2\right)\right) 1_{\|X\|_2 > tS}\right]$$

$$= \int_0^1 \frac{2cnt^2}{u^3} e^{-cnt^2/u^2} \Pr\left(\frac{\|X\|_2}{S} > u \vee t\right)$$

$$\leq \int_0^{T/\sqrt{n}} \frac{2cnt^2}{u^3} e^{-cnt^2/u^2} + \int_{T/\sqrt{n}}^\infty \frac{2cnt^2}{u^3} e^{-cnt^2/u^2} e^{-cnu} du. \tag{9}$$

The first summand in (9) is equal to $e^{-cn^2t^2/T^2}$. Noticing that the maximum of $e^{-cnt^2/u^2}e^{-cnu}$ occurs at $u = 2^{1/3}t^{2/3}$ and is equal to $e^{-c'nt^{2/3}}$, we get that the second summand in (9) is dominated by

$$e^{-c'nt^{2/3}} \int_{T/\sqrt{n}}^\infty \frac{2cnt^2}{u^3} du = \frac{cn^2t^2}{T^2} e^{-c'nt^{2/3}} \leq C'e^{-c''nt^{2/3}}$$

for the relevant range of t (i.e., t larger than an absolute constant times $1/n$). Summarizing, we get that (with different absolute constants c, C)

$$\Pr\left(\frac{\|X\|_2}{S}\left|\frac{S}{n} - 1\right| > t\right) \leq Ce^{-cn^2t^2} + Ce^{-cnt^{2/3}} \tag{10}$$

and it is easy to see that this is dominated by $Ce^{-c\alpha(n,t)}$.

Proof of (8). First note that by Theorem 2.3, $\Pr(\|Y\|_2 < c\sqrt{n}) < Ce^{-c\sqrt{n}}$ for an appropriate C and that $\sqrt{n}$ is larger than a constant times $\alpha(n,t)$ in the range in question. Thus,

$$\Pr\left(\|X\|_2 - \|Y\|_2 > nt\right) = \Pr\left(\|X\|_2^2 > \|Y\|_2^2 + 2\|Y\|_2 nt + n^2 t^2\right)$$
$$\leq Ce^{-c\alpha(n,t)} + \Pr\left(\|X\|_2^2 > \|Y\|_2^2 + cn^{3/2}t\right) \tag{11}$$

and it is enough to prove that

$$\Pr\left(\sum_{i=1}^n (X_i^2 - Y_i^2) > cn^{3/2}t\right) \leq Ce^{-c\alpha(n,t)}. \tag{12}$$

To prove (12) note that for each p the left hand side of (12) is dominated by an absolute constant times $\dfrac{\mathbb{E}|\sum \epsilon_i X_i|^p}{(n^{3/2}t)^p}$. Apply now the result of Hitczenko, Montgomery-Smith and Oleszkiewicz [HMO, Th. 4.2] to the variables X_i^2 and get that, for each p, the left hand side of (12) is dominated by

$$\left(\frac{C(n^{1/p}(2p)^2 \wedge p^{1/2}n^{1/2})}{n^{3/2}t}\right)^p. \tag{13}$$

For t larger than $n^{-5/6}$ take $p = c'n^{3/4}t^{1/2}$ for c' small with respect to C. Then (13) is dominated by $C'e^{-c''n^{3/4}t^{1/2}}$. For t smaller than $n^{-5/6}$ but larger than a constant times $1/n$ take $p = c'n^2t^2$ and get that (13) is dominated by $C'e^{-c''n^2t^2}$. $\qquad\square$

Remarks. 1. The use of [HMO] was suggested to us by S. Kwapien. The result of [HMO] was considerably generalized by Latała [Lat]. Originally we had another (more direct but also more special to the variables in question) proof of (12): After appropriately truncating the X_i's evaluate $\exp(t\sum \epsilon_i X_i^2 1_{|X_i|<a})$ by carefully evaluating the terms in the Taylor expansion. This follows closely [Bou, Lemma 1].

2. Is the statement of Lemma 5.1 best possible? for $t > Cn^{-1/2}$ this was proved in [SZ]. For $n^{-5/6} < t < Cn^{-1/2}$ this is still the case. This easily follows from the following three facts: a. The proof of (7) in this case gives an estimate, (10) which is of better order of magnitude than $e^{-c\alpha(n,t)}$.
b. The inequalities in (11) can be inverted (with different constants of course).
c.

$$2\Pr\left(|\sum_{i=1}^n \epsilon_i X_i^2| > cn^{3/2}t\right) \geq \Pr\left(X_1^2 > cn^{3/2}t\right) = e^{-n^{3/4}t^{1/2}}.$$

For $n^{-1} < t < n^{-5/6}$ one can show the right lower bound for the quantity in the left end sides of (11) but it doesn't seem to combine nicely with upper bound for (7).

References

[BL]   Bobkov S., Ledoux M. (1997) Poincaré's inequalities and Talagrand's concentration phenomenon for the exponential distribution. Probab. Theory Related Fields 107:383–400

[Bou]   Bourgain J. (1999) Random points in isotropic convex sets. In: Convex Geometric Analysis (Berkeley, CA, 1996), Math. Sci. Res. Inst. Publ., 34, Cambridge Univ. Press, Cambridge, 53–58

[GM]   Gromov M., Milman V.D. (1983) A topological application of the isoperimetric inequality. Amer. J. Math. 105:843–854

[HMO]   Hitczenko P., Montgomery-Smith S.J., Oleszkiewicz K. (1997) Moment inequalities for sums of certain independent symmetric random variables. Studia Math. 123:15–42

[Lat]   Latała R. (1997) Estimation of moments of sums of independent real random variables. Ann. Probab. 25:1502–1513

[LO]   Latała R., Oleszkiewicz K. Between Sobolev and Poincare. This volume

[Led]   Ledoux M. (1997) Concentration of measure and logarithmic Sobolev inequalities. Preprint

[Mau]   Maurey B. (1991) Some deviation inequalities. Geom. Funct. Anal. 1:188–197

[MS]   Milman V.D., Schechtman G. (1986) Asymptotic theory of finite-dimensional normed spaces. Lecture Notes in Mathematics, 1200, Springer-Verlag, Berlin-New York

[SZ]   Schechtman G., Zinn J. (1990) On the volume of the intersection of two L_p^n balls. Proc. A.M.S. 110:217–224

[ScSc] Schechtman G., Schmuckenschläger M. (1991) Another remark on the volume of the intersection of two L_p^n balls. Geometric Aspects of Functional Analysis (1989–90), Lecture Notes in Math., 1469, Springer, 174–178

[Tal]   Talagrand M. (1991) A new isoperimetric inequality and the concentration of measure phenomenon. Geometric Aspects of Functional Analysis (1989–90), Lecture Notes in Math., 1469, Springer, 94–124

Shannon's Entropy Power Inequality via Restricted Minkowski Sums

S.J. Szarek[1] and D. Voiculescu[2]

[1] Department of Mathematics, Case Western Reserve University, Cleveland, OH 44106-7058, U.S.A. and Université Pierre et Marie Curie, Equipe d'Analyse, Bte 186, 4, Place Jussieu, 75252 Paris, France
[2] Department of Mathematics, University of California, Berkeley, CA 94720-3840, U.S.A.

1 Introduction and Preliminaries

If X is an $\mathbb{R}^n$-valued random variable whose distribution μ_X is absolutely continuous with respect to the Lebesgue measure λ_n and f is the corresponding density, the *entropy* of X is defined via $h(X) := \int_{\mathbb{R}^n} f \log \frac{1}{f} d\lambda_n$. One of the fundamental results of Information Theory (see, e.g., [SW]) is the Shannon's *Entropy Power Inequality*, which affirms that if X, Y are two such variables which are independent, then

$$\exp\left(2h(X)/n\right) + \exp\left(2h(Y)/n\right) \leq \exp\left(2h(X+Y)/n\right). \tag{1}$$

Shannon's original variational argument seems incomplete, but there exist (at least) two other proofs of (1) due to Stam ([S], 1959) and Lieb ([L], 1978); see [CT] or [DCT] for more background and history. The purpose of this note is to present a new proof of the classical Entropy Power Inequality in the spirit of [SV], where its noncommutative (free) analogue was shown. Our proof is conceptually related to Lieb's argument as it uses a rearrangement inequality from [BLL], belonging to the same circle of ideas as [L].

While having one more proof of a classical fact may be perceived as being of limited value, the present argument appears to have the advantage of being much more direct than the other ones. Additionally, we hope that the more geometric approach may shed some new light on the noncommutative theory, where even the most appropriate definitions of concepts, particularly in the multivariate case, haven't been determined, cf. the series of papers [V] and their references.

As in [SV], our argument is based on a geometric result resembling formally the classical Brunn-Minkowski inequality. Let A, B be subsets of a

---

Both authors were supported in part by grants from the National Science Foundation. This research was partially carried out by the second named author for the Clay Mathematics Institute.

vector space and $\Theta \subset A \times B$. We will call

$$A +_\Theta B := \{x + y : (x, y) \in \Theta\}$$

the restricted (to Θ) sum of A and B. We then have (below and in what follows, all sets and functions are assumed to be measurable and $\lambda = \lambda_n$ is the Lebesgue measure in the appropriate dimension which may vary between occurrences)

Lemma 1. *For any* $\varepsilon > 0$ *there exists* $\delta > 0$ *such that if* $n \in \mathbb{N}$, $A, B \subset \mathbb{R}^n$ *and* $\Theta \subset A \times B \subset \mathbb{R}^{2n}$ *verify*

$$\lambda(\Theta) \geq (1 - \delta)^n \lambda(A) \lambda(B),$$

then

$$\lambda(A +_\Theta B)^{2/n} \geq (1 - \varepsilon) \left(\lambda(A)^{2/n} + \lambda(B)^{2/n} \right). \tag{2}$$

Remarks.

(i) The Lemma above is slightly different from the version stated in [SV]. Its analogue in that paper (Theorem 1.2) asserts a stronger inequality $\lambda(A +_\Theta B)^{2/n} \geq \lambda(A)^{2/n} + \lambda(B)^{2/n}$ under a stronger hypothesis: $(1 - \delta)^n$ replaced by $1 - \delta$, where – in cases of interest – $\delta > 0$ can be chosen independently of n, A, B and Θ. However, the present variant follows easily, with a "nearly" optimal dependence $\varepsilon = O(\delta^{1/2})$, from Corollary 1.5 and Remark 1.6 in [SV], or can be directly derived from the rearrangement inequality of [BLL].

(ii) A formally stronger "restricted Prékopa-Leindler inequality" was proved in [B]; it is quite likely that it can be used to prove (1) in an even more direct way.

We will also need the following elementary fact, closely related to the traditional information-theoretic definition of entropy as a measure of the volume of the "effective support" of a "large" sample of X. To make the exposition more clear, we shall concentrate on the scalar case, which contains all the ingredients of the general setting (see the remark at the end of this section).

Let $X_1, X_2, \ldots, X_N, \ldots$ be a sequence of independent copies of a real random variable X (with density f, as before; only variables with density need to be considered in this context) and denote by $\mathcal{P}$ the underlying probability measure. Given $N \in \mathbb{N}$, let $F = F_N : \mathbb{R}^N \to \mathbb{R}_+$ be the joint density of $X_1, X_2, \ldots, X_N$ with respect to λ_N; of course $F(x_1, x_2, \ldots, x_N) = f(x_1) f(x_2) \ldots f(x_N)$. We then have

Proposition 2. *There exist two positive sequences* (α_k) *and* (β_k) *(depending on X) converging to 0 such that if we set, for $N \in \mathbb{N}$,*

$$V_N = V_N(X) := \left\{ e^{N(-h(X)-\alpha_N)} \leq F_N \leq e^{N(-h(X)+\alpha_N)} \right\},$$

then

$$\mathcal{P}\left((X_1, X_2, \ldots, X_N) \in V_N\right) \equiv \int_{V_N} F_N d\lambda_N > 1 - \beta_N.$$

V_N is closely related to the set of "typical sequences" from Information Theory (cf. [SW], sections 7 and 21). For completeness, we include a simple proof of the Proposition in the Appendix, even though neither the result nor the gist of the argument are new. Of course we could have simplified the statement by requiring that $(\alpha_k) = (\beta_k)$; however, as shall be clear from what follows, the roles of these two sequences are quite different. In particular, for our purposes it would have been enough to have (β_k) just "sufficiently small", e.g., $\leq 1/4$ for large k.

An immediate consequence of the Proposition is that

$$(1 - \beta_N)e^{N(h(X)-\alpha_N)} < \lambda(V_N) \leq e^{N(h(X)+\alpha_N)} \tag{3}$$

and hence, as $N \to \infty$,

$$\frac{\log \lambda(V_N(X))}{N} \to h(X) \quad and \quad \lambda(V_N(X))^{2/N} \to \exp(2h(X)). \tag{4}$$

2   The Proof

The idea of the rest of the proof is now as follows. Let (X_k) and (Y_k) be sequences of *jointly* independent copies of X and Y respectively. Given N, the set $V_N(X+Y)$ of typical sequences $X_1+Y_1, X_2+Y_2, \ldots, X_N+Y_N$ is *roughly* the same as the Minkowski sum of $V_N(X)$ and $V_N(Y)$, the sets of typical sequences $X_1, X_2, \ldots, X_N$ and $Y_1, Y_2, \ldots, Y_N$ respectively. Accordingly, by the inequality (2) for restricted Minkowski sums ("restricted" because of the qualification "roughly" above) we have *approximately* $\lambda(V_N(X+Y))^{2/N} \geq \lambda(V_N(X))^{2/N} + \lambda(V_N(Y))^{2/N}$ ("approximately" because of the $1 - \varepsilon$ factor in (2)) and (1) follows by letting $N \to \infty$ and using the second limit relation in (4).

To make this sketch precise, we apply Lemma 1 with $A := V_N(X)$, $B := V_N(Y)$ and

$$\Theta := \left\{ (a, b) \in A \times B : a + b \in V_N(X+Y) \right\}.$$

Since, by definition, $A +_\Theta B \subset C := V_N(X+Y)$, leaving for a moment aside the issue of the exact values of $\varepsilon = \varepsilon_N$ and $\delta = \delta_N$ that intervene, we get from Lemma 1 that

$$\lambda\left(V_N(X+Y)\right)^{2/N} \geq (1 - \varepsilon_N)\left(\lambda(V_N(X))^{2/N} + \lambda(V_N(Y))^{2/N}\right). \tag{5}$$

We claim that with our choice of A, B and Θ one has

$$\left(\frac{\lambda_{2N}(\Theta)}{\lambda_N(A) \cdot \lambda_N(B)}\right)^{1/N} \to 1$$

as $N \to \infty$ and so, when applying Lemma 1, one may have choose $\delta = \delta_N$ so that (as $N \to \infty$) $\delta_N \to 0$, hence $\varepsilon_N \to 0$, and so, by the argument sketched earlier, the inequality (5) becomes in the limit (1).

To prove our claim $\delta_N \to 0$ we denote $\mathbf{X} := (X_1, X_2, \ldots, X_N)$, $\mathbf{Y} := (Y_1, Y_2, \ldots, Y_N)$ and observe that

$$
\begin{aligned}
1 - \beta_N &\leq \mathcal{P}(\mathbf{X} + \mathbf{Y} \in C) \\
&\leq \mathcal{P}(\mathbf{X} \in A, \mathbf{Y} \in B \& \mathbf{X} + \mathbf{Y} \in C) + \mathcal{P}((\mathbf{X}, \mathbf{Y}) \notin A \times B) \\
&= \mathcal{P}((\mathbf{X}, \mathbf{Y}) \in \Theta) + (1 - \mathcal{P}(\mathbf{X} \in A)\mathcal{P}(\mathbf{Y} \in B)) \\
&\leq \mathcal{P}((\mathbf{X}, \mathbf{Y}) \in \Theta) + 1 - (1 - \beta_N)^2 .
\end{aligned}
$$

by the definitions of A, B, C and the estimates on the corresponding probabilities given by Proposition 2. (We note that even though *a priori* the sequences $(\alpha_k), (\beta_k)$ depend on the random variable in question, they can be chosen to verify the Proposition for the random variables X, Y and $X + Y$ *simultaneously*.) By simple calculation, the above implies

$$1 - 3\beta_N + \beta_N^2 \leq \mathcal{P}((\mathbf{X}, \mathbf{Y}) \in \Theta) \equiv \int_\Theta G_N d\lambda,$$

where G_N is the density of $(\mathbf{X}, \mathbf{Y})$ on $\mathbb{R}^{2N}$, necessarily equal to the product of the densities of $\mathbf{X}$ and $\mathbf{Y}$. Using the upper bounds on the latter densities implicit in the hypothesis of Proposition 2 (where the sets $V_N(\cdot)$ were defined; recall that $\Theta \subset A \times B \equiv V_N(X) \times V_N(Y)$) we deduce

$$\int_\Theta G_N d\lambda \leq \lambda(\Theta) \cdot e^{N(-h(X)+\alpha_N)} \cdot e^{N(-h(Y)+\alpha_N)} .$$

On the other hand, by (3),

$$\lambda(A) \cdot \lambda(B) \leq e^{N(h(X)+\alpha_N)} \cdot e^{N(h(Y)+\alpha_N)} .$$

Combining the last three inequalities we obtain

$$
\begin{aligned}
\frac{\lambda(\Theta)}{\lambda(A) \cdot \lambda(B)} &\geq \frac{(1 - 3\beta_N + \beta_N^2) \cdot e^{N(h(X)-\alpha_N)} \cdot e^{N(h(Y)-\alpha_N)}}{e^{N(h(X)+\alpha_N)} \cdot e^{N(h(Y)+\alpha_N)}} \\
&= (1 - 3\beta_N + \beta_N^2)e^{-4N\alpha_N} ,
\end{aligned}
$$

whence

$$\left(\frac{\lambda(\Theta)}{\lambda(A) \cdot \lambda(B)}\right)^{1/N} \geq (1 - 3\beta_N + \beta_N^2)^{1/N} e^{-4\alpha_N} \to 1$$

when $N \to \infty$, as required.

Remark. The proof extends immediately to the multivariate case. Note that if X is $\mathbb{R}^n$-valued, then the corresponding density F_N "lives" on $\mathbb{R}^{nN}$, and so an application of Lemma 1 yields exponents $2/(Nn)$ and results in an additional n in $\exp(2h(X)/n)$.

3   Appendix: The Proof of Proposition 2

As mentioned earlier, Proposition 2 and its immediate consequences are closely related to the traditional information-theoretic definition of entropy, which is as follows. In the notation of the Proposition, let $N \in \mathbb{N}$ and let $V' = V'_N \subset \mathbb{R}^N$ be any smallest (volumewise) set verifying $\int_{V'_N} F_N = 1/2$; we then set $h(X) := \lim_{n \to \infty} \log \lambda_N (V'_N)/N$. The fact that the limit exists, that it remains unchanged if we replace $1/2$ by some other $\beta \in (0,1)$ as well as the equivalence of the two definitions can be easily deduced, e.g., ¿from the argument below, or formally from the assertion of the Proposition.

To prove the Proposition, consider the expression

$$\frac{\log F_N}{N} \equiv \frac{\sum_{j=1}^N \log f(x_j)}{N}$$

and think of it as a random variable on $\mathbb{R}^\infty$ endowed with the product measure $\mathcal{P} := \otimes_{j=1}^\infty \mu_X$ (where $d\mu_X \equiv f d\lambda_1$ is the law of X); it becomes then $\frac{1}{N} \sum_{j=1}^N \log f(X_j)$. By the law of large numbers, as $N \to \infty$, the last sequence converges ($\mathcal{P}$-a.e., or in probability with respect to $\mathcal{P}$), to the expected value of $\log f(X)$ which in turn equals $\int_\mathbb{R} \log f d\mu_X = \int_\mathbb{R} f \log f d\lambda_1 \equiv -h(X)$. In particular, there exists a positive sequence (α_k) converging to 0 (depending on X) such that,

$$\mathcal{P} \left(-h(X) - \alpha_N \leq \frac{1}{N} \sum_{j=1}^N \log f(X_j) \leq -h(X) + \alpha_N \right) \to 1$$

as $N \to \infty$, which is just a rephrasing of the assertion of the Proposition.

References

[B]   Barthe F. (1999) Restricted Prékopa-Leindler inequality. Pacific J. Math. 189(2):211–222

[BLL]   Brascamp H.J., Lieb E.H., Luttinger J.M. (1974) A general rearrangement inequality for multiple integrals. J. Funct. Analysis 17:227–237

[CT]   Cover, T.M., Thomas, J.A. (1991) Elements of information theory. Wiley Series in Telecommunications. John Wiley & Sons, Inc., New York.

[DCT]   Dembo A., Cover T.M., Thomas J.A. (1991) Information theoretic inequalities. IEEE Transactions on Information Theory 37(6):1501–1518

[L]   Lieb E.H. (1978) Proof of an entropy conjecture of Wehrl. Commun. Math. Phys. 62:35–41

[SW]   Shannon C.E., Weaver W. (1963) The Mathematical Theory of Communications. University of Illinois Press

[S]   Stam A.J. (1959) Some inequalities satisfied by the quantities of information of Fisher and Shannon. Information and Control 2:101–112

[SV]   Szarek S.J., Voiculescu D. (1996) Volumes of restricted Minkowski sums and the free analogue of the entropy power inequality. Commun. Math. Phys. 178:563–570

[V]   Voiculescu D. (1993) The analogues of entropy and of Fisher's information measure in free probability theory, I. Commun. Math. Phys. 155:71–92; (1994) *ibidem*, II. Invent. Math. 118:411–440; (1996) *ibidem*, III, The absence of Cartan subalgebras. Geom. Funct. Anal. 6(1):172–199; (1997) *ibidem*, IV, Maximum entropy and freeness. In: Free Probability Theory (Waterloo, ON, 1995), Fields Inst. Commun., 12, Amer. Math. Soc., 293–302; (1998) *ibidem*, V, Noncommutative Hilbert transforms. Invent. Math. 132(1):189–227

Notes on an Inequality by Pisier for Functions on the Discrete Cube

R. Wagner

School of Mathematical Sciences, Tel Aviv University, Tel Aviv 69978, Israel

1 Introduction

We study functions from the discrete cube $\{-1,1\}^n$ to a linear normed space B. For such functions f define $D_i f(x)$ to be the vector $\frac{f(x)-f(x^i)}{2}$, where x^i has the same coordinates as x except at the i-th place. Define

$$|\nabla f|_p(x) = \left(\int_{y_i \in \{-1,1\}^n} \| \sum_{i=1}^n y_i D_i f(x) \|_B^p \right)^{\frac{1}{p}}.$$

In [P] Pisier proved that for every $1 \leq p \leq \infty$ and every $f : \{-1,1\}^n \to B$ one has

$$\|f(x) - Ef\|_{L_p(B)} \leq C \log n \left\| |\nabla f|_p(x) \right\|_{L_p(B)}.$$

In [T] Talagrand showed that for $B = \mathbb{R}$ the logarithmic factor can be removed, whereas in general it is sharp with dependence on $p < \infty$. We will close here the final gap in this inequality, by showing that the logarithmic factor can be dropped for $p = \infty$. The argument we employ is an elementary counting argument. We will then turn to revise Pisier's proof of his inequality above. The proof we present uses the same mechanism, but applies it differently.

2 The Main Result

Theorem 1. *Let $f : \{-1,1\}^n \to B$, where B is a normed linear space. Then*

$$\max_{x,y \in \{-1,1\}^n} \|f(x) - f(y)\|_B \leq 2 \max_{x \in \{-1,1\}^n} |\nabla f|_\infty(x).$$

Proof. Choose two points, x and y, where the left-hand maximum is attained. Without loss of generality we can assume these are diametrically opposite points in the discrete cube. Otherwise we will restrict to the subcube where x and y are diametrically opposed, and prove the inequality in this context; then the "unconditional" nature of $|\nabla f|_\infty(x)$ allows to return the right-hand term to the full-cube context without decreasing it.

Denote $x = a_0 = (-1,-1,\ldots)$ and $y = a_n = (1,1,\ldots)$. We will be interested in all *monotone* paths of neighbouring vertices $a_0, a_1, \ldots, a_n$, that is paths where a_{i+1} is achieved from a_i by changing one of its -1's into a 1. Now

$$f(x) - f(y) = \frac{1}{n!} \sum_{\substack{\{a_0,\dots,a_n\} \\ \text{a monotone} \\ \text{path}}} \left(\sum_{i=1}^{n} f(a_i) - f(a_{i-1}) \right); \qquad (1)$$

The normalising factor $n!$ is just the number of monotone paths.

We would like to express (1) using the following objects:

$$\delta f(x) = \sum_{\substack{a \text{ is a} \\ \text{neighbour of } x}} y_a \frac{f(x) - f(a)}{2},$$

where $y_a = 1$ if a has more 1's than x, and $y_a = -1$ if a has less 1's than x. It is clear that $\|\delta f(x)\|_B \leq |\nabla f|_\infty(x)$. Thus, our claim is proved if we rearrange (1) in the form

$$\frac{2}{n!} \sum_{i=1}^{n!} \delta f(x_i),$$

where x_i is an enumeration of the vertices in $\{-1, 1\}^n$ with repetitions. This rearrangement will be obtained heuristically. Consider the sum

$$\frac{2}{n!} \sum_{x \in \{-1,1\}^n} b_x \delta f(x). \qquad (2)$$

We will find conditions on the integers b_x which force (1) and (2) to be equal.

Take a point a_k with k coordinates equal to 1, and a point a_{k+1} obtained from a_k by changing one of its -1's to a 1. The term $f(a_k) - f(a_{k+1})$ appears in (1) exactly $k!(n - k - 1)!$ times, because there are exactly that many monotone paths going through (a_k, a_{k+1}). That same term appears $b_{a_k} + b_{a_{k+1}}$ times in (2), because it is part of both $\delta f(a_k)$ and $\delta f(a_{k+1})$. We conclude that (1) and (2) are equal iff

$$b_{a_k} + b_{a_{k+1}} = k!(n - k - 1)!.$$

Now a monotone path, as well as a $\delta f(x)$, both have exactly n terms of the form $f(a_k) - f(a_{k+1})$. Since we are matching these latter terms bijectively between (1) and (2), we have exactly as many paths in (1) as δf's in (2). This equality affirms proper normalisation in (2), namely $\sum b_x = n!$. Adding the constriction that the b_{a_k}'s be positive will allow the triangle inequality to bound (2) by $2|\nabla f|_\infty(x)$.

That integers satisfying our demands of the b_{a_k}'s do in fact exist is easily verified; Actually, our b_{a_k}'s will depend only on k. The main observation behind our construction is that the sequence $k!(n - k - 1)!$ decreases until k reaches $(n - 1)/2$, and then climbs back symmetrically.

Consider the case n odd (the even case is similar). Set

$$b_{\frac{n-1}{2}} = \frac{1}{2}\left(\frac{n-1}{2}\right)!\left(\frac{n-1}{2}\right)! = b_{\frac{n+1}{2}}.$$

We can define inductively $b_{k-1} = (k-1)!(n-k)! - b_k$. An induction going backwards from $k = \frac{n-1}{2}$ will prove that

$$k!(n-k-1)! > b_k > 0.$$

Indeed, suppose b_k satisfies this premise. We have $b_{k-1} = (k-1)!(n-k)! - b_k$, which implies

$$(k-1)!(n-k)! > b_{k-1} > (k-1)!(n-k)! - k!(n-k-1)! > 0.$$

The argument for $k > \frac{n}{2}$ is symmetric. $\qquad\qquad\qquad\qquad\qquad\qquad\square$

Remark 2. This method seems to be applicable to other highly symmetric regular graphs. The constant 2 should in general be replaced by a number close to the ratio $2\frac{\text{diameter}}{\text{degree}}$. An example where this holds (by an easy argument following our scheme) is the crosspolytope.

3   Pisier's Proof Revisited

We present a proof of Pisier's inequality as stated in the introduction. our proof repeats the original proof's mechanism. Here, however, this mechanism is applied to a different operator (the identity). This approach takes the weight off the lower estimate, and may clarify the picture for the upper estimate. The price we pay is the conversion of a convexity argument into a more intense combinatorial consideration.

The proof requires some Walsh analysis. A function $f : \{-1,1\}^n \to B$ can be decomposed uniquely into $\sum_{A \subseteq \{1,\ldots,n\}} f_A w_A(x)$, where $f_A \in B$, $w_A(x) = \prod_{i \in A} x_i$, and the functions $\{w_A(x)\}_{A \subseteq \{1,\ldots,n\}}$ form an orthonormal basis in the L_2 norm. It is important to observe that the operator D_i in this context is simply an orthogonal projection on those Walsh functions $w_A(x)$ for which $i \in A$.

Proof. The argument is a duality argument. Let $g : \{-1,1\}^n \to B^*$ such that $\|g\|_q = 1$ and

$$\|f\|_p = \int_{x \in \{-1,1\}^n} f(x) \cdot g(x)\, dx$$

(we assume, without loss of generality that $f_\emptyset = Ef = 0$). Our goal is to rewrite this expression in the form

$$\int_{x,y \in \{-1,1\}^n} \varphi(x,y) \cdot \sum_{i=1}^{n} y_i D_i f(x)\, dx dy.$$

Bounding $\|\varphi(x,y)\|_q$ by $C \log n$ will complete the argument.

Since we assume $f_\emptyset = 0$, orthonormality of the Walsh system implies:

$$\int_{x \in \{-1,1\}^n} g(x) \cdot f(x) dx = \int_{x \in \{-1,1\}^n} (g - g_\emptyset)(x) \cdot f(x) dx. \qquad (3)$$

Now note that the identity can be decomposed into

$$I = R \sum_{i=1}^{n} D_i^* D_i,$$

where R is the following selfadjoint operator

$$R\left(\sum_{\substack{A \subseteq \{1,\ldots,n\} \\ |A| \geq 1}} f_A w_A(x) \right) = \sum_{\substack{A \subseteq \{1,\ldots,n\} \\ |A| \geq 1}} \frac{1}{|A|} f_A w_A(x).$$

Applying this decomposition, we turn (3) into

$$\int_{x \in \{-1,1\}^n} (g - g_\emptyset)(x) \cdot \left(R \sum_{i=1}^{n} D_i^* D_i \right) f(x) dx.$$

Conjugating and transferring the operators to the left we get

$$\sum_{i=1}^{n} \int_{x \in \{-1,1\}^n} D_i R(g - g_\emptyset)(x) \cdot D_i f(x) dx,$$

and finally, decoupling obtains

$$\int_{x,y \in \{-1,1\}^n} \left(\sum_{i=1}^{n} y_i D_i R(g - g_\emptyset)(x) \right) \cdot \left(\sum_{i=1}^{n} y_i D_i f(x) \right) dx dy.$$

Observe that all terms in the Walsh decomposition of the right-hand function have the form $y_i w_A(x)$, where $i \in A$. Orthonormality implies we can replace the left-hand function by any $\varphi(x,y)$, which coincides with the left-hand function on those Walsh terms.

Our candidate will be:

$$\varphi(x,y) = \sum_{k=1}^{n} \frac{1}{k} \frac{1}{\binom{n}{k}} \sum_{\substack{B \subseteq \{1,\ldots,n\} \\ |B|=k}} g(x_1^B, \ldots, x_n^B),$$

where $x_i^B = x_i$ if $i \notin B$, and $x_i^B = x_i y_i$ if $i \in B$ (the reader interested in motivation should check out the cases $n = 2$ and 3; there it is visible that φ is obtained by 'completing' the relevant terms from the original left hand function into a sum of copies of g). Note that the sequence $\{x_i^B\}_{i=1}^{n}$ is a sequence of independent uniform variables. This and the fact that the inner sum has exactly $\binom{n}{k}$ terms allow the triangle inequality to conclude

$$\|\varphi\|_q \leq \sum_{k=1}^{n} \frac{1}{k} \|g\|_q \leq 1 + \log n.$$

Therefore, the proof is done, once we verify that φ can indeed replace the original left-hand function.

Let's compare the relevant Walsh terms in the two functions. Let $i \in A$. On the one hand,

$$\int_{x,y\in\{-1,1\}^n} \left(\sum_{j=1}^{n} y_j D_j R(g - g_\emptyset)(x)\right) \cdot (y_i w_A(x)) dx dy =$$

$$\int_{x\in\{-1,1\}^n} (D_i R(g - g_\emptyset)(x)) \cdot w_A(x) dx = \frac{1}{|A|} g_A.$$

On the other hand,

$$\int_{x,y\in\{-1,1\}^n} \varphi(x, y) \cdot y_i w_A(x) dx dy = \sum_{k=1}^{n} \frac{1}{k} \frac{1}{\binom{n}{k}} \sum_{\substack{B \subseteq \{1,\ldots,n\} \\ |B|=k \\ A\cap B=\{i\}}} g_A.$$

To complete the proof, all we need to show is that

$$\sum_{k=1}^{n} \frac{1}{k} \frac{1}{\binom{n}{k}} |\{B \subseteq \{1,\ldots,n\}; |B| = k, A \cap B = \{i\}\}| = \frac{1}{|A|}.$$

Indeed, the left-hand sum equals

$$\sum_{k=1}^{n} \frac{1}{k} \frac{1}{\binom{n}{k}} \binom{n - |A|}{k - 1},$$

which can be regrouped as

$$\frac{(n - |A|)!(|A| - 1)!}{n!} \sum_{k=1}^{n} \binom{n - k}{|A| - 1}.$$

A simple combinatorial identity yields

$$\frac{(n - |A|)!(|A| - 1)!}{n!} \binom{n}{|A|},$$

which in turn equals the desired $\frac{1}{|A|}$. $\square$

References

[P]   Pisier G. (1986) Probabilistic methods in the geometry of Banach spaces. In: Probability and Analysis (Varenna, Italy, 1985), Lecture Notes in Math., 1206, Springer Verlag, 167–241

[T]   Talagrand M. (1993) Isoperimetry, logarithmic Sobolev inequalities on the discrete cube, and Margulis' graph connectivity theorem. GAFA 3:295–314

More on Embedding Subspaces of L_p into ℓ_p^N, $0 < p < 1$

A. Zvavitch

Department of Theoretical Mathematics, The Weizmann Institute of Science, Rehovot 76100, Israel

Abstract. It is shown that n-dimensional subspace of L_p, $0 < p < 1$, $(1+\varepsilon)$-embeds in l_p^N, whenever $N \geq c(\varepsilon,p)n(\log n)(\log\log n)^2$.

1 Introduction

Let X be a n-dimensional subspace of L_p, $0 < p < \infty$, and let $\varepsilon > 0$. What is the smallest $N = N(n, \varepsilon)$ such that there is a subspace Y of l_p^N for which $d(X, Y) \leq 1 + \varepsilon$, where $d(X, Y)$ is the Banach-Mazur distance?

In this paper we consider only the case $0 < p < 1$. In this case, it is shown in [S-Z] that $N(n, \varepsilon)$ grows in n at most like $n(\log n)^3$. The tools used in [S-Z] are random choice argument, Lewis' theorem in the non-convex setting and finally Dudley's theorem for bounding of the supremum of a Gaussian process. Here instead of using Dudley's theorem we will use the majorizing measures technique, developed by M. Talagrand.

M. Talagrand [T2] used this method to find an estimate for $N(n, \varepsilon)$, for $1 < p < 2$, which improved a previous estimate of [B-L-M] by reducing the power of the $\log n$ factor. We simplify a bit the methods used in [T2], and we improve the result of [S-Z] by proving the following theorem.

Theorem 1.1 *Let* $0 < p < 1$ *then*

$$N(n, \varepsilon) \leq C(p)\varepsilon^{-2}n(\log n)(\log\log \varepsilon^{-2}n + \log \varepsilon^{-1})^2.$$

Throughout this paper C denotes absolute constant whose value may change from line to line, $C(p), c(p)$ denote constants depending only on the argument p.

The proof is by the method of random selection of coordinates (for more details on this method see e.g. [J-S 2]). We may assume without loss of generality that X lies in a finite (but large) dimensional $L_p^M(\mu)$ space, where μ is a probability measure on $\{1, \ldots, M\}$. We would like to show that the restriction operator to a set of relatively few of the M coordinates is a $(1+\varepsilon)$-isometry on X. We do it iteratively, reducing the size of M by a factor of at least $3/4$ each time. To apply a random choice argument we first of all need

Supported by the Israel Science Foundation.

to change the "location" of X in $L_p^M(\mu)$. This is done using the following version of Lewis' theorem (see Theorem 2.1 in [S-Z] and Proposition 2.2(i) in [J-S 1]).

Theorem 1.2 *Let X be an n dimensional subspace of $L_p^M(\mu)$, $0 < p < \infty$. Then, one can find an integer $K \leq 3M/2$, a probability measure λ on $\{1,\ldots,K\}$ and a subspace F of $L_p(\lambda)$, isometric to X, which admits a basis $(h_i)_{i \leq n}$ orthogonal in $L_2(\lambda)$ such that $\sum_{i=1}^n h_i^2 \equiv n$. Moreover, for each $i \leq K$ we have $\lambda(\{i\}) \leq 2/K$.*

Next we will change the density in such a way that a new density ν will have the property $1/2K \leq \nu(\{i\}) \leq 2/K$. We define

$$\nu(\{i\}) = \frac{\lambda(\{i\}) + \frac{1}{K}}{2}$$

then ν is a probability measure on $\{1,\ldots,K\}$ and we can define an isometry $I : L_p^K(\lambda) \to L_p^K(\nu)$ by

$$I(f)(i) = f(i)\left(\frac{\lambda(\{i\})}{\nu(\{i\})}\right)^{1/p}.$$

Define $F_1 = I(F)$, then it is shown in [S-Z], that there is $K_1 \leq K/2$ and a subspace Y of $L_p^{K_1}(\nu)$, such that $d(F_1, Y) \leq 1 + c(p)A_{F_1}$, where

$$A_{F_1} = \mathbb{E} \sup_{x \in F_1, \|x\|_p \leq 1} \left|\sum_{i=1}^K \nu(\{i\})g_i|x(i)|^p\right|,$$

and the sequence $\{g_i\}_{i=1}^K$ denote here a sequence of independent standard Gaussian variables.
In Section 3 below, we will prove that

$$A_{F_1} \leq C(p)\left(\frac{n}{K}\log n \left[\log\frac{K}{n} + \log\log K\right]^2\right)^{\frac{1}{2}}. \tag{1}$$

Finally we can replace the bound given in Proposition 4.1 in [S-Z] by (1) and we follow Section 5 in [S-Z] to finish the proof of Theorem 1.

2 Entropy Estimates

Let $B_r(X) = \{x \in X : \|x\|_r \leq 1\}$. By $N(B_r(X), tB_s(X))$ we denote the metric entropy of $B_r(X)$ with respect to $tB_s(X)$, i.e., the minimal number of translates of $tB_s(X)$ needed to cover $B_r(X)$. We shall use the following result from [S-Z] (Proposition 3.1 there).

Lemma 2.1 *Let $0 < p < 2$, and let F be as in Theorem 1.2. Then, for any $t > 1$,*

$$\log N(B_p(F), tB_2(F)) \leq C(p)n(\log n)t^{-2p/(2-p)}.$$

Now we would like to show that the same estimate (up to a constant) is valid for F_1.

$$\|I(f)\|_{2,\nu}^2 = \sum_{i=1}^{K} \lambda(\{i\})|f(i)|^2 \left(\frac{2\lambda(\{i\})}{\lambda(\{i\}) + \frac{1}{K}}\right)^{2/p-1} \leq 2^{2/p-1}\|f\|_{2,\lambda}^2$$

and then

$$N(B_p(F_1), tB_2(F_1)) \leq N(B_p(F), 2^{1/p-1/2}tB_2(F)),$$

consequently

$$N(B_p(F_1), 2^{1/p-1/2}tB_2(F_1)) \leq C(p)n(\log n)t^{-2p/(2-p)}. \tag{2}$$

Let us define $B_p = B_p(F_1)$ and $B_2 = B_2(F_1)$.

Lemma 2.2

$$\log N(B_p + tB_2, 3tB_2) \leq C(p)n(\log n)t^{-2p/(2-p)}. \tag{3}$$

It is easy to see that $N(B_p + tB_2, 3tB_2) \leq \log N(B_p, tB_2)$ and then the lemma follows from (2).

3   A Bound on the Gaussian Process

As we explained in the introduction Theorem 1.1 will be proved once we establish the following theorem.

Theorem 3.1 *Let $0 < p < 1$, let g_i, $i = 1, \ldots, K$, be independent standard Gaussian random variables and assume $1/2K \leq \nu(\{i\}) \leq 2/K$. Then*

$$\mathbb{E} \sup_{x \in B_p} \left| \sum_{i=1}^{K} \nu(\{i\})g_i|x(i)|^p \right| \leq C(p) \left(\frac{n}{K} \log n \left[\log \frac{K}{n} + \log\log K\right]^2\right)^{\frac{1}{2}}. \tag{4}$$

To prove this theorem, we will use the majorizing measure bound (for details see [T2],[T3])

$$\mathbb{E} \sup_{x \in B_p} \left| \sum_{i=1}^{K} \nu(\{i\})g_i|x(i)|^p \right| \leq C\gamma(B_p, d), \tag{5}$$

where

$$\gamma(B_p, d) = \inf_{\mu} \sup_{x \in B_p} \int_0^{\infty} \sqrt{\log \frac{1}{\mu(B(x,\varepsilon))}} \, d\varepsilon. \tag{6}$$

Here the infimum is taken over all probability measures μ on B_p and $B(x, \varepsilon)$ is the d-ball where

$$d(y, z) = \Big(\sum_{i=1}^{K} [\nu(\{i\})(|y(i)|^p - |z(i)|^p)]^2 \Big)^{1/2}, \quad y, z \in B_p. \tag{7}$$

Define

$$\delta(y, z) = \Big(\sum_{i=1}^{K} \nu^2(\{i\})|y(i) - z(i)|^{2p} \Big)^{1/2}, \quad y, z \in B_p, \tag{8}$$

and let us show that δ is a metric. If $p < 1/2$ then $2p < 1$ and

$$\delta^2(y, z) \leq \delta^2(y, x) + \delta^2(x, z), \tag{9}$$

which implies

$$\delta(y, z) \leq \delta(y, x) + \delta(x, z). \tag{10}$$

If $1 > p \geq 1/2$ then $2p \geq 1$ and

$$\delta^{1/p}(y, z) \leq \delta^{1/p}(y, x) + \delta^{1/p}(x, z), \tag{11}$$

which again implies

$$\delta(y, z) \leq \delta(y, x) + \delta(x, z). \tag{12}$$

Moreover, $d(y, z) \leq \delta(y, z)$ for any $y, z \in B_p$, which clearly implies that $\gamma(B_p, d) \leq \gamma(B_p, \delta)$. Our goal now is to bound $\gamma(B_p, \delta)$.

Consider a metric space (T, ρ) and fix an integer $s \geq 1$. Following Talagrand [T2] we next define the notion of an **s-tree** .

Recall that a tree of subsets of T is a collection $\mathcal{F}$ of subsets of T with the property that for all $A, B \in \mathcal{F}$, either $A \cap B = \emptyset$, or $A \subset B$, or $B \subset A$. We say that B is a **son** of A if $B \subset A$ and

$$C \in F, \quad B \subset C \subset A \implies C = B \text{ or } C = A. \tag{13}$$

A **branch** of $\mathcal{F}$ is a sequence $A_1 \supset A_2 \supset \ldots$ such that A_{k+1} is a son for A_k. A branch is **maximal** if it is not contained in a longer branch. To each $A \in \mathcal{F}$ we denote by $N(A)$ the number of sons of A. Set $r = 8$. A tree $\mathcal{F}$ called an **s-tree** if we can associate to each $A \in \mathcal{F}$ an integer $n(A)$ in manner that

1. $\operatorname{diam}(A) \leq r^{-n(A)}$.
2. If $B \subset A$ then

$$n(B) \geq n(A) + s. \tag{14}$$

3. If B, B' are two distinct sons of A then

$$\rho(B, B') \geq \frac{1}{2}r^{-n(A)-1}.\qquad(15)$$

The **size** $\gamma(\mathcal{F}, \rho)$ of $\mathcal{F}$ is the infimum over all possible maximal branches of

$$\sum_{k \geq 1} r^{-n(A_k)}\sqrt{\log N(A_k)}.\qquad(16)$$

Theorem 3.2 (Talagrand [T2]) *For some universal constant C and for $s \geq 1$, any metric space (T, ρ), with $\gamma(T, \rho) < \infty$, contains an s-tree satisfying*

$$\gamma(\mathcal{F}, \rho) \geq \frac{1}{Cs}\gamma(T, \rho).\qquad(17)$$

Consider a parameter s to be determined later. We would like to bound the size of an s-tree $\mathcal{F}$ contained in (B_p, δ) and show that

$$\gamma(\mathcal{F}, \delta) \leq \sqrt{\frac{n \log n}{K}},\qquad(18)$$

for

$$s \leq C(p)\left[\log\frac{K}{n} + \log\log K\right],\qquad(19)$$

thus establish (4).

Let n_1 to be the smallest integer such that for some $A \in \mathcal{F}$ we have $n(A) = n_1$. We pick A_1 such that $n(A_1) = n_1$. Starting with A_1 we will construct a branch $(A_k)_{k \geq 1}$ of $\mathcal{F}$, disjoint subsets I_k of $J = \{1, \ldots, K\}$ and non negative numbers $(a_k)_{k \geq 1}$ such that

$$\forall y \in A_{k+1}, \quad \sum_{i \in I_k} \nu(\{i\})|y(i)|^p \geq a_k,\qquad(20)$$

and

$$r^{-n_k}\sqrt{\log N_k} \leq \min\left\{C(p)\left(\frac{1}{\log K} + a_k\right)\sqrt{\frac{n \log n}{K}}, \quad C(p)r^{-n_k}\sqrt{n}\right\},\qquad(21)$$

where $n_k = n(A_k)$, $N_k = N(A_k)$.
Observe that necessarily $\sum a_k \leq 1$.
Once (21) is established, to bound $\gamma(\mathcal{F}, \delta)$, we simply use that by definition

$$\gamma(\mathcal{F}, \delta) \leq \sum_{k \geq 1} r^{-n_k}\sqrt{\log N_k}.\qquad(22)$$

Consider the smallest integer k_0 with $N_{k_0} \geq 2$. Fix some $k_1 > k_0$ then

$$\sum_{k \geq 1} r^{-n_k} \sqrt{\log N_k} = \sum_{k \geq k_0} r^{-n_k} \sqrt{\log N_k}$$

$$\leq C(p)\Big(\frac{k_1 - k_0}{\log K} + 1\Big)\sqrt{\frac{n \log n}{K}} + C(p) r^{-n_{k_1}} \sqrt{n}.$$

It is easy to see that the δ-diameter of B_p is at most 4. Since $N_{k_0} \geq 2$, $n_{k_0} > 1$; thus, since $n_{k_1} \geq n_{k_0} + (k_1 - k_0)s$, if $k_1 - k_0 = \log K$, we have $r^{-n_{k_1}} \leq 1/\sqrt{K}$, so that in particular, we see that

$$\gamma(\mathcal{F}, \delta) \leq C(p)\sqrt{\frac{n \log n}{K}}. \tag{23}$$

We now turn to the construction. To make the construction possible, we will also require that

$$\nu(I_k) \leq (\bar{c}K)^{-1} r^{2n_k}, \tag{24}$$

where $\bar{c} = \bar{c}(p) \in (0, 1]$ is a constant to be chosen later.
We set $I_0 = \emptyset$. Assuming that A_k, I_{k-1} have been constructed, we construct A_{k+1}, I_k.

Using standard volume considerations in the n-dimensional space $(F_1, \| * \|_{2p, \nu^2})$, we observe that $\log N_k \leq C(p)n$ so that

$$r^{-n_k} \sqrt{\log N_k} \leq C(p) r^{-n_k} \sqrt{n}, \tag{25}$$

and the second term in (21) does not cause any problem.

Case 1. If

$$r^{-n_k} \sqrt{\log N_k} \leq \frac{1}{\log K} \sqrt{\frac{n \log n}{K}}, \tag{26}$$

then set $I_k = \emptyset, a_k = 0$, and we chose A_{k+1} to be any of the sons of A_k.

Case 2. We have

$$r^{-n_k} \sqrt{\log N_k} \geq \frac{1}{\log K} \sqrt{\frac{n \log n}{K}}, \tag{27}$$

(i.e. the number of sons is not too small).
We denote as $(X_l)_{l \leq N_k}$ the sons of A_k. For each $l \leq N_k$, we pick $x_l \in X_l$. We set $J_k = \bigcup_{l \leq k-1} I_l$, then

$$\nu(J_k) \leq 2(\bar{c}K)^{-1} r^{2n_{k-1}}. \tag{28}$$

We set $\eta_k = \bar{c}Kr^{-2n_k}$ and we consider

$$S = \sup_{l \leq N_k} \sum_{J \setminus J_k} \nu(\{i\})|x_l(i)|^p 1_{\{|x_l(i)|^p \geq \eta_k\}}. \tag{29}$$

The **Main Claim** in Case 2 is that

$$c(p)\frac{r^{-n_k}\sqrt{\log N_k}}{\sqrt{\frac{n \log n}{K}}} \leq S. \tag{30}$$

Before proving the "Main Claim", we show how to conclude the construction. We select $l \leq N_k$ that achieves the sup in S. We set $A_{k+1} = X_l$ and

$$I_k = \{i \leq K; i \notin J_k; |x_l(i)|^p \geq \eta_k\}. \tag{31}$$

Note that

$$\nu(I_k) \leq \eta_k^{-1} = (\bar{c}K)^{-1}r^{2n_k}.$$

We also set $a_k = S/2$. Consider $y \in A_{k+1}$ (we need to prove that $\sum_{i \in I_k} \nu(\{i\})|y(i)|^p \geq a_k$). Then

$$\delta(y, x_l) \leq r^{-n_{k+1}} \leq r^{-n_k-s} \tag{32}$$

since $\mathcal{F}$ is an s-tree. Thus

$$\sum_{i \in I_k} \nu(\{i\})||y(i)|^p - |x_l(i)|^p| \leq \sum_{i \in I_k} \nu(\{i\})|y(i) - x_l(i)|^p \leq \tag{33}$$

(by Cauchy-Schwartz)

$$\left(\sum_{i \in I_k} \nu^2(\{i\})|y(i) - x_l(i)|^{2p}\right)^{1/2}(\text{card}\{i : i \in I_k\})^{1/2} \leq \tag{34}$$

(by $\nu(\{i\}) \geq \frac{1}{2k}$ and $\nu(I_k) \leq (\bar{c}K)^{-1}r^{2n_k}$)

$$\left(\sum_{i \in I_k} \nu^2(\{i\})|y(i) - x_l(i)|^{2p}\right)^{1/2}(2K\nu(I_k))^{1/2} \leq C(p)r^{-s}. \tag{35}$$

Now by (27) and the "Main Claim" we have

$$S \geq c(p)\frac{1}{\log K}, \tag{36}$$

and we get

$$\sum_{i \in I_k} \nu(\{i\})||y(i)|^p - |x_l(i)|^p| \leq \frac{S}{2}, \tag{37}$$

as soon as

$$r^s \geq C(p)(\log K), \tag{38}$$

i.e. $s \geq C(p) \log \log K$.
Finally,

$$\sum_{i \in I_k} \nu(\{i\})|y(i)|^p \geq \frac{S}{2} = a_k. \tag{39}$$

Proof of the "Main Claim". We have that

$$\forall l, l' \leq N_k, \quad l \neq l', \quad \frac{r^{-n_k-1}}{2} \leq \delta(x_l, x_{l'}) \leq r^{-n_k}, \tag{40}$$

define $\mathcal{X} = \{x_l\}_{l \leq N_k}$.

Proposition 3.3 *If $r^{2s} \geq C(p)\frac{K\log^2 K}{n\log n}$ ($s \geq C(p)\left(\log \frac{K}{n} + \log\log K\right)$), then we can find a subset L of $\{1, \ldots, N_k\}$ with*

$$\log \operatorname{card} L \geq \frac{1}{2} \log N_k \tag{41}$$

and such that for all $l, l' \in L, l \neq l'$:

$$\sum_{i \in J \setminus J_k} \nu^2(\{i\})|x_l(i) - x_{l'}(i)|^{2p} \geq \frac{r^{-2n_k-2}}{8}. \tag{42}$$

Proof. Consider

$$H = \left\{x \in B_p : \sum_{i \in J_k} \nu^2(\{i\})|x(i)|^{2p} \leq \frac{r^{-2n_k-2}}{32}\right\}.$$

Fix $l_0 \leq N_k$, from $\delta^2(x_l, x_{l_0}) \leq \frac{r^{-2n_k-2}}{32}32r^2$, we get that

$$x_l - x_{l_0} \in (32r^2)^{1/2p}H, \text{ for all } l \leq N_k,$$

i.e. $\mathcal{X} \subset x_{l_0} + (32r^2)^{1/2p}H$.
By standard volume considerations we get

$$\log N(\mathcal{X}, H) \leq \log((8r^2)^{1/2p}H, H)$$
$$\leq C(p)\operatorname{card}J_k \leq C(p)r^{2n_k-1} \leq C(p)r^{-2s}r^{2n_k} \leq$$

(by inequality (27) and bound on s from the statement of the proposition)

$$C(p)r^{-2s}\frac{K\log^2 K}{n\log n}\log N_k \leq \frac{1}{2}\log N_k,$$

i.e. $\log N(\mathcal{X}, H) \le \frac{1}{2} \log N_k$ or $N(\mathcal{X}, H) \le \sqrt{N_k}$.

Next we use the "pigeon hole principle": we can cover $\mathcal{X}$ by Q translates of the set H where $Q \le \sqrt{N_k}$. One of these sets must contain at least $\sqrt{N_k}$ points of $\mathcal{X}$, let L to be the set of indexes of these points. Finally, applying the triangle inequality we get that

$$\sum_{i \in J_k} \nu^2(\{i\})|x_l(i) - x_{l'}(i)|^{2p} \le \frac{r^{-2n_k-2}}{8},\tag{43}$$

for all $l, l' \in L$, $l \ne l'$, and (42) easily follows from (43) and (40). $\square$

Proposition 3.4 *Let L satisfy the conditions of Proposition 3.3 and $l, l' \in L$, $l \ne l'$. Then*

$$\|x_l - x_{l'}\|_2^{2p/(2-p)} \ge C \frac{Kr^{-2n_k}}{S^q},\tag{44}$$

where $q = 2(1-p)/(2-p)$.

Proof. Fix $l, l' \in L, l \ne l'$, then

$$\sum_{J \setminus J_k} \nu^2(\{i\})|x_l(i) - x_{l'}(i)|^{2p} 1_{\{|x_l(i)|^p \le \eta_k, |x_{l'}(i)|^p \le \eta_k\}} \le\tag{45}$$

(by $\nu(\{i\}) \le \frac{2}{K}$ and $|x_l(i) - x_{l'}(i)|^p \le |x_l(i)|^p + |x_{l'}(i)|^p$)

$$\frac{2}{K} 2\eta_k \sum_{J \setminus J_k} \nu(\{i\})|x_l(i) - x_{l'}(i)|^p \le \frac{8(\bar{c}Kr^{-2n_k})}{K} \le \frac{r^{-2n_k-2}}{16},\tag{46}$$

as soon as $\bar{c} \le (128r^2)^{-1}$. By the properties of L,

$$\sum_{i \in J \setminus J_k} \nu^2(\{i\})|x_l(i) - x_{l'}(i)|^{2p} 1_{\{|x_l(i)|^p > \eta_k \text{ or } |x_{l'}(i)|^p > \eta_k\}} \ge \frac{r^{-2n_k-2}}{16}.\tag{47}$$

On the other hand Hölder inequality implies (see e.g. [S-Z] proof of Proposition 4.1),

$$\sum_{i \in J \setminus J_k} \nu^2(\{i\})|x_l(i) - x_{l'}(i)|^{2p} 1_{\{|x_l(i)|^p > \eta_k \text{ or } |x_{l'}(i)|^p > \eta_k\}} \le$$

$$6K^{-1}\|x_l - x_{l'}\|_2^{\frac{2p}{2-p}} \left(\sum_{i \in J \setminus J_k} \nu(\{i\})|x_l(i) - x_{l'}(i)|^p 1_{\{|x_l(i)|^p > \eta_k \text{ or } |x_{l'}(i)|^p > \eta_k\}} \right)^q.$$

But $|x_l(i) - x_{l'}(i)|^p \le |x_l(i)|^p + |x_{l'}(i)|^p$, and it is easy to see that

$$(|x_l(i)|^p + |x_{l'}(i)|^p) 1_{\{|x_l(i)|^p > \eta_k \text{ or } |x_{l'}(i)|^p > \eta_k\}}$$
$$\le 2(|x_l(i)|^p 1_{\{|x_l(i)|^p \ge \eta_k\}} + |x_{l'}(i)|^p 1_{\{|x_{l'}(i)|^p \ge \eta_k\}}),\tag{48}$$

and by the definition of S (29),

$$\sum_{i \in J \setminus J_k} \nu(\{i\})|x_l(i) - x_{l'}(i)|^p 1_{\{|x_l(i)|^p > \eta_k \text{ or } |x_{l'}(i)|^p > \eta_k\}} \leq 4S. \qquad (49)$$

$\square$

Fix some $l_0 \in L$ and consider the points $z_l = x_l - x_{l_0}$. Then $z_l \in 2^{1/p} B_p$ and

$$\|z_l - z_{l'}\|_2^{2p/(2-p)} \geq C \frac{K r^{-2n_k}}{S^q}. \qquad (50)$$

Applying the entropy estimate from Lemma 2.1 we get

$$\log N_k \leq C(p) \frac{n \log n}{K} r^{2n_k} S^q, \qquad (51)$$

or

$$c(p) \frac{r^{-n_k} \sqrt{\log N_k}}{\sqrt{\frac{n \log n}{K}}} \leq S^{q/2}. \qquad (52)$$

This estimation is weaker then the "Main Claim" (because $q/2 = \frac{1-p}{2-p} < 1$), however, from (27) and (51) we can get the following very useful lower bound on S:

$$\frac{c(p)}{\log K} \leq S^{\frac{1-p}{2-p}}, \qquad (53)$$

and finally

$$\left(\frac{c(p)}{\log K} \right)^{\frac{2-p}{1-p}} \leq S. \qquad (54)$$

Proposition 3.5 If $r^s \geq \left(\frac{\log K}{c(p)} \right)^{\frac{2-p}{1-p}}$ $(s \geq C(p) \log \log K)$, then

$$\sum_{i \in J_k} \nu(\{i\})|z_l(i)|^p \leq S, \qquad (55)$$

for every $l \in L$.

Proof.

$$\sum_{i \in J_k} \nu(\{i\})|x_l(i) - x_{l_0}(i)|^p$$

$$\leq \left(\sum_{i \in J_k} \nu^2(\{i\})|x_l(i) - x_{l_0}(i)|^{2p} \right)^{1/2} \left(2K\nu(J_k) \right)^{1/2}$$

$$\leq C(p) r^{-n_k + n_{k-1}} \leq C(p) r^{-s} \leq \left(\frac{\log K}{c(p)} \right)^{-\frac{2-p}{1-p}} \leq S. \qquad (56)$$

$\square$

We are now ready to conclude the proof of the "Main Claim". For $l \in L$ we represent z_l as

$$\{z_l(i)\}_{i=1}^K = \{z_l(i)\}_{i \in J_k} + \{z_l(i)\}_{i \notin J_k, |z_l(i)|^p \geq 2\eta_k} + \{z_l(i)\}_{i \notin J_k, |z_l(i)|^p < 2\eta_k}. \tag{57}$$

Applying Proposition 3.5 we get $\{z_l(i)\}_{i \in J_k} \in S^{1/p} B_p$ and from (48) and the definition of S (29) we get that

$$\{z_l(i)\}_{i \notin J_k, |z_l(i)|^p \geq 2\eta_k} \in (4S)^{1/p} B_p. \tag{58}$$

Using $\|x\|_2^2 \leq \|x\|_\infty^{2-p} \|x\|_p^p$, we get

$$\{z_l(i)\}_{i \notin J_k, |z_l(i)|^p < 2\eta_k} \in 2(2\eta_k)^{\frac{2-p}{2p}} B_2. \tag{59}$$

Combining, we get

$$z_l \in C(p) \left(S^{1/p} B_p + \eta_k^{\frac{2-p}{2p}} B_2 \right). \tag{60}$$

By (50) and (60),

$$\log N_k \leq C \log N \left(S^{1/p} B_p + \eta_k^{\frac{2-p}{2p}} B_2, \ c(p) \frac{(Kr^{-2n_k})^{\frac{2-p}{2p}}}{S^{\frac{1-p}{p}}} B_2 \right)$$

$$= C \log N \left(B_p + (\eta_k)^{\frac{2-p}{2p}} S^{-1/p} B_2, \ c(p) \frac{(Kr^{-2n_k})^{\frac{2-p}{2p}}}{S^{\frac{2-p}{p}}} B_2 \right). \tag{61}$$

To improve upon (51) we want to find a good bound for (61), which is done by applying Lemma 2.2. For that we need

$$3(\eta_k)^{\frac{2-p}{2p}} S^{-1/p} \leq c(p) \frac{(Kr^{-2n_k})^{\frac{2-p}{2p}}}{S^{\frac{2-p}{p}}}, \tag{62}$$

but, by definition, $\eta_k = \bar{c} K r^{-2n_k}$ and $S \leq 1$, so for (62) to be true we just have to take $\bar{c} \leq c(p)$.
Finally, we apply Lemma 2.2 and get

$$\log N_k \leq C(p) \frac{n \log n}{K} r^{2n_k} S^2, \tag{63}$$

which establish the "Main Claim" (30). $\square$

References

[B-L-M] Bourgain J., Lindenstrauss J., Milman V. (1989) Approximation of zonoids by zonotopes. Acta. Math. 162:73–141

[J-S 1]   Johnson W.B., Schechtman G. (1994) Computing p-summing norms with few vectors. Israel Journal of Math. 87:19–31

[J-S 2]   Johnson W.B., Schechtman G. Finite dimensional L_p spaces, $1 \leq p < \infty$. In preparation

[S-Z]   Schechtman G., Zvavitch A. Embedding subspaces of L_p into ℓ_p^N, $0 < p < 1$. Math. Nachr, to appear

[T1]   Talagrand M. (1990) Embedding subspaces of L_1 into L_1^N. Proc. Amer. Math. Soc. 108:363–369

[T2]   Talagrand M. (1995) Embedding subspaces of L_p in L_p^N. In: Geometric Aspects of Functional Analysis (Israel, 1992-1994), Oper. Theory Adv. Appl., 77, Birkhauser, Basel, 311–325

[T3]   Talagrand M. (1996) Majorizing measures: the generic chaining. Ann. of Probability 24(3):1049–1103

Israel Seminar on Geometric Aspects of Functional Analysis (GAFA) 1996-2000

SEMINAR TALKS

November 8, 1996

1. **B. Maurey** (Paris) Gowers' dichotomy and the solution of the homogeneous space problem
2. **S. Bobkov** (Syktyvkar/Bielefeld) Discrete Poincare-type inequalities (joint work with F. Goetze)
3. **S. Alesker** (Tel Aviv) Polynomial rotation invariant valuations

November 12, 1996

S. Bobkov (Bielefeld) Logarithmic Sobolev inequalities and concentration of measure

November 19, 1996

A. Litvak (Tel Aviv) Unitary averages of norms and quasi-norms

November 26, 1996

G. Schechtman (Rehovot) Embedding l_p^k subspaces of l_p^n a partial review

November 29, 1996

1. **I. Benyamini** (Rehovot) Some recent results in percolation
2. **O. Schramm** (Rehovot) Expanding graphs contain expanding trees
3. **E. Gluskin** (Tel Aviv) Kashin's approach to the Menshov's correction theorem

February 21, 1997

1. **L. Aizenberg** (Ramat Gan) The duality in complex analysis
2. **W. Kaup** (Tubingen) On the applications of bounded symmetric domains in Functional Analysis
3. **V. Zahariuta** (Rostov-on-Don/Istanbul) Hadamard-type inequalities for harmonic functions and their application to approximations

March 7, 1997

1. **S. Kwapien** (Warsaw) Some maximal inequalities for sums of independent random variables
2. **A. Vershik** (St. Petersburg) Weak convergences and invariant measures in the theory of big groups

March 28, 1997

1. **Y. Shalom** (Jerusalem) Property T of Kazhdan: some applications and new quantitative results

    2. **B. Tsirelson** (Tel Aviv) Triple points: from non-Brownian filtrations to harmonic measures

    3. **V. Bergelson** (Haifa/Columbus, Ohio) Polynomials, IP-convergence, Ramsey theory

April 4, 1997

    1. **P.M. Gruber** (Vienna) Stability of Fejes Toth's moment lemma and its applications

    2. **A. Pelczynski** (Warsaw) The weak type problem of canonical projections of anisotropic Sobolev space: the solution for second order smoothnesses

    3. **V. Milman** (Tel Aviv) Averaging of norms by rotations (joint work with Litvak and Schechtman)

November 7, 1997

    1. **N. Alon** (Tel Aviv) Nearly orthogonal vectors and Shannon capacity

    2. **E. Bombieri** (Princeton, New Jersey) Diophantine equations in low dimension (as part of the Sackler's lecture series)

November 14, 1997

    1. **J. Lindenstrauss** (Jerusalem) Nonlinear quotients (joint work with Bates, Johnson, Preiss and Schechtman)

    2. **M. Gromov** (Bures-sur-Yvette/New York) Infinite dimensional dynamical systems

December 19, 1997

    1. **V. Milman** (Tel Aviv) Asymptotic formulas and isotropicy positions in convex geometry

    2. **A. Pogorelov** (Kharkov) On foundations of geometry (framework Blumenthal distinguished lectures on geometry)

January 9, 1998

    1. **A. Giannopoulos** (Iraklion) Isotropic surface area measures (joint work with Papadimitrakis)

    2. **K. Oleszkiewicz** (Warsaw) Comparison of moments of sums of independent random vectors via Poincare-type inequality

February 23, 1998

    1. **D. Preiss** (London) Some new facts on null sets in Banach spaces

    2. **S. Reisner** (Haifa) Some inequalities in convex geometry

November 20, 1998

    1. **V. Palamodov** (Tel Aviv) Reconstruction of a function from its arc means

    2. **G. Kalai** (Jerusalem) Boolean functions: sensitivity, Fourier expansion, concentration and application to percolation

December 11, 1998

    1. **L. Pastur** (Paris) Some mathematical aspects of the spin glass theory

2. **R. Raz** (Rehovot) Quantum communication complexity and some properties of S^n

December 25, 1998

1. **Y. Yomdin** (Rehovot) A tractable problem on the way (hopefully) to counting limit cycles of polynomial vector fields
2. **E. Gordon** (Nizhnii Novgorod) Finite dimensional approximations of pseudodifferential operators on the locally compact Abelian groups

January 8, 1999

1. **A. Giannopoulos** (Iraklion) Isotropic positions of convex bodies
2. **B. Tsirelson** (Tel Aviv) Trees, not cubes: hypercontractivity, coziness, and noise stability

January 29, 1999

1. **V. Matsaev and M. Sodin** (Tel Aviv) Variations on the theme of M. Riesz and Kolmogorov
2. **D. Romik** (Tel Aviv) The central limit problem for convex bodies (after K. Ball, Anttila and Perissinaki)

April 16, 1999

1. **V. Pestov** (Wellington) Fixed points, amenability, and measure concentration
2. **E. Glasner** (Tel Aviv) On Polish monothetic groups
3. **V. Milman** (Tel Aviv) Concentration phenomenon without metric: new view

May 7, 1999

1. **B. Mityagin** (Columbus, Ohio) Eigenvalues of periodic Schroedinger operator with complex potential
2. **M. Sodin** (Tel Aviv) Zeros of Gaussian entire functions: the mean, rigidity, and large deviations

May 28, 1999

1. **E. Matouskova** (Prague) Measure-null sets in infinite dimensional Banach spaces
2. **I. Benjamini** (Rehovot) Graph indexed random walks

August 19, 1999

J. Bourgain (Princeton, New Jersey) Lyapounov exponents and quasi periodic localization

December 10, 1999

1. **A. Pajor** (Marne-la-Vallee) On width of lattice-point-free convex body
2. **S. Alesker** (Tel Aviv) Affirmative solution of P. McMullen's conjecture on translation invariant valuations

December 24, 1999

1. **B. Rubin** (Jerusalem) Arithmetical properties of generalized Minkowski-Funk transforms and small denominators

2. **V. Bergelson** (Columbus, Ohio) A generalization of Khintchine's recurrence theorem

March 17, 2000

F. Barthe (Marne-la-Vallee) Spherical models in isoperimetry

In the framework of the educational seminar:

March 28, 2000

R. Schneider (Freiburg) Stability versions of inequalities and their applications

April 11, 2000

G. Pisier (College Station, Texas/Paris) Topics on non-commutative analysis in L_p

PIMS* Workshop on Geometric Functional Analysis and Conference on Convex Geometric Analysis

(June–July 1999)

WORKSHOP PROGRAM
(Organizers: V. Milman and N. Tomczak-Jaegermann)

Monday, June 28

1. **C. Schütt** (Kiel) Random polytops
2. **R. Latala** (Atlanta, Georgia) and **K. Oleszkiewicz** (Warsaw) S-conjecture

Tuesday, June 29

1. **G. Schechtman** (Rehovot) Graph indexed random walks
2. **F. Barthe** (Marne-la-Vallee) Brascamp-Lieb inequality and measure transportation

Monday, July 5

1. **M. Rudelson** (Columbia, Missouri) Lipschitz embeddings of Levy families
2. **H. Rosenthal** (Austin, Texas) M-approximate identities for X in Y with $X \subset Y$

Tuesday, July 6

1. **T. Schlumprecht** (College Station, Texas) Asymptotic structures of Banach spaces
2. **G. Androulakis** (College Station, Texas) Candidates for prime spaces
3. **S. Argyros** (Athens) An example of non-separable no ℓ_1, c_0 or reflexive
4. **S. Bobkov** (Syktyvkar) Sobolev inequalities

Wednesday, July 7

1. **S. Bobkov** (Syktyvkar) On log-Sobolev inequalities
2. **M. Meyer** (Marne-la-Vallee) Maximal volume sections

Thursday, July 8

1. **S. Argyros** (Athens) An example of non-separable no ℓ_1, c_0 or reflexive (continued)
2. **M. Rudelson** (Columbia, Missouri) Levy families
3. **A. Pelczynski** (Warsaw) Ellipsoidal sections of convex bodies
4. **G. Schechtman** (Rehovot) Lewis' change of density revisited

---

* The Pacific Institute for the Mathematical Sciences, Vancouver

Friday, July 9

1. **A. Pajor** (Marne-la-Vallee) Questions on non-symmetry
2. **A. Koldobsky** (Columbia, Missouri) Functional analysis approach to intersection bodies
3. **J. Lindenstrauss** (Jerusalem) A simple proof of James' characterization of ω compact sets (in the separable case)
4. **W. Banaszczyk** (Lódź) Balancing vectors in $\mathbb{R}^n$

Monday, July 12

1. **H. Koenig** (Kiel) Variants of Knintchin inequality
2. **F. Barthe** (Marne-la-Vallee) Isoperimetric inequalities of Gaussian type

Tuesday, July 13

1. **K. Oleszkiewicz** (Warsaw) Between Sobolev and Poincaré
2. **Y. Gordon** (Haifa) Volume ratios and local theory

Wednesday, July 14

1. **P. Mankiewicz** (Warsaw) Convexified distance between random quotients l_1^n
2. **S. Reisner** (Haifa) Dropping a vertex or a facet from a convex polytope

Thursday, July 15

1. **O. Guédon** (Paris) Random methods in geometry of Schatten classes
2. **A. Litvak** (Edmonton, Alberta) Random projections of convex bodies and distances between them

CONFERENCE PROGRAM
(Scientific Committee: J. Bourgain, T. Gowers, M. Gromov, G. Pisier, N. Tomczak-Jaegermann)

Wednesday, June 30

1. **T. Gowers** (Cambridge) Szemeredi's theorem
2. **A. Giannopoulos** (Iraklion) Isotropic positions of convex bodies
3. **V. Milman** (Tel Aviv) Concentration phenomenon without metric; phase transitions in the asymptotic geometric analysis
4. **J. Bourgain** (Princeton, New Jersey) Non-perturbative localization and semi-algebraic sets
5. **S. Alesker** (Paris/Tel Aviv) Continuous valuations on convex sets

Thursday, July 1

1. **G. Pisier** (College Station, Texas/Paris) Similarity degree and random matrices
2. **P. Milman** (Toronto) All you ever wanted to know about resolution of singularities
3. **S. Bobkov** (Syktyvkar) Concentration phenomena in probability
4. **T. Odell** (Austin, Texas) and **T. Schlumprecht** (College Station, Texas) Asymptotic properties of Banach spaces under renormings
5. **N. Tomczak-Jaegermann** (Edmonton, Alberta) Asymptotic finite-dimensional structure of infinite-dimensional Banach spaces

Friday, July 2

1. **N. Ghoussoub** (Vancouver, British Columbia) Liouville theorems and the De Giorgi and Gibbons conjectures
2. **M. Gromov** (Bur-sur-Yvette/New York)
3. **M. Talagrand** (Columbus, Ohio) Spin glasses and combinatorial optimization

Saturday, July 3

1. **J. Lindenstrauss** (Jerusalem) Exceptional sets in Banach spaces
2. **W. Johnson** (College Station, Texas) Affine approximation of Lipschitz functions
3. **N. Kalton** (Columbia, Missouri) Local theory of Banach spaces and approximation of polynomials

Workshop on Geometric Convex Analysis (March 2000)

(Organized by V. Milman, Y. Gordon, J. Lindenstrauss and G. Schechtman
Supported by I.S.F. and Minkowski Center for Geometry)

Wednesday, March 15

1. **E. Werner** (Clevelend, Ohio) Santalo's regions and connected inequalities
2. **Y. Gordon** (Haifa) Relations between normed ideal theory, volume estimates and local theory
3. **J. Zinn** (College Station, Texas) Concentration on the ℓ_p^n ball
4. **R. Villa** (Seville) Concentration of volume in classical spaces
5. **C. Schuett** (Kiel) Random and best approximation of convex bodies by polytopes

Thursday, March 16

1. **D. Burago** (University Park, Pennsylvania) A few remarks on surfaces in normed spaces
2. **A. Litvak** (Haifa) Projections of convex bodies

Wednesday, March 22

1. **L. Pastur** (Paris) Selfaveraging property of the free energy and related problems of many random variables
2. **A. Pajor** (Marne-la-Vallee) Discussion on lattice width vs. geometric width
3. **R. Schneider** (Freiburg) On intersection bodies and projection bodies

Thursday, March 23

1. **M. Ludwig** (Vienna) Invariant valuations on the space of convex bodies
2. **S. Alesker** (Tel Aviv) Solution of McMullen's conjecture and further applications
3. **A. Giannopoulos** (Iraklion) Representation of log-concave measures by discrete sets of points

Friday, March 24

1. **R. Wagner** (Tel Aviv) Random graph - homomorphism (after J.Kahn)
2. **F. Barthe** (Marne-la-Vallee) Spherical models in isoperimetry
3. **S. Argyros** (Athens) Recent results concerning mixed Tsirelson spaces

Workshop on Geometric Convex Analysis (March 2000)

Technion Institute of Advanced Studies in Mathematics
(Organized by Y. Gordon and A. Litvak)

Sunday, March 19

1. **F. Barthe** (Marne-la-Vallee) Isoperimetry for product measures, comparison with the Gaussian case
2. **R. Villa** (Seville) Concentration of the distance for finite dimensional normed spaces
3. **A. Zvavitch** (Rehovot) Embeddings of subspaces of L_p $(0 < p < 1)$ in ℓ_p^N via majorizing measures
4. **D. Cordero** (Marne-la-Vallee, France) Prekopa-Leindler inequalities in manifolds

Monday, March 20

1. **M. Zippin** (Jerusalem) A local minimality property of translation invariant projections on $L_p(G)$
2. **A. Naor** (Jerusalem) Isomorphic embeddings of ℓ_p^m in ℓ_p^n
3. **O. Guedon** (Paris) Euclidean projections of a p-convex body
4. **A. Tsolomitis** (Samos) John's theorem for an arbitrary pair of convex bodies
5. **S. Reich** and **A. Zaslavski** (Haifa) The set of divergent descent methods in a Banach space is sigma-porous
6. **G. Schechtman** (Rehovot) Affine approximation of Lipschitz maps between Banach spaces

Printing: Weihert-Druck GmbH, Darmstadt
Binding: Buchbinderei Schäffer, Grünstadt

Lecture Notes in Mathematics

For information about Vols. 1–1560
please contact your bookseller or Springer-Verlag

Vol. 1706: S. Yu. Pilyugin, Shadowing in Dynamical Systems. XVII, 271 pages. 1999.

Vol. 1707: R. Pytlak, Numerical Methods for Optimal Control Problems with State Constraints. XV, 215 pages. 1999.

Vol. 1708: K. Zuo, Representations of Fundamental Groups of Algebraic Varieties. VII, 139 pages. 1999.

Vol. 1709: J. Azéma, M. Émery, M. Ledoux, M. Yor (Eds), Séminaire de Probabilités XXXIII. VIII, 418 pages. 1999.

Vol. 1710: M. Koecher, The Minnesota Notes on Jordan Algebras and Their Applications. IX, 173 pages. 1999.

Vol. 1711: W. Ricker, Operator Algebras Generated by Commuting Projections: A Vector Measure Approach. XVII, 159 pages. 1999.

Vol. 1712: N. Schwartz, J. J. Madden, Semi-algebraic Function Rings and Reflectors of Partially Ordered Rings. XI, 279 pages. 1999.

Vol. 1713: F. Bethuel, G. Huisken, S. Müller, K. Steffen, Calculus of Variations and Geometric Evolution Problems. Cetraro, 1996. Editors: S. Hildebrandt, M. Struwe. VII, 293 pages. 1999.

Vol. 1714: O. Diekmann, R. Durrett, K. P. Hadeler, P. K. Maini, H. L. Smith, Mathematics Inspired by Biology. Martina Franca, 1997. Editors: V. Capasso, O. Diekmann. VII, 268 pages. 1999.

Vol. 1715: N. V. Krylov, M. Röckner, J. Zabczyk, Stochastic PDE's and Kolmogorov Equations in Infinite Dimensions. Cetraro, 1998. Editor: G. Da Prato. VIII, 239 pages. 1999.

Vol. 1716: J. Coates, R. Greenberg, K. A. Ribet, K. Rubin, Arithmetic Theory of Elliptic Curves. Cetraro, 1997. Editor: C. Viola. VIII, 260 pages. 1999.

Vol. 1717: J. Bertoin, F. Martinelli, Y. Peres, Lectures on Probability Theory and Statistics. Saint-Flour, 1997. Editor: P. Bernard. IX, 291 pages. 1999.

Vol. 1718: A. Eberle, Uniqueness and Non-Uniqueness of Semigroups Generated by Singular Diffusion Operators. VIII, 262 pages. 1999.

Vol. 1719: K. R. Meyer, Periodic Solutions of the N-Body Problem. IX, 144 pages. 1999.

Vol. 1720: D. Elworthy, Y. Le Jan, X-M. Li, On the Geometry of Diffusion Operators and Stochastic Flows. IV, 118 pages. 1999.

Vol. 1721: A. Iarrobino, V. Kanev, Power Sums, Gorenstein Algebras, and Determinantal Loci. XXVII, 345 pages. 1999.

Vol. 1722: R. McCutcheon, Elemental Methods in Ergodic Ramsey Theory. VI, 160 pages. 1999.

Vol. 1723: J. P. Croisille, C. Lebeau, Diffraction by an Immersed Elastic Wedge. VI, 134 pages. 1999.

Vol. 1724: V. N. Kolokoltsov, Semiclassical Analysis for Diffusions and Stochastic Processes. VIII, 347 pages. 2000.

Vol. 1725: D. A. Wolf-Gladrow, Lattice-Gas Cellular Automata and Lattice Boltzmann Models. IX, 308 pages. 2000.

Vol. 1726: V. Marić, Regular Variation and Differential Equations. X, 127 pages. 2000.

Vol. 1727: P. Kravanja, M. Van Barel, Computing the Zeros of Analytic Functions. VII, 111 pages. 2000.

Vol. 1728: K. Gatermann, Computer Algebra Methods for Equivariant Dynamical Systems. XV, 153 pages. 2000.

Vol. 1729: J. Azéma, M. Émery, M. Ledoux, M. Yor, Séminaire de Probabilités XXXIV. VI, 431 pages. 2000.

Vol. 1730: S. Graf, H. Luschgy, Foundations of Quantization for Probability Distributions. X, 230 pages. 2000.

Vol. 1731: T. Hsu, Quilts: Central Extensions, Braid Actions, and Finite Groups,. XII, 185 pages. 2000.

Vol. 1732: K. Keller, Invariant Factors, Julia Equivalences and the (Abstract) Mandelbrot Set. X, 206 pages. 2000.

Vol. 1733: K. Ritter, Average-Case Analysis of NumericalProblems. IX, 254 pages. 2000.

Vol. 1734: M. Espedal, A. Fasano, A. Mikelić, Filtration in Porous Media and Industrial Applications. Cetraro 1998. Editor: A. Fasano. 2000.

Vol. 1735: D. Yafaev, Scattering Theory: Some Old and New Problems. XVI, 169 pages. 2000.

Vol. 1736: B. O. Turesson, Nonlinear Potential Theory and Weighted Sobolev Spaces. XIV, 173 pages. 2000.

Vol. 1737: S. Wakabayashi, Classical Microlocal Analysis in the Space of Hyperfunctions. VIII, 367 pages. 2000.

Vol. 1738: M. Emery, A. Nemirovski, D. Voiculescu, Lectures on Probability Theory and Statistics. XI, 356 pages. 2000.

Vol. 1739: R. Burkard, P. Deuflhard, A. Jameson, J.-L. Lions, G. Strang, Computational Mathematics Driven by Industrial Problems. Martina Franca, 1999. Editors: V. Capasso, H. Engl, J. Periaux. VII, 418 pages. 2000.

Vol. 1740: B. Kawohl, O. Pironneau, L. Tartar, J.-P. Zolesio, Optimal Shape Design. Troia 1999. Editors: A. Cellina, A. Ornelas. IX, 388 pages. 2000.

Vol. 1741: E. Lombardi, Oscillatory Integrals and Phenomena Beyond all Algebraic Orders. XV, 413 pages. 2000.

Vol. 1742: A. Unterberger, Quantization and Non-holomorphic Modular Forms. VIII, 253 pages. 2000.

Vol. 1743: L. Habermann, Riemannian Metrics of Constant Mass and Moduli Spaces of Conformal Structures. XII, 116 pages. 2000.

Vol. 1744: M. Kunze, Non-Smooth Dynamical Systems. X, 228 pages. 2000.

Vol. 1745: V. D. Milman, G. Schechtman, Geometric Aspects of Functional Analysis. VIII, 289 pages. 2000.

Vol. 1746: A. Degtyarev, I. Itenberg, V. Kharlamov, Real Enriques Surfaces. XVI, 259 pages. 2000.

4. Lecture Notes are printed by photo-offset from the master-copy delivered in camera-ready form by the authors. Springer-Verlag provides technical instructions for the preparation of manuscripts. Macro packages in T_EX, L^AT_EX2e, $L^AT_EX2.09$ are available from Springer's web-pages at

http://www.springer.de/math/authors/b-tex.html.

Careful preparation of the manuscripts will help keep production time short and ensure satisfactory appearance of the finished book.

The actual production of a Lecture Notes volume takes approximately 12 weeks.

5. Authors receive a total of 50 free copies of their volume, but no royalties. They are entitled to a discount of 33.3 % on the price of Springer books purchase for their personal use, if ordering directly from Springer-Verlag.

Commitment to publish is made by letter of intent rather than by signing a formal contract. Springer-Verlag secures the copyright for each volume. Authors are free to reuse material contained in their LNM volumes in later publications: A brief written (or e-mail) request for formal permission is sufficient.

Addresses:

Professor F. Takens, Mathematisch Instituut,
Rijksuniversiteit Groningen, Postbus 800,
9700 AV Groningen, The Netherlands
E-mail: F.Takens@math.rug.nl

Professor B. Teissier
Université Paris 7
UFR de Mathématiques
Equipe Géométrie et Dynamique
Case 7012
2 place Jussieu
75251 Paris Cedex 05
E-mail: Teissier@math.jussieu.fr

Springer-Verlag, Mathematics Editorial, Tiergartenstr. 17,
D-69121 Heidelberg, Germany,
Tel.: *49 (6221) 487-701
Fax: *49 (6221) 487-355
E-mail: lnm@Springer.de